W9-BTI-450

The Beta-Adrenergic Receptors

The Receptors

Series Editor
David B. Bylund
University of Nebraska Medical Center, Omaha, NE

Board of Editors

S. J. Enna
Nova Pharmaceuticals
Baltimore, Maryland

Bruce S. McEwen
Rockefeller University
New York, New York

Morley D. Hollenberg
University of Calgary
Calgary, Alberta, Canada

Solomon H. Snyder
Johns Hopkins University
Baltimore, Maryland

The Beta-Adrenergic Receptors

Edited by

John P. Perkins

The University of Texas,
Southwestern Medical Center at Dallas,
Dallas, Texas

Withdrawn
University of Waterloo

Humana Press
Clifton, New Jersey

Library of Congress Cataloging-in-Publication Data

The Beta-adrenergic receptors / edited by John P. Perkins.
    p. cm. — (The Receptors)
  Includes index.
  ISBN 0-89603-173-X
  1. Beta adrenoceptors. I. Perkins, John P. (John Phillip)
II. Series.
  [DNLM: 1. Receptors, Adrenergic, Beta—physiology. WL 102.8
B562]
QP364.7.B48   1991
612.8'042—dc20
DNLM/DLC
for Library of Congress   90-15616
                                                CIP

© 1991 The Humana Press Inc.
Crescent Manor
PO Box 2148
Clifton, NJ 07015

Printed in the United States of America

Preface

Prior to 1974, the β-adrenergic receptors were known only indirectly as entities that responded to drugs in a selective manner to mediate a variety of physiologically important responses. During the intervening years, our view of β-adrenergic receptors has changed dramatically. The availability of high affinity ^{125}I-labeled radioligands selective for these receptors presaged an explosion of experimentation utilizing direct binding assays to establish the biochemical properties of the receptor protein. In the opening chapter, Stadel and Lefkowitz describe this development and its impact on our understanding of the molecular basis of β-adrenergic receptor function.

The availability of well-characterized receptor ligands, coupled with the development of efficient methods for detergent solubilization, formed the basis of receptor purification using affinity chromatography. The related technique of photoaffinity labeling provided a means to estimate the molecular mass of these receptors. The availability of substantial amounts of purified $β_2$-adrenergic receptor allowed determination of segments of its amino acid sequence. This information led to the production of polynucleotide probes and eventually to cloning of the receptor gene and determination of the complete primary sequence of the receptor protein. Caron and Lefkowitz review the investigations leading to this major development and discuss the methods involved. They analyze our current perception of the relation of receptor function to its structure and discuss the general features of the G protein-interacting receptor family, of which the β-adrenergic receptors are prototypes.

Perkins, Hausdorff, and Lefkowitz examine the current status of ligand-induced desensitization of β-adrenergic receptor function. The role of receptor phosphorylation is discussed in terms of the involvement of cyclic AMP-dependent protein kinase and the newly discovered β-adrenergic receptor kinase (βARK). Receptor endocytosis is examined as a mechanism for ligand-induced βAR sequestration and a novel mechanism of receptor down-regulation is considered.

Elliott Ross next discusses in detail the methods and theories of membrane protein solubilization, purification, and reconstitution in a functional state, in defined phospholipid vesicles. His treatment focuses on the β-adrenergic receptor, but provides general insight into this important methodology for the analysis of structure/function relationships of membrane proteins.

The intact β-adrenergic receptor has routinely been found to be a poor antigen. However, knowledge of the amino acid sequence of the receptor provided the means for production of sequence-specific antibodies. The early frustrations and eventual success of the search for antibodies to the β_2-adrenergic receptor is next described by Malbon, Moxham, and Brandwein. They critically review the techniques involved and discuss the insights gained into receptor secondary structure using immunological methods.

Barry Wolfe analyzes the information gained about the localization of β-adrenergic receptors in tissues and on cells using autoradiographic techniques. He provides a critical analysis of the techniques and the limits to conclusions that can be drawn from such data.

The chapters by Stiles and Insel, respectively, review our understanding of the regulation of β-adrenergic receptor function and/or expression by other hormones and neurotransmitters and the involvement of altered receptor function or expression in disease states.

Many of the important new insights into the structure of β-adrenergic receptors and the molecular bases of their various modes of action are revealed to be generally applicable to other members of this class of receptor molecules.

John P. Perkins

Contents

Contributors

Harvey J. Brandwein • Genetic Diagnostics Corporation, Great Neck, New York

Marc Caron • Department of Cell Biology, Duke University Medical Center, Durham, North Carolina

Paul A. Insel • Departments of Pharmacology and Medicine, University of California-San Diego, La Jolla, California

Robert J. Lefkowitz • Howard Hughes Medical Institute, Departments of Medicine and Biochemistry, Duke University Medical Center, Durham, North Carolina

Craig C. Malbon • Department of Pharmacological Sciences, School of Medicine, Health Sciences Center, State University of New York, Stony Brook, New York

Cary P. Moxham • Department of Pharmacological Sciences, School of Medicine, Health Sciences Center, State University of New York, Stony Brook, New York

John P. Perkins • Department of Pharmacology, University of Texas, Southwestern Medical Center, Dallas, Texas

Elliott M. Ross • Department of Pharmacology, University of Texas, Southwestern Medical Center, Dallas, Texas

Jeffrey M. Stadel • Department of Molecular Pharmacology, Smith Kline and French Laboratories, Philadelphia, Pennsylvania

Gary L. Stiles • Departments of Medicine (Cardiology) and Biochemistry, Duke University Medical Center, Durham, North Carolina

Barry B. Wolfe • Department of Pharmacology, Georgetown University School of Medicine, Washington, DC

CHAPTER 1

Beta-Adrenergic Receptors

*Identification and Characterization
by Radioligand Binding Studies*

*Jeffrey M. Stadel
and Robert J. Lefkowitz*

1. Introduction

The catecholamines epinephrine and norepinephrine evoke specific beta-adrenergic responses in a variety of tissues. Examples of processes modulated by these agonists are chronotropic and inotropic cardiac responses, relaxation of smooth muscle, and lipolysis in adipose tissue. The facts that beta-adrenergic responses are limited to specific tissues and that there exist stereospecific constraints, i.e., the naturally occurring (–)-isomers of the catecholamines are more potent than the (+)-isomers, imply a recognition system based on stereo-complementarity (Gilbert and Greenberg, 1984). These observations, based on adrenergic responses, reinforce one of the underlying tenets of pharmacology and therapeutics: The specific actions of hormones and neurotransmitters result from high-affinity, stereospecific interactions with tissues. The concept of an entity or substance that recognizes and discriminates on the basis of geometric properties of hormones or drugs has been evolving for more than a century (Langley, 1905; Dale,

The β-Adrenergic Receptors Ed.: J. P. Perkins © 1991 The Humana Press Inc.

1906). This proposed moiety has been functionally designated "receptor." Receptors are defined by their ability to recognize hormones or drugs of a specific class through direct binding interactions and, of equal importance, translate the binding event into a biological response. In the case of beta-adrenergic receptors, stereospecific interactions with epinephrine or norepinephrine and synthetic agonists initiate characteristic biological effects. Beta-adrenergic biological effects can be discriminated from other catecholamine-dependent responses based on the potency series for catecholamine agonists (Ahlquist, 1948). Beta-adrenergic responses are stimulated by isoproterenol more potently than by epinephrine or norepinephrine. Evaluation of a large volume of data generated in the study of beta-adrenergic pharmacology suggests a further division of the responses into subtypes termed beta-1 and beta-2 (Lands et al., 1967). This distinction is based on the relative potency of the naturally occurring catecholamines, epinephrine, and norepinephrine. Beta-1 responses are equally sensitive to these two agonists; beta-2 responses are more potently stimulated by epinephrine. Beta-1 responses appear to be initiated by the neurotransmitter, norepinephrine, in innervated tissues, whereas beta-2 responses are most likely triggered by the circulating hormone epinephrine (O'Donnell and Wanstall, 1987).

The beta-adrenergic system has in many ways served as the premier model system in which to investigate the processes by which external stimuli regulate cellular behavior. This is the result of many factors. The recognition of the widespread distribution of beta-adrenergic responses throughout the body prompted investigation by both pharmacologists and physiologists. These studies pointed to potential therapeutic benefits of modulating beta-adrenergic responses, thereby stimulating research interest in the pharmaceutical industry. This led to the development of high-affinity, highly specific antagonists of beta-adrenergic responses, such as propranolol, alprenolol, and pindolol. These antagonist agents have played an important role in the investigation of the molecular mechanisms of hormone and drug action.

Another property of the beta-adrenergic system that has attracted the interest of both pharmacologists and biochemists is the ability to measure an agonist-dependent response in vitro, in broken cell preparations. The pioneering work of Sutherland and his colleagues showed in rat liver plasma membranes that beta-adrenergic agonists induced the production of a factor that stimulated the activity of cytosolic enzymes involved in glycogen metabolism (Rall et al., 1957; Robison et al., 1971). The soluble factor was identified as cyclic AMP. Cyclic AMP is produced by the membrane-bound enzyme, adenylate cyclase, from substrate ATP (Sutherland et al., 1963). In membrane preparations, the enzyme's activity could be stimulated by the addition of beta-adrenergic agonists. These observations gave rise to the second messenger hypothesis (Robison et al., 1971). This proposal states that cells sense external stimuli, such as hormones and neurotransmitters, by specific binding interactions of these agents at the level of the plasma membrane. The information inherent in these interactions is transduced across the membrane, resulting in the activation of adenylate cyclase, and thus, an increased intracellular concentration of cyclic AMP. This process culminates in the activation of cyclic AMP-dependent protein kinase (Walsh et al., 1968; Krebs and Beavo, 1979), which regulates the activity of key rate-limiting enzymes by reversible phosphorylation. The specificity of hormonal signaling is, therefore, twofold. First, only target tissues possessing the specific recognition factors or receptors are sensitive to the stimuli. The second level of specificity resides in the substrates for the cyclic AMP-dependent protein kinase that are programmed for expression during differentiation. It is noteworthy that, in all cases, beta-adrenergic receptors, regardless of subtype, couple to adenylate cyclase to stimulate the synthesis of cyclic AMP. This tightly coupled receptor–effector complex allows the investigation of beta-adrenergic receptors in isolated membrane fractions under carefully controlled conditions. These studies have supplied a wealth of information pertaining to the mechanisms of transmembrane signaling.

This chapter will focus on the direct identification and characterization of beta-adrenergic receptors using radioligand binding techniques. The data generated from radioligand binding studies have contributed significant insights, leading to the development of a molecular model for the communication between agonist-occupied receptors and adenylate cyclase.

2. Radioligands for Beta-Adrenergic Receptors

Classical pharmacological studies both in vivo and in vitro indicated the existence of beta-adrenergic receptors. The development of synthetic agonists and antagonists generated a large body of structure–activity data leading to the description of beta-adrenergic receptor subtypes. In order to translate the "receptor concept" into a biochemical reality, more direct experimental approaches were required. To study the biochemical properties of receptors, a reliable, direct assay was needed. The development of radioligand binding techniques was an important breakthrough in bringing about a better understanding of the nature of hormone and drug receptors. The feasibility of the radioligand approach was demonstrated in the late 1960s and early 1970s, when procedures were defined for the covalent incorporation of radionuclides into hormone structures without significantly altering the biological activity or specificity of the hormones (Roth, 1973). Radiolabeling of hormones provides the necessary sensitivity to detect the low concentrations of receptors on cell surfaces. These procedures were first applied to peptide hormones, such as ACTH (Lefkowitz et al., 1970) and angiotensin (Lyn and Goodfriend, 1970).

The first attempts to utilize this technology for direct identification of beta-adrenergic receptors led to the synthesis of $[^3H]$catecholamines as ligands (Lefkowitz et al., 1976a). Unfortunately, the binding sites investigated in these early studies did not display the characteristics expected of beta-adrenergic receptors and most likely represented catecholamine uptake sites or metabolizing enzymes. It became apparent that high-affinity antagonists might make better radioligands. In 1974 three groups, working independently, used this insight to develop radioligands for beta-adrenergic receptors (*see* Table 1). The initial radioligands were (–)$[^3H]$dihydroalprenolol ($[^3H]$ DHA) (Lefkowitz et al., 1974); (+/–)$[^3H]$propranolol (Atlas et al., 1974), and (+/–)$[^{125}I]$iodohydroxybenzylpindolol (^{125}I-HYP) (Aurbach et al., 1974). Of these three, $[^3H]$DHA and ^{125}I-HYP have been used most extensively to characterize beta-adrenergic receptors because of their low nonspecific background, i.e., nonreceptor binding. Since

Table 1
Radioligands for Beta-Adrenergic Receptors

Radioligand	Structure	Specific Activity (Ci/m mol)	K_D
Antagonists			
(−) [³H]-dihydroalprenolol ([³H]-DHA)		120	.5-2 nM
(±) [¹²⁵I]-iodohydroxybenzylpindolol (¹²⁵I -HYP)		2,200	10-1000 pM
(±) [³H]-propranolol ([³H]-PRO)		30	1-10 nM
(−) [¹²⁵I]-iodocyanopindolol (¹²⁵I -CYP)		2,200	10-50 pM
(−) [¹²⁵I]-iodopindolol (¹²⁵I -PIN)		2,200	50-200 pM
(±) [³H]-CGP-12177 ([³H]-CGP)		60	100-500 pM
Agonists			
(±) [³H]-hydroxybenzyl isoproterenol ([³H]-HBI)		20	1-10 nM
(−) [³H]-epinephrine ([³H]-EPI)		85	80 nM
(±) [³H]-isoproterenol ([³H]-ISO)		15	100 nM

these original radioligands were introduced, new ones have been synthesized. (–)[^{125}I]Iodocyanopindolol (^{125}I-CYP) (Hoyer et al., 1982) and (–)[^{125}I]iodopindolol (^{125}I-PIN) (Barovsky and Brooker, 1980) possess high affinity for beta-adrenergic receptors and appear to have very low nonspecific interactions.

A characteristic common to the five radioligands described above is their hydrophobic nature. This property limits their usefulness in experiments employing intact cells because of the potential permeation of the radioligand into the cell. A hydrophilic antagonist radioligand, (+/–)[^{3}H]CGP-12177, has been synthesized (Staehelin and Hertel, 1983). Investigations of its binding to membrane preparations and to whole cells indicate that it binds exclusively to beta-adrenergic receptors in the plasma membrane on the cell surface.

The utility of these antagonist radioligands is based on their high affinity and specificity for beta-adrenergic receptors. They also can be radiolabeled to a high specific activity (50–2000 Ci/mmol). This is a prerequisite for the identification of beta-adrenergic receptors, which often exist in very minute quantities on target cells. If the receptor concentration is a limiting factor, then the high specific activity ^{125}I-labeled compounds are the ligands of choice. Although all of these radioligands are specific for beta-adrenergic receptors they do not appear to discriminate between beta-receptor subtypes. However, it should be noted that [^{125}I]iodopindolol can bind to serotonin receptors in brain preparations.

Radiolabeled antagonist ligands have proven to be valuable tools for identifying beta-adrenergic receptors and characterizing their properties. However, one of the goals in studying receptors is to understand how agonist binding triggers a biological response. In the case of the beta-adrenergic receptors, this question can be simplified to examining agonist stimulation of adenylate cyclase activity. These types of questions fostered the search for a radiolabeled agonist ligand. Initial efforts with [^{3}H]epinephrine and [^{3}H]norepinephrine were not successful because the high concentrations of the radioligand employed in the experiments led to an unacceptably high background, which masked true receptor binding. To date only one radiolabeled agonist ligand has been developed and employed to any extent,

(+/−)[^{3}H]-hydroxybenzylisoproterenol ([^{3}H]HBI) (Lefkowitz and Williams, 1977). This radioligand has been an important tool in characterizing properties of beta-2 adrenergic receptors, since the ligand is approx 30-fold selective for this receptor subtype. In general, [^{3}H]HBI has not been routinely used to quantify or characterize beta-adrenergic receptors because of its limitations of specific radioactivity (5–20 Ci/mmol), receptor affinity, subtype specificity, and complex interactions with the beta-adrenergic receptor *(see below)*. However, [^{3}H]HBI has provided novel information as to how agonist binding to receptors promotes coupling to adenylate cyclase.

Following the validation of radioligand binding methods to measure beta-adrenergic receptors, a logical extension of the technology was the development of radiolabeled affinity ligands for covalent labeling of beta-adrenergic receptors (Stadel, 1985; Lefkowitz et al., 1983; Rashidfaigi and Ruoho, 1982; Burgermeister et al., 1982). Once again, antagonist compounds were used because of their high affinity and specificity. Both chemically active and photoactivated substituents have been introduced into antagonist structures. Photoaffinity labels have the advantage that, following equilibration with receptors, excess label can be washed away before the covalent insertion is initiated by irradiation. This usually provides for a relatively low background of nonspecific labeling. The disadvantage of photoaffinity labels is their low efficiency of covalent incorporation (5–20%). It is necessary to determine that the labeling of a protein or proteins can be inhibited by unlabeled ligands of appropriate pharmacological specificity. The addition of protease inhibitors throughout the procedure may simplify the labeling pattern. Covalent labeling of beta-adrenergic receptors with affinity labels has provided valuable information about the structure of the receptor and about the processes regulating receptors.

3. Radioligand Binding Methodology

Following the synthesis of radioligands for beta-adrenergic receptors, there was a virtual explosion of data. These receptors were identified in many tissues from multiple species. Although the radioligands are relatively easy to use, some caution must be exercised in validating and interpreting data. Over the past ten years, several books and articles have been published to guide investigators in the theory and proper usage of radioligands (Williams and Lefkowitz, 1978; Limbird, 1986; Weiland and Molinoff, 1981; Burgisser and Lefkowitz, 1984). The reader is referred to these published works and to the primary references cited in this chapter for in-depth discussions of theory and applications. Several practical issues are discussed below.

One of the preliminary objectives of a radioligand binding experiment is to establish a reliable signal. Optimal incubation conditions are often determined empirically. For beta-adrenergic receptors, antagonist binding in general is less sensitive to the composition of the incubation medium than is agonist binding. Buffers, ions, pH, and other modulators, e.g., nucleotides, may all need to be optimized for the receptor preparation to be examined. If binding is to be tested on intact cells, isotonic conditions must be maintained and a Ringers solution or tissue culture medium is recommended. If at all possible the same incubation conditions should be maintained for the measuring of receptor binding by radioligands as are used to measure a physiological response, e.g., adenylate cyclase activity.

All radioligand binding experiments require the ability to efficiently separate the free radioligand from that associated with the biological preparation. The choice of method will depend to some extent on whether binding experiments are performed on intact cells or on membrane preparations. There are basically three choices: equilibrium dialysis, centrifugation, and vacuum filtration. Equilibrium dialysis is time consuming and laborious, although accurate and thermodynamically correct. Centrifugation is most applicable to whole cell binding, but may lead to high backgrounds caused by trapping of the radioligand and is unsuitable for kinetic experiments. Vacuum filtration is currently the most popular method because of the ease of sample handling, low background, and equal applicability to equilibrium and kinetic experiments. The choice of filter material is important. Glass fiber filters have consistently given the lowest backgrounds for beta-adrenergic radioligands although a filter blank should be examined initially to determine how much radioligand is adsorbed. Depending on the kinetics of the dissociation of the radioligand from the receptor, the receptor–ligand complex trapped by vacuum filtration may only approx the true steady-state level. The filters are often washed with ice-cold buffer to retard ligand dissociation and allow reliable determinations of receptor properties.

Additional separation techniques were developed for the direct assay of beta-adrenergic receptors solubilized from their membrane environment. Two procedures in addition to equilibrium dialysis have been successfully employed to assay beta-adrenergic receptors in solution. The ligand–receptor complex can be precipitated in the presence of polyethylene glycol and the precipitate collected by either centrifugation or vacuum filtration (Homcy et al., 1983). An alternative method is to separate the receptor–radioligand complex from the free radioligand by gel filtration (Caron and Lefkowitz, 1976). Both methods permit reliable assays for soluble beta-adrenergic receptors.

The total amount of radioactivity associated with the biological preparation is not usually equivalent to receptor binding. Specific receptor binding is defined as the total associated radioligand minus the radioactivity measured in the presence of an excess of unlabeled competing ligand. The competing compound is usually present at a concentration 100- to 1000-fold excess over its dissociation constant (K_d) for the receptor. The remaining radioactivity is nonspecifically adsorbed or trapped during the binding procedures. Whenever possible, nonspecific binding should be defined by more than one competing agent. For beta-adrenergic receptor binding, nonspecific binding is usually <20% of the total associated radioactivity.

When a workable specific receptor signal is established, the next imperative is to determine that binding to the physiologically relevant receptor is being measured. A set of criteria (Williams and Lefkowitz, 1978; Limbird, 1986) has been established that needs to be satisfied to determine the relevance of the radioligand binding site as a receptor. These criteria derive from the definition of a receptor as a moiety that (1) specifically recognizes the appropriate agonists and antagonists through high affinity binding interactions and (2) initiates a defined biological response. Thus, one relies on a wealth of pharmacological data to validate radioligand binding experiments. The basic criteria for establishing a valid radioligand binding assay are as follows:

1. The kinetics of radioligand binding should approximate the kinetics of the compound's effect on the biological response, after making appropriate adjustment for receptor and radioligand concentrations.
2. The specific binding of increasing concentrations of the radioligand should reach saturation, reflecting a finite receptor number.
3. The concentration range over which the radioligand binds to the receptor should reflect the concentration range over which the ligand initiates or inhibits a biological response.
4. The receptor sites occupied by the radioligand should demonstrate the appropriate specificity of the biological response. For beta-adrenergic receptors this means that the ability of agonist or antagonists to compete for radioligand binding should reflect the potency of these compounds to evoke or antagonize adenylate cyclase activity. In addition, the (–)-stereoisomers of adrenergic agents should be more potent in competing for the receptor than the corresponding (+)-enantiomers.

In order to determine if these criteria are satisfied two types of experiments are generally employed; kinetic and equilibrium binding studies. Kinetic experiments provide several types of data. First, they allow comparison of the association and dissociation rate of the radioligand with the known kinetic parameters of the biological response, thus satisfying the first criterion. Second, the experimentally determined rate constants can be used to independently calculate a dissociation constant for the radioligand by using the formula: $K_d = k_2/k_1$ for $R + L = RL$, where R is the receptor, L is the radioligand, k_1 and k_2 are the kinetic rate constants, and K_d is the equilibrium dissociation constant. Third, the kinetic data, determined at several radioligand concentrations, will establish the appropriate incubation times to perform equilibrium experiments.

Equilibrium experiments can be divided into two types: saturation and competition. Saturation experiments characterize the receptor as to its affinity for the radioligand and the total number of receptor sites

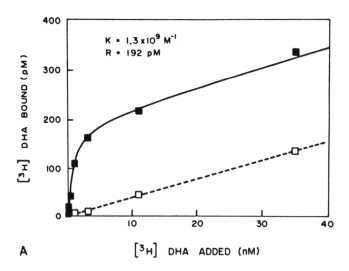

Fig. 1. Saturation binding of the antagonist [³H]DHA to frog erythrocyte beta-adrenergic receptors. The binding experiment was performed in the absence (■) and presence (-) of $10^{-5}M$ propranolol. The curves through the data points represent the best fit of the data by computer-aided nonlinear least-squares curve fitting. The best fit was obtained with a model for one class of binding sites with an affinity constant $K = 1.3 \times 10^9 M^{-1}$ and a receptor concentration $R = 192$ pM (from DeLean et al., 1980).

in the biological preparation. This type of experiment is illustrated in Fig. 1. These experiments establish the existence of specific receptors in tissues of interest and, in addition, provide information as to the subcellular localization of the receptors. This type of experiment can be used to quantify changes in receptor density and/or affinity that may by related to physiological or pathophysiological regulation.

Alternatively, radioligands are used in competition binding experiments. In this type of study, the radioligand is employed at a single fixed concentration while the dose-dependent inhibition of its binding is determined by increasing concentrations of unlabeled compounds. These studies help determine the specificity, including stereospecificity, of the receptor for a broad spectrum of hormones and drugs.

Under certain experimental circumstances, equilibrium may not be achievable or may not be desirable, because the properties of the receptor may be changing during the incubation period. The equilibrium dissociation constant (K_d) for an unlabeled ligand can be estimated from competition binding experiments conducted under initial velocity conditions, as long as the competing ligand comes rapidly to equilibrium with the receptor during the assay (Toews et al., 1983; Motulsky and Mahan, 1984). This approach has been employed to assay beta-adrenergic receptors on intact cells *(see below)*.

Analysis of saturation binding experiments yields the affinity of the receptor for the radioligand (K_d) and the total number of receptor binding sites (B_{max}). Until recently, standard data analysis usually involved the linear transformation of the binding data. The linear transformation, known as Scatchard analysis (Scatchard, 1949) or more correctly, Rosenthal analysis (Rosenthal, 1967), has the advantage that true saturation need not be obtained in order to determine receptor parameters. However, there are pitfalls in this type of analysis. These include the use of too narrow a range of radioligand concentrations to define accurately the saturation and propagation of data scatter and binding artifacts through the transformation calculations (Klotz, 1983; Burgisser, 1984). Computer-aided data analysis using nonlinear least squares curve-fitting programs is an alternative approach and the method of choice (Hancock et al., 1980; Burgisser and Lefkowitz, 1984). The computer programs use a mathematical framework based on the Law of Mass Action to model primary binding data.

Computer programs also have aided in the quantitative analysis of competition binding data. These programs are able to accommodate multiple independent binding sites and determine the affinity constants and relative proportions of each (Hancock et al., 1980; Kent et al., 1979). Through the use of beta-adrenergic receptor subtype-specific ligands in competition binding studies, it has been possible to quantify the relative proportions of the receptor subtypes in a variety of tissues (Hancock et al., 1980; DeLean et al., 1982). Table 2 shows the beta-adrenergic receptor subtype proportion in selected rat tissues.

Table 2
Distribution of Beta-Adrenergic Receptor Subtypes
in Rat Tissues Determined by Radioligand Binding Techniques[a]

Tissue	% Beta-1	% Beta-2
Lung	20	80
Heart		
Left ventricle	90	10
Uterus		
Estrogen dominated	20	80
Progesterone dominated	0	100
Brain		
Cerebral cortex	65	35
Cerebellum	0	100
Limbic forebrain	55	45
Spleen	35	65
Erythrocyte	0	100

[a]Compiled from Minneman et al., 1981; Nahorski, 1981; and data from authors' laboratories.

4. Radioligand Binding to Beta-Adrenergic Receptors on Whole Cells

To understand better how catecholamines modulate the physiological activity of tissues, radioligand binding techniques have been applied to intact cells. These studies permit a comparison of beta-adrenergic receptor binding properties and a physiological response such as cyclic AMP accumulation. As the number of beta-adrenergic receptors on the surface of cells is usually quite low (500–5000 sites/cell), radioiodinated ligands of high specific radioactivity are the best choice. Antagonist radioligands have the advantage of high specificity and are not likely to interact with catecholamine uptake mechanisms or catabolic enzymes. Although direct agonist binding to whole cells would be anticipated to provide important and useful data, the current agonist radioligands are of insufficient specific activity and affinity to be of general utility for this purpose.

Initial studies of beta-adrenergic receptors on intact cells were both gratifying and troubling. It was quickly shown that radiolabeled antagonists bound with high specificity to relevant receptors, providing a reliable signal. Antagonist interactions with beta-adrenergic receptors, deduced from competition binding studies, were as would be predicted from their inhibitory effect on agonist-stimulated cyclic AMP accumulation in terms of both kinetics and potency. In contrast, investigations of agonist interactions with receptors on intact cells generated results that were somewhat surprising. Under equilibrium conditions, the affinity of the beta-adrenergic receptors for the agonist, as determined from competition binding experiments, was generally much lower than would be predicted from the potency (K_{act}) of that agonist to stimulate cyclic AMP synthesis (Terasaki and Brooker, 1978). Specific uptake into the cells or degradation of the agonist were eliminated as possible explanations for the data. The results were interpreted as evidence for very-high-efficiency coupling between the cell surface receptors and the effector enzyme, adenylate cyclase. It was proposed that in certain cell types maximal cyclic AMP accumulation was elicited when less than 1% of the beta-adrenergic receptors were occupied by agonist.

Recently the question of agonist binding to beta-adrenergic receptors on intact cells has been reinvestigated by comparing binding interactions in equilibrium and initial rate experiments (Pittman and Molinoff, 1980; Toews et al., 1983; Insel et al., 1983; Toews and Perkins, 1984; Hoyer et al., 1984). Initial rate experiments yield a valid estimate of K_d so long as the compound competing for the radioligand rapidly equilibrates with the receptor. These experiments showed that beta-adrenergic receptors display a distinctly different affinity for agonists during the first minute of the incubation compared to the affinity determined at equilibrium (Fig. 2). In all cases, the apparent affinity of the beta-adrenergic receptor for agonist decreased with time. Since agonist-dependent activation of adenylate cyclase proceeds immediately upon agonist binding, these results provide an explanation for the discrepancy between K_{act} and apparent K_d during equilibrium experiments.

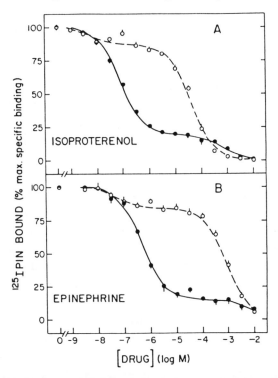

Fig. 2. Agonist competition binding for [^{125}I]PIN to intact 1321Nl cells. The cells were incubated with [^{125}I]PIN and varying concentrations of isoproterenol (A) or epinephrine (B) at 37°C for either 15 s (●) or 60 min (○) (from Toews et al., 1983).

The rapidly induced change in beta-adrenergic receptor affinity following agonist binding appears to be caused by a redistribution of the receptors within the cell (Toews et al., 1984; Harden et al., 1980; Stadel et al., 1983). Through the use of agonists and hydrophilic antagonists in competition binding experiments, it was demonstrated that a portion of the beta-adrenergic receptors was not readily accessible to these ligands, even though the entire receptor population was labeled by the more hydrophobic radioligands. Preexposure of cells to agonists, but not antagonists, altered the distribution of beta-adrenergic receptors, increasing the fraction of receptors in the less accessible compartment. It is possible that agonist-promoted redistribution of the receptors reflects their internalization, accounting for the decreased affinity of the receptor for agonist and other hydrophilic ligands; but the exact localization of the receptors remains to be established (Harden et al., 1980; Mahan et al., 1985). However, the rapid alteration in beta-adrenergic receptor binding properties is clearly associated with the process of agonist-induced desensitization (*see* Chapter 3).

The potential changes in beta-adrenergic receptor binding characteristics during the time-course of incubation in radioligand binding experiments complicate data interpretation. The generally available hydrophobic radioligands bind to both intracellular and cell surface receptors. The definition of a binding equilibrium in whole cells is tenuous, since regulatory mechanisms affect both beta-adrenergic receptor number and affinity. It has been suggested that radioligand binding studies of intact cells may best be conducted at low temperatures (0–4°C) and/or in the presence of metabolic inhibitors to block receptor processing (Limbird, 1986; Motulsky et al., 1985). These nonphysiological conditions, however, may also affect beta-adrenergic receptor properties.

Although radioligand binding experiments using intact cells are complicated, they provide valuable information about the regulatory processes governing receptor binding characteristics and tissue sensitivity to agonist. However, experimental conditions need to be explicitly defined and the results cautiously interpreted.

5. Radioligand Binding to Beta-Adrenergic Receptors in Membranes

To eliminate many of the problems associated with radioligand binding to beta-adrenergic receptors on intact cells, investigators turned to studies of receptors in partially purified plasma membrane preparations. Binding experiments conducted using receptors in membrane fractions have several advantages:

1. The regulatory processes described for intact cell binding studies are not observed when membranes are used.

2. In assays using membranes, the beta-adrenergic receptor concentration is high enough to allow the ready use of tritiated as well as iodinated radioligands, making it possible to employ both agonists and antagonists.
3. As the membrane preparations also contain adenylate cyclase, it is possible to investigate receptor binding properties and an immediate coupled response under controlled conditions.

Therefore, the vast majority of the literature concerning radioligand binding to beta-adrenergic receptors derives from investigations using isolated membrane preparations.

Early studies of beta-adrenergic receptors in membranes showed that the receptors are located predominantly, if not exclusively, in the plasma membrane of responsive cells. This is also true for the closely associated enzyme adenylate cyclase. A partially purified membrane preparation can contain up to 2 pmol of beta-adrenergic receptors/ mg of membrane protein. Proteases and chemical reagents directed toward specific amino acid side chains eliminated the specific binding of the radioligand and established that beta-adrenergic receptors were proteins (Lefkowitz et al., 1976b). Since beta-adrenergic receptors appear to be intrinsic membrane proteins, lipid-perturbing agents, such as phospholipases and amphotericin B, also diminished radioligand binding, suggesting that the lipid environment of the receptor also influences its activity (Lefkowitz et al., 1976b). In most cases the lipid-perturbing compound simultaneously inhibited beta-adrenergic receptor binding and agonist stimulation of adenylate cyclase activity. However, one membrane perturbant, filipin, attenuated agonist-dependent adenylate cyclase activity without inhibiting the binding of the radioligand to its receptor (Limbird and Lefkowitz, 1976). This result indicated that ligand binding to beta-adrenergic receptors could be "uncoupled" from adenylate cyclase stimulation and that the two activities could be differentially modulated.

One of the first major applications of radioligand binding techniques to membrane-bound beta-adrenergic receptors was the investigation of interaction of unlabeled adrenergic compounds. A wide spectrum of both agonists and antagonists was tested with beta-adrenergic receptors in competition binding experiments, and their effects on adenylate cyclase stimulation were compared. These studies established that agonists and antagonists compete for the same receptor site. In addition, these investigations showed that the efficacy of an agonist in stimulating adenylate cyclase activity is not simply determined by its binding affinity to the beta-adrenergic receptor, but rather involves additional interactions not triggered by antagonists. Structure–activity studies confirmed previous pharmacological data showing that the affinity of an adrenergic ligand for the receptor is primarily determined by its stereoconfiguration and by the substituents on the amino nitrogen (Lefkowitz et al., 1976b). Chemical substitution on the aromatic ring, however, had a much greater effect on the stimulation of the cyclase than on the affinity of the compound for the receptor.

As drugs were synthesized that appeared to be selective for one or the other of the beta-adrenergic receptor subtypes, they were also tested in radioligand competition binding assays. The most widely used radioligands for beta-adrenergic receptors appear not to discriminate between the receptor subtypes. These studies provided direct evidence for two beta-adrenergic receptor subtypes and clearly showed that both receptor subtypes can coexist in a single tissue (Hancock et al., 1980). The quantification of beta-adrenergic receptor subtypes from radioligand binding experiments has been greatly aided by the development of nonlinear least squares curve-fitting computer programs (DeLean et al., 1982).

In addition to identifying and quantifying beta-adrenergic receptors in a variety of cell types, radioligand binding studies have shed light on the molecular mechanisms by which agonist binding to receptors results in the stimulation of adenylate cyclase activity. The basic hypothesis underlying these studies was that agonists must interact with beta-adrenergic receptors in a manner that is fundamentally different from that of competing antagonists. Radioligand binding techniques provided an opportunity to examine the differences directly at the receptor level. A comparison of agonist and antagonist binding vs radiolabeled antagonist in competition binding experiments using membrane fractions as the source of beta-adrenergic receptors revealed several such differences. For example, the apparent affinity of beta-adrenergic receptors for agonists, but not antagonists, was increased when the temperature of the binding assay was decreased (Pike and Lefkowitz, 1978). These observations were extended to agonist binding to both beta-1 and beta-2 receptors in mammalian tissues (Weiland et al., 1980). The effects of temperature on agonist binding to beta-adrenergic receptors may reflect a conformational change in the receptor and/or its interaction with other components of the membrane. The fact that agonist-occupied beta-adrenergic receptors are susceptible to inactivation by sulfhydryl reagents, although antagonist-occupied receptors are not (Bottari et al., 1979; Vauquelin and Maguire, 1980; Korner et al., 1982), supports the notion that agonist binding produces a unique conformational change in the receptor, thereby promoting its association with additional elements of the adenylate cyclase (Korner et al., 1982).

Detailed analysis of competition binding data pointed out another unique property of agonist–receptor binding. The competition curves for agonists were shallow, indicating complex interactions with the receptor (Fig. 3). These experiments also revealed that guanine nucleotides modulate agonist binding to beta-adrenergic receptors by reducing the affinity of the receptor for agonist (Maguire et al., 1976; Lefkowitz et al., 1976c). In contrast, the binding of an antagonist displayed a relatively "steep" competition curve (pseudo Hill coefficient = 1) consistent with a single homogeneous class of beta-adrenergic receptor sites, and this binding was unaffected by the presence of nucleotides (Fig. 3).

The rationale for investigating the effects of guanine nucleotides on receptor binding properties stems from early observations made on glucagon-sensitive adenylate cyclase in rat liver membranes (Rodbell et al., 1971a,b). These studies showed that guanine nucleotides were required for hormonal activation of adenylate cyclase but, unexpectedly, guanine nucleotides decreased the binding of ^{125}I-glucagon to its receptor. The beta-adrenergic system provided an opportunity to explore this anomalous relationship because of the existence of well-defined agonist and antagonist compounds. The inclusion of guanine nucleotides, such as GTP or its nonhydrolyzible analog Gpp(NH)p, in

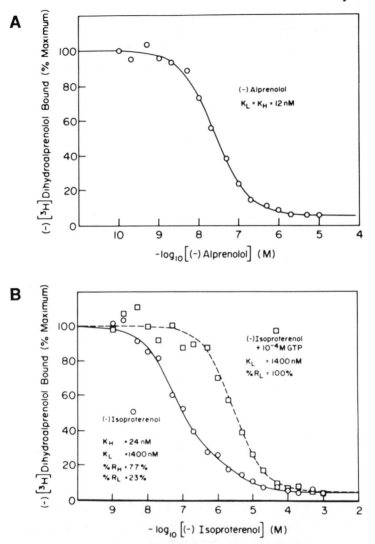

Fig. 3. Computer modeling of competition binding data of alprenolol for
[³H]DHA **(A)** and isoproterenol for [³H]DHA **(B)** in frog erythrocyte membranes. The competition binding curve of the antagonist alprenolol for
[³H]DHA is adequately modeled to a homogeneous class of binding sites.
Competition binding of the agonist isoproterenol for [³H]DHA was performed
in the absence (O) and presence (-) of GTP. The curve in the absence of
nucleotide was significantly ($p < 0.001$) better fit by a model for two binding
states of the beta-adrenergic receptor. *See* text for details (from Kent et al., 1979).

beta-adrenergic receptor binding assays had two effects on agonist competition curves. First, the position of the curve moved to the right, indicating a lower apparent affinity of the receptor for the agonist. Second, the agonist competition curve also steepened to become parallel to antagonist competition curves, suggesting a single, homogeneous receptor population. The addition of guanine nucleotides to the binding assays did not affect the binding of the radiolabeled antagonist, nor did it affect the position or shape of the competition curves generated by unlabeled antagonists. Thus agonists, but not antagonists, form a high-affinity complex with beta-adrenergic receptors and this complex is modulated by guanine nucleotides.

The formation of a high-affinity, nucleotide-sensitive complex between agonist and the beta-adrenergic receptor requires the presence of divalent cations during the binding assay. Millimolar concentrations of Mg^{2+} or Mn^{2+} are necessary for agonist high-affinity binding (Bird and Maguire, 1978; Williams et al., 1978). These same cations also support adenylate cyclase activity. Monovalent cations cannot substitute for these divalent metal ions. The regulation of beta-adrenergic receptor affinity by divalent cations is specific for agonist binding since antagonist binding is unaffected by the presence or absence of metal ions in the binding assay.

The guanine nucleotide-dependent shift to lower receptor affinity for agonists was observed for a series of beta-adrenergic agonists (Lefkowitz et al.,1976c). The magnitude of the change in position of the competition curves, induced by guanine nucleotides as determined by the difference in IC_{50} for the agonist, correlated with the intrinsic activity of the agonist to stimulate adenylate cyclase. The correlation suggested a close relationship between guanine nucleotide regulation of receptor affinity and the ability of an agonist to activate the cyclase.

The application of computer-aided analysis to the competition binding data helped to clarify the differences between agonist and antagonist binding to the beta-adrenergic receptor. Computer modeling techniques applied to complex agonist competition binding curves were similar to those used to quantify receptor subtypes from competition experiments as described above. The agonist competition curves generated in the absence of exogenously added nucleotide were best fitted by a model for two binding states of the receptor (Kent et al., 1979). The analysis permitted the determination of the specific dissociation constants for both high- and low-affinity receptor states (K_H, K_L) and the relative proportion of the total beta-adrenergic receptor population in each state (R_H, R_L) (*see* Fig. 3). The two-state model was found to be statistically appropriate for all agonists tested. Within a series of full and partial agonists, a significant correlation between agonist-intrinsic activity and the ratio of the dissociation constants of the agonist for both receptor states (K_H, K_L) was established (Kent et al., 1979). Intrinsic activity could also be significantly correlated with the proportion of the receptors in the high affinity state (R_H).

Computer modeling of agonist competition curves generated in the presence of guanine nucleotides defined a dissociation constant for agonist binding to a single class of beta-adrenergic receptors that was indistinguishable from the low-affinity dissociation constant (K_L) determined in the absence of nucleotide (Kent et al., 1979). The relative proportion of the high- and low-affinity states of the receptor (R_H, R_L) was dependent on the concentration of guanine nucleotide in the binding assay. The observations that guanine nucleotides mediate a transition of the agonist-induced high-affinity form of the receptor to its low-affinity state, without a similar effect on antagonist binding, and that partial agonists promote the formation of differing proportions of the two affinity states is strong evidence that the high- and low-affinity states of the agonist-occupied receptor are interconvertible. Thus, quantitative analysis of radioligand binding data provided evidence for the importance of agonist high-affinity binding in receptor–adenylate cyclase coupling and new insights into the mechanism of receptor regulation by guanine nucleotides.

The correlation established between guanine nucleotide regulation of agonist–receptor binding and adenylate cyclase activation provides a comparative index of the coupling efficiency between receptors and adenylate cyclase. Competition binding studies of beta-adrenergic receptors in membranes derived from agonist-desensitized tissues suggest that an impaired ability of agonists to promote a high-affinity complex with the receptor contributes to the attenuated adenylate cyclase responsiveness of these tissues (Harden, 1983; Lefkowitz et al., 1983; Stiles et al., 1984).

Direct characterization of agonist interactions with the beta-adrenergic receptor became possible with the development of a radiolabeled agonist ligand, [^{3}H]HBI (Lefkowitz and Williams, 1977). High-affinity [^{3}H]HBI binding to the beta-adrenergic receptor was characterized by a very slow dissociation rate (Williams and Lefkowitz, 1977). As shown in Fig. 4, the dissociation of the radioligand from the receptor was not affected by the presence of a competing adrenergic compound, but the radioligand could be rapidly and completely released in the presence of guanine nucleotide. These results indicated that [^{3}H]HBI bound only to the high-affinity, nucleotide sensitive form of the receptor, and explained why [^{3}H]HBI labeled fewer receptor sites than antagonist radioligands (Wessels et al., 1978; DeLean et al., 1980). From these results it was evident that the radiolabeled agonist would be useful only to investigate beta-receptors in a membrane preparation freed of contaminating nucleotides. The effect of guanine nucleotides in promoting the low-affinity form of the receptor appears to result from the destabilization of the agonist–receptor high-affinity complex, resulting in the rapid release of the agonist.

Since guanine nucleotides are required for hormonal stimulation of adenylate cyclase, it was of interest to determine the role of the agonist–receptor high-affinity complex in cyclase activation. Because the complex dissociates very slowly, it was possible to isolate this intermediate in washed membranes preexposed to agonist (Stadel et al., 1980). Nucleotide-dependent adenylate cyclase activity was significantly higher in membranes preincubated with the agonist than in membranes exposed to antagonist. Thus, the formation of the high-affinity complex of the agonist with the receptor facilitates the activation of adenylate cyclase by regulatory guanine nucleotides.

A computer-aided comparison of the ability of several mechanistic models to fit and reproduce radioligand binding data in the presence and absence of guanine nucleotides led to the proposal of a ternary complex model to explain agonist-specific interactions with beta-adrenergic receptors (DeLean et al., 1980). According to this model, agonist (H) initially binds to the receptor (R) to form a low-affinity binary complex HR. Antagonist binds to the receptor by a similar bimolecular reaction. However, agonist binding uniquely promotes

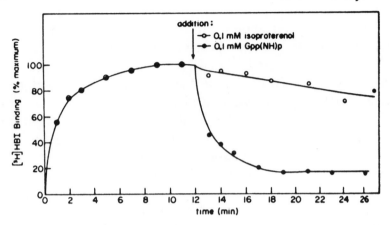

Fig. 4. Effect of added guanine nucleotide on the binding of the agonist [³H]HBI to frog erythrocyte membranes. [³H]HBI was incubated with frog erythrocyte membranes for the indicated time intervals. At 12 min of incubation, either 0.1 mM isoproterenol (O) or 0.1 mM 5'-guanylimidodiphosphate (Gpp[NH]p) (●) were added to the equilibrated mixture (from Williams and Lefkowitz, 1977).

the association of the receptor with an additional membrane component (X) to form a ternary complex, HRX. The ternary complex facilitates the activation of the effector adenylate cyclase (E) in the presence of nucleotide. These interactions are represented in the model shown in Fig. 5. The modeling requires that the stoichiometry of R and X is nearly 1:1. The association of HR and X can be characterized by an affinity constant, L. The intrinsic activity of an agonist correlates with the predicted value of L. This correlation is entirely consistent with the previous relationships established between intrinsic activity and the agonist-receptor high-affinity state.

Although the modeling was not dependent on the identity of the additional membrane component, several lines of evidence suggested that the stimulatory guanine nucleotide regulatory protein, G_s, was equivalent to X. First, agonist binding to the receptor is uniquely regulated by guanine nucleotides. Computer modeling of radioligand binding data indicated that guanine nucleotide-dependent regulation resulted from the inhibition of the ability of agonists to stabilize a complex between HR and X (DeLean et al., 1980). Biochemical

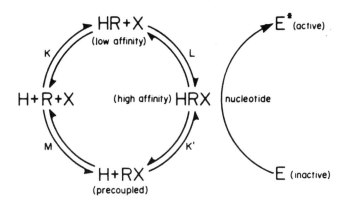

Fig. 5. Schematic diagram of the ternary complex model. The model involves the interactions of hormone (H), receptor (R), an additional membrane component (X), and an effector (E). *See* text for details (from DeLean et al., 1980).

experiments using Mn^{2+} to uncouple agonist-occupied receptors from adenylate cyclase stimulation or using sulfhydryl-specific reagents to inactivate adenylate cyclase catalytic activity demonstrated that a functional cyclase enzyme was not required for agonist high-affinity nucleotide-sensitive binding (Limbird et al., 1979; Howlett et al., 1978; Stadel and Lefkowitz, 1979). However, an additional sulfhydryl-containing component was identified as being necessary for agonist high-affinity binding (Stadel and Lefkowitz, 1979). The concentration range over which this additional component was sensitive to sulfhydryl reagents was consistent with its being G_s (Ross et al., 1978). Finally, a variant of the S49 murine lymphoma cell line, cyc⁻, which lacks a functional G_s component, but possesses beta-adrenergic receptors, does not demonstrate agonist high-affinity binding in competition experiments, nor does it have agonist-stimulated adenylate cyclase activity (Sternweis and Gilman, 1979). Both defects can be corrected by reconstitution of membranes from cyc⁻ cells with an exogenous G_s protein (Sternweis and Gilman, 1979; Northup et al., 1980). Following the reconstitution, agonist competition curves become shallow and are affected by guanine nucleotides; agonists now stimulate the cyclase. Therefore, multiple lines of investigation suggest that agonist high-affinity binding is the result of a relatively stable association of the agonist occupied receptor with G_s.

6. Radioligand Binding to Soluble Beta-Adrenergic Receptors

One of the goals of receptor research is better understanding of the relationship between receptor structure and function. Complete characterization of the biochemical properties of beta-adrenergic receptors requires the purification of sufficient quantities of the receptor protein for biochemical analysis. Since the beta-adrenergic receptor is an intrinsic membrane protein, a major hurdle to be overcome was the solubilization of the receptor from the plasma membrane through the use of detergents. Many different detergents were screened, but the plant glycoside digitonin was found to be uniquely capable of extracting the receptor in an active state, i.e., able to bind ligands subsequent to solubilization (Caron and Lefkowitz, 1976). Other detergents, such as lubrol and triton, could also solubilize beta-adrenergic receptors, but only if the receptors were prelabeled with a radioligand. The properties of the digitonin-solubilized receptors were characterized in radioligand binding experiments (Caron and Lefkowitz, 1976). The affinity of the soluble receptor for antagonist radioligands was equal to that observed in particulate preparations, and the pharmacological specificity, including stereoselectivity, for a series of agonists and antagonists was comparable to that of the membrane-bound receptor. It was concluded that the beta-adrenergic receptor had indeed been solubilized and that digitonin did not severely alter its binding properties.

One important difference between soluble and membrane-associated beta-adrenergic receptors is the absence of the agonist-induced high-affinity state of the receptor (Caron and Lefkowitz, 1976). Agonist competition binding curves for the receptor in detergent solution are uniformly of low-affinity and are unaffected by guanine nucleotides. These results indicate that the soluble receptors are uncoupled from G_s and the other components of adenylate cyclase. Although adenylate cyclase activity in the detergent extracts is not stimulated by agonists, both the enzyme's activity and receptor binding can be measured. Fractionation of these two activities either by gel filtration (Limbird and Lefkowitz, 1977) or by sucrose density centrifugation (Haga et al., 1977) show that they exist on two different macromolecules.

The observation that beta-adrenergic receptors retain their ability to bind ligands following solubilization by digitonin led directly to complete purification of the receptors. Radioligand binding techniques provided a sensitive and quantitative assay to follow receptor activity throughout the steps of purification. Since beta-adrenergic receptors are usually present in such low concentrations in tissues, affinity chromatography procedures were a necessary part of the purification scheme. Affinity chromatography using immobilized antagonist ligands affords up to 1000-fold purification of the receptor in a single step (Caron et al., 1979). Using affinity chromatography in conjunction with other conventional techniques, beta-adrenergic receptors have been successfully purified from a variety of mammalian (Benovic et al., 1984; Cubero and Malbon, 1984) and nonmammalian (Shorr et al., 1981,1982) tissues as described in detail in Chapter 2.

Studies of soluble beta-adrenergic receptors have also shed additional light on the unique interactions of agonists and receptors. Detergent extracts of membranes prelabeled with antagonist or agonist radioligands were analyzed by gel filtration chromatography (Limbird and Lefkowitz, 1978; Limbird et al., 1979). The agonist-prelabeled receptor resolved from the antagonist-prelabeled receptor and eluted from the column earlier, thus indicating an increased apparent molecular size (Fig. 6). This observation is consistent with other data suggesting that agonist binding to beta-adrenergic receptors promotes the receptors association with G_s. Even though soluble receptors appear to be uncoupled from G_s, by prelabeling the membrane receptors with agonist, it was then possible to solubilize the high-affinity agonist–receptor complex. This soluble complex was shown to retain its sensitivity to guanine nucleotides that promoted the rapid dissociation of the radiolabeled agonist from the receptor (Stadel et al., 1981).

Direct evidence of the molecular composition of the agonist-high-affinity receptor complex was obtained by specifically labeling G_s prior

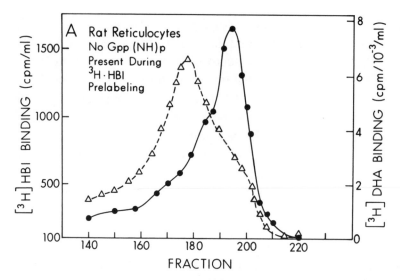

Fig. 6. Gel-exclusion chromatography of beta-adrenergic receptors solubilized from rat reticulocyte membranes with digitonin after prelabeling with
agonist [^{3}H]HBI (Δ) or antagonist [^{3}H]DHA (●). The column matrix was
AcA34 (from Limbird et al., 1980).

to solubilization. This was accomplished using cholera toxin, which
catalyzes the specific covalent incorporation of ADP-ribose from
substrate NAD$^+$, into the structure of G_s (Gill, 1977). By using
[^{32}P]NAD$^+$, it was possible to tag G_s with a radiolabeled marker. It
was shown that the ^{32}P-labeled G_s specifically associated with the
agonist-occupied beta-adrenergic receptor and not with the receptor
prebound with an antagonist (Limbird et al., 1980; Stadel et al., 1981).
Following destabilization of the agonist-promoted ternary complex
by guanine nucleotides, the released G_s can go on to stimulate
adenylate cyclase activity in reconstitution experiments with an
exogenous source of adenylate cyclase catalytic unit (Stadel et al.,
1981). These experiments provide biochemical evidence for an agonist-
promoted ternary complex of HRG_s and for agonist activation of
adenylate cyclase, which occurs by facilitation of guanine nucleotide
binding to G_s through a direct interaction of this regulatory protein
with the receptor.

7. Radioligand Binding to Reconstituted Beta-Adrenergic Receptors

Reconstitution experiments have played a crucial role in defining the components of the beta-adrenergic receptor–adenylate cyclase complex and how they interact. The initial reconstitution studies did not attempt to use purified components, but instead used fusion techniques to produce hybrid cells from cells containing complementary activities. One cell type contained beta-adrenergic receptors and a chemically inactivated adenylate cyclase. This donor cell was fused to an acceptor cell devoid of beta-adrenergic receptors but possessing an active adenylate cyclase. Stimulation of the adenylate cyclase by beta-adrenergic agonists was observed in the fused hybrid cells (Orly and Schramm, 1976). Such experiments were among the first to indicate that receptors and adenylate cyclase are distinct components. They also suggested a highly conserved nature of adenylate cyclase coupled receptors, since components from evolutionarily divergent sources communicate efficiently. Similar experiments helped to define the genetic lesions in S49 variants that are defective in adenylate cyclase responsiveness (Schwarzmeier and Gilman, 1977).

Cell fusion techniques were further exploited using beta-adrenergic receptors from different tissue sources to establish that beta-1 and beta-2 receptor subtype specificity is an intrinsic property of the receptor moiety (Pike et al., 1979). The beta-adrenergic receptor subtype expressed in hybrid cells, as determined by radioligand binding and adenylate cyclase assays, was always the same as the receptor subtype of the donor cells. Agonist-intrinsic activity, on the other hand, appears to involve additional components of the adenylate cyclase complex distal to the receptor (Pike et al., 1979; Kaslow et al., 1979).

Experimental conditions were eventually established for the incorporation of crude solubilized beta-adrenergic receptors into lipid vesicles (Eimerl et al., 1980; Fleming and Ross, 1980; Citri and Schramm, 1980; Cerione et al., 1983). The vesicle-associated receptors showed the appropriate specificity for adrenergic compounds in radioligand binding assays. To protect against inactivation of the receptor during solubilization using deoxycholate as detergent, the

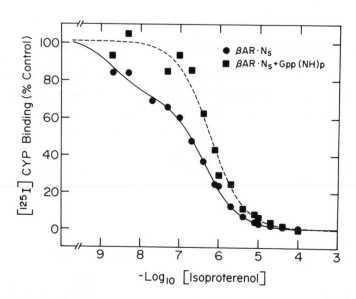

Fig. 7. Competition binding of isoproterenol for [^{125}I]CYP to phospholipid vesicles containing pure beta-adrenergic receptor and pure G_s ($N_s = G_s$). The reconstituted vesicles were assayed in the absence (●) and presence (■) of 5'-guanylimidodiphosphate (from Cerione et al., 1984).

receptor could be stabilized by prelabeling with adrenergic ligands. Deoxycholate appears to mimic G_s in stabilizing a high affinity, slowly dissociable agonist–receptor complex (Neufeld et al., 1983). The reconstituted beta-adrenergic receptor was shown to retain functional activity, as determined by its fruitful interactions with crude preparations of G_s in the reconstitution system (Citri and Schramm, 1980, 1982; Pedersen and Ross, 1982).

As the components of the beta-adrenergic receptor–adenylate cyclase complex have been purified to homogeneity, they have been characterized by coreconstitution in defined lipid vesicles. These important experiments showed that pure beta-adrenergic receptor and pure G_s can interact in the presence of an agonist to promote a high-affinity, nucleotide-sensitive ternary complex (Fig. 7) (Cerione et al., 1984). Recently, agonist-stimulated adenylate cyclase activity has been reconstituted in a vesicle using purified beta-adrenergic receptor,

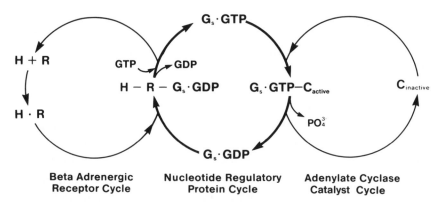

Fig. 8. Schematic model of hormonal activation of adenylate cyclase involving agonist (H), receptor (R), stimulatory nucleotide regulatory protein (G_s), and enzyme catalytic unit (C). *See* text for details (from Stadel and Lefkowitz, 1983).

G_s, and catalytic unit of adenylate cyclase (May et al., 1985). These studies emphasize the bifunctional nature of each of these components. For the beta-adrenergic receptor, which has been identified as a single polypeptide, this means recognizing adrenergic ligands through specific binding interactions and translating this binding event into a fruitful interaction with G_s, leading to adenylate cyclase stimulation. The reconstitution of the beta-adrenergic receptor with additional components of transmembrane signaling systems is described in detail (Ross, Chapter 4).

8. Molecular Model of Beta-Adrenergic Receptor Coupling to Adenylate Cyclase

Radioligand binding studies of beta-adrenergic receptors, along with concurrent investigations of the guanine nucleotide regulatory proteins and adenylate cyclase, have provided sufficient information to formulate a unifying model that explains how agonist binding to receptors is translated across the plasma membrane into increased synthesis of the second messenger cyclic AMP (Fig. 8) (Cassel and Selinger, 1978; Ross and Gilman, 1980; Stadel et al., 1982; Lefkowitz et al., 1983). In this model, the initial binding of agonist and receptor is followed by the formation of a high-affinity ternary complex, HRG_s. The ternary complex facilitates the exchange of GDP for GTP on G_s. The binding of a guanine nucleotide triphosphate has two effects: First, it destabilizes the ternary complex, leading to a release of the agonist to free the receptor; second, it activates G_s to associate with the catalytic unit of adenylate cyclase. The binding of G_s to the catalytic unit stimulates the conversion of ATP to cyclic AMP. This activated state is maintained so long as GTP remains bound to G_s. However, G_s has an intrinsic GTPase activity that cleaves the GTP to GDP, resulting in the dissociation of the regulatory protein from the catalytic unit of the cyclase. The free catalytic unit returns to basal state and the liberated G_s-GDP is available to interact with the receptor in the presence of agonist to reinitiate the response. The salient features of this model are:

1. Agonist binding promotes the formation of a ternary complex that facilitates nucleotide exchange on G_s.
2. G_s conveys information from the receptor to the adenylate cyclase catalytic unit, stimulating cyclic AMP synthesis.
3. The intrinsic GTPase activity of G_s is the turn-off mechanism that returns the system to its basal state.

It should be stressed that this simplified model does not explicitly include the dissociation of G_s into its subunits, which appears to occur as a consequence of R–G_s interactions. Current reconstitution experiments with pure components of the transmembrane signaling pathway will no doubt extend and refine this working model.

References

Ahlquist, R. P. (1948) A study of the adrenotropic receptors. *Am. J. Physiol.* **153,** 585–600.

Atlas, D., Steer, M. L., and Levitzki, A. (1974) Stereospecific binding of propranolol and catecholamines to the beta-adrenergic receptor. *Proc. Natl. Acad. Sci. USA* **71,** 4246–4248.

Aurbach, G. D., Fedak, S. A., Woodard, C. J., Palmer, J. S., Hauser, D., and Troxler, F. (1974) The beta-adrenergic receptor: Stereospecific interaction of an iodinated beta-blocking agent with a high affinity site. *Science* **186,** 1223,1224.

Barovsky, K. and Brooker, G. (1980) ^{125}I-Iodopindolol, a new highly selective radioiodinated beta-adrenergic receptor antagonist: Measurement of beta-receptors on intact rat astrocytoma cells. *J. Cyclic Nucl. Res.* **6**, 297–307.

Benovic, J. L., Shorr, R. G. L., Caron, M. G., and Lefkowitz, R. J. (1984) The mammalian beta-2 adrenergic receptor: Purification and characterization. *Biochemistry* **23**, 4510–4518.

Bird, S. J. and Maguire, M. B. (1978) The agonist-specific effect of magnesium ion on the binding by beta-adrenergic receptor in S49 lymphoma cells: Interaction of GTP and magnesium in adenylate cyclase activation. *J. Biol. Chem.* **253**, 8826–8829.

Bottari, S., Vauquelin, G., Darien, O., Klutchko, C., and Strosberg, A. D. (1979) The beta-adrenergic receptor of turkey erythrocyte membranes: Conformation modification by beta-adrenergic agonists. *Biochem. Biophys. Res. Commun.* **86**, 1311–1318.

Burgermeister, W., Hekman, M., and Helmreich, F. J. M. (1982) Photoaffinity labeling of the beta-adrenergic receptor with azide derivatives of iodocyano-pindolol. *J. Biol. Chem.* **257**, 5306–5311.

Burgisser, E. (1984) Radioligand-receptor binding studies: What's wrong with the Scatchard analysis. *Trends Pharmacol. Sci.* **5**, 142–144.

Burgisser, E. and Lefkowitz, R. J. (1984) Beta-adrenergic receptors, in *Brain Receptor Methodologies Part A -General Methods and Concepts* (Marango, P. J., Campbell, I., and Cohen, R. M., eds.), Academic, Orlando, Florida, pp. 229–253.

Caron, M. G. and Lefkowitz, R.J . (1976) Solubilization and characterization of the beta-adrenergic bindings sites of frog erythrocytes. *J. Biol. Chem.* **251**, 2374–2384.

Caron, M. G., Srinivasan, Y., Pitha, J. Kociolek, K., and Lefkowitz, R. J. (1979) Affinity chromatography of the beta-adrenergic receptor. *J. Biol. Chem.* **254**, 2923–2927.

Cassel, D. and Selinger, Z. (1978) Mechanism of adenylate cyclase activation through the beta-adrenergic receptor: Catecholamine-induced displacement of bound GDP by GTP. *Proc. Natl. Acad. Sci. USA* **75**, 4155–4159.

Cerione, R. A., Codina, J. Benovic, J. L., Lefkowitz, R. J., Birnbaumer, L., and Caron, M. G. (1984) The mammalian beta-2 adrenergic receptor: Reconstitution of functional interactions between pure receptor and pure stimulatory nucleotide binding protein of the adenylate cyclase system. *Biochemistry* **23**, 4519–4525.

Cerione, R. A., Strulovici, B., Benovic, J. L., Strader, C. D., Caron, M. G., and Lefkowitz, R. J. (1983) Reconstitution of beta-adrenergic receptors in lipid vesicles: Affinity chromatography-purified receptors confer catecholamine responsiveness on a heterologous adenylate cyclase system. *Proc. Natl. Acad. Sci. USA* **80**, 4899–4903.

Citri, Y. and Schramm, M., (1980) Resolution, reconstitution, and kinetics of the primary action of a hormone receptor. *Nature* **287**, 297–300.

Citri, Y. and Schramm, M. (1982) Probing of the coupling site of the beta-adrenergic receptor. *J. Biol. Chem.* **257**, 13257–13262.

Cubero, A. and Malbon, C. C. (1984) The fat cell beta-adrenergic receptor: Purification and characterization of a mammalian beta-1 adrenergic receptor. *J. Biol. Chem.* **259**, 1344–1350.

Dale, H. H. (1906) On some physiological actions of ergot. *J. Physiol. (Lond.)* **34**, 165–206.

DeLean, A., Hancock, A. A., and Lefkowitz, R. J. (1982) Validation and statistical analysis of a computer modeling method for quantitative analysis of radioligand binding data for a mixture of pharmacological receptor subtypes. *Mol. Pharmacol.* **21**, 5–16.

DeLean, A., Stadel, J. M., and Lefkowitz, R. J. (1980) A ternary complex model explains the agonist-specific binding properties of the adenylate cyclase-coupled beta-adrenergic receptor. *J. Biol. Chem.* **255**, 7108–7117.

Eimerl, S., Neufeld, G., Korner, M., and Schramm, M. (1980) Functional implantation of a solubilized beta-adrenergic receptor in the membrane of a cell. *Proc. Natl. Acad. Sci. USA* **77**, 760–764.

Fleming, J. W. and Ross, E. M. (1980) Reconstitution of beta-adrenergic receptors into phospholipid vesicles: Restoration of [^{125}I]-iodohydroxybenzylpindolol binding to digitonin-solubilized receptors. *J. Cyclic Nucleotide Res.* **6**, 407–419.

Gilbert, S. F. and Greenberg, J. P. (1984) Intellectual traditions in the life sciences II: Stereocomplementarity. *Perspect. Biol. Med.* **28**, 18–34.

Gill, D. M. (1977) The mechanism of action of choleratoxin. *Adv. Cyclic Nucleotide Res.* **8**, 85–118.

Haga, T., Haga, K., and Gilman, A. G. (1977) Hydrodynamic properties of the beta-adrenergic receptors and adenylate cyclase from wild type and variant S49 lymphoma cells. *J. Biol. Chem.* **252**, 5776–5782.

Hancock, A. A., DeLean, A., and Lefkowitz, R. J. (1980) Quantitative resolution of beta-adrenergic subtypes by selective ligand binding: Application of a computerized model fitting technique. *Mol. Pharmacol.* **16**, 1–9.

Harden, T. K. (1983) Agonist-induced desensitization of the beta- adrenergic receptor-linked adenylate cyclase. *Pharmacol. Rev.* **35**, 5–32.

Harden, T. K., Cotton, C. U., Waldo, G. L., Lutton, J. K., and Perkins, J. P. (1980) Catecholamine-induced alteration in the sedimentation behavior of membrane-bound beta-adrenergic receptors. *Science* **210**, 441–443.

Homcy, C. J., Rockson, S. G., Countaway, J., and Egan, D. A. (1983) Purification and characterization of the mammalian beta-2 adrenergic receptor. *Biochemistry* **22**, 660–668.

Howlett, A. C., Van Arsdale, P. M., and Gilman, A. G. (1978) Efficiency of coupling between the beta-adrenergic receptor and adenylate cyclase. *Mol. Pharmacol.* **14**, 531–539.

Hoyer, D., Engle, G., and Berthold, R. (1982) Binding characteristics of (+)-, (+/–) and (–)- [^{125}Iodo]cyanopindolol to guinea pig left ventricle membranes. *Naunyn-Schmiedebergs Arch. Pharmacol.* **318**, 319–329.

Hoyer, D., Reynolds, E. E., and Molinoff, P. B. (1984) Agonist-induced changes in the properties of beta-adrenergic receptors on intact S49 lymphoma cells: Time dependent changes in the affinity of the receptor for agonists. *Mol. Pharmacol.* **25**, 209–218.

Insel, P. A., Mahan, L. C., Motulsky, A. J., Stoolman, L. M., and Koachman, A. M. (1983) Time-dependent decreases in binding affinity of agonists for beta-adrenergic receptors of intact S49 lymphoma cells: A mechanism of desensitization. *J. Biol. Chem.* **258**, 13597–13605.

Kaslow, H. R., Farfel, Z., Johnson, G. L., and Bourne, H. R. (1979) Adenylate cyclase assembled in vitro: Cholera toxin substrates determine different patterns of regulation by isoproternol and guanosine 5'-triphosphate. *Mol. Pharmacol.* **15**, 472–479.

Kent, R. S., DeLean, A., and Lefkowitz, R. J. (1979) A quantitative analysis of beta-adrenergic receptor interactions: Resolution of high- and low-affinity states of the receptor by computer modeling of ligand binding data. *Mol. Pharmacol.* **17**, 14–23.

Klotz, I. M. (1983) Ligand-receptor interactions: What we can and cannot learn from binding measurements? *Trends Pharmacol. Sci.* **4**, 253–255.

Korner, M., Gilon, G., and Schramm, M. (1982) Locking of hormone in the beta-adrenergic receptor by attack on a sulfhydryl in an associated component. *J. Biol. Chem.* **257**, 3389–3396.

Krebs, E. G. and Beavo, J. (1979) Phosphorylation-dephosphorylation of enzymes. *Annu. Rev. Biochem.* **48**, 923–960.

Lands, A. M., Arnold, A., McAuliff, J. P., Luduena, F. P., and Braun, T. G. (1967) Differentiation of receptor systems activated by sympathomimetic amines. *Nature* **214**, 597, 598.

Langley, J. N. (1905) On the reaction of cells and nerve endings to certain poisons, in regards the reaction of striated muscle to nicotine and curari. *J. Physiol. (Lond.)* **33**, 374–413.

Lefkowitz, R. J. and Williams, L. T. (1977) Catecholamine binding to the beta-adrenergic receptor. *Proc. Natl. Acad. Sci. USA* **74**, 515–519.

Lefkowitz, R. J., Stadel, J. M., and Caron, M. G. (1983) Adenylate cyclase coupled beta-adrenergic receptors: Structure and mechanisms of activation and desensitization. *Annu. Rev. Biochem.* **52**, 159–186.

Lefkowitz, R. J., Roth, J., Pricer, W., and Pastan, I. (1970) ACTH receptors: Specific binding of ACTH-[^{125}I] and its relationship to adenyl cyclase. *Proc. Natl. Acad. Sci. USA* **65**, 745–753.

Lefkowitz, R. J., Mukherjee, C., Coverstone, M., and Caron, M. G. (1974) Stereospecific [^{3}H](–)alprenolol binding sites, beta-adrenergic receptors and adenyl cyclase. *Biochem. Biophys. Res. Commun.* **60**, 703–709.

Stadel and Lefkowitz

Lefkowitz, R. J., Limbird, L. E., Mukherjee, C., and Caron, M. G. (1976a) The beta-adrenergic receptor and adenylate cyclase. *Biochem. Biophys. Acta* **457**, 1–39.

Lefkowitz, R. J., Mukherjee, C., Limbird, L. E., Caron, M. G., Williams, L. T., Alexander, R. W., Mickey, J. V., and Tate, R. (1976b) Regulation of adenylate cyclase coupled beta-adrenergic receptors. *Recent Prog. Horm. Res.* **32**, 597–632.

Lefkowitz, R. J., Mullikin, D., and Caron, M. G. (1976c) Regulation of beta-adrenergic receptors by guanyl-5'-ylimidophosphate and other purine nucleotides. *J. Biol. Chem.* **252**, 799–802.

Limbird, L. E. (1986) *Cell Surface Receptors: A Short Course on Theory and Methods* (Martinies Nijhoff, Boston).

Limbird, L. E. and Lefkowitz, R. J. (1976) Adenylate cyclase coupled beta-adrenergic receptors: Effects of membrane lipid-perturbing agents on receptor binding and enzyme stimulation by catecholamines. *Mol. Pharmacol.* **12**, 559–564.

Limbird, L. E. and Lefkowitz, R. J. (1977) Resolution of beta-adrenergic receptor binding and adenylate cyclase activity by gel exclusion chromatography. *J. Biol. Chem.* **252**, 779–781.

Limbird, L. E. and Lefkowitz, R. J. (1978) Agonist-induced increase in apparent beta-adrenergic receptor size. *Proc. Natl. Acad. Sci. USA* **75**, 228–232.

Limbird, L. E., Hickey, A. R., and Lefkowitz, R.J. (1979) Unique uncoupling of the frog erythrocyte adenylate cyclase system by manganese. *J. Biol. Chem.* **254**, 2677–2683.

Limbird, L. E., Gill, D. M., and Lefkowitz, R. J. (1980) Agonist-promoted coupling of the beta-adrenergic receptor with the guanine nucleotide regulatory protein of the adenylate cyclase system. *Proc. Natl. Acad. Sci. USA* **77**, 775–779.

Lyn, S. Y. and Goodfriend, T. L. (1970) Angiotensin receptors. *Am. J. Physiol.* **218**, 1319–1328.

Maguire, M. E., Van Arsdale, P. M., and Gilman, A. G. (1976) An agonist-specific effect of guanine nucleotides on binding to the beta-adrenergic receptor. *Mol. Pharmacol.* **12**, 335–339.

Mahan, L. C., Motulsky, H. J., and Insel, P. A. (1985) Do agonists promote rapid internalization of beta-adrenergic receptors? *Proc. Natl. Acad. Sci. USA* **82**, 6566–6570.

May, D. C., Ross, E. M., Gilman, A. G., and Smigel, M. D. (1985) Reconstitution of catecholamine-stimulated adenylate cyclase activity using three purified proteins. *J. Biol. Chem.* **260**, 15829–15833.

Minneman, K. P., Pittman, R. N., and Molinoff, P. B. (1981) Beta-adrenergic receptor subtypes: Properties, distribution and regulation. *Ann. Rev. Neurosci.* **4**, 419–461.

Motulsky, H. J. and Mahan, L. C. (1984) The kinetics of competitive radioligand binding predicted by the law of mass action. *Mol. Pharmacol.* **25**, 1–9.

Motulsky, H. J., Mahan, L. C., and Insel, P. A. (1985) Radioligands, agonists and membrane receptors on intact cells: Data analysis in a bind. *Trends Pharmacol. Sci.* **6,** 317–319.

Nahorski, S. R. (1981) Identification and significance of beta-adrenergic subtypes, in *Towards Understanding Receptors* (Lamble, J. W., ed.), Elsevier/North Holland Biomedical, Amsterdam, pp. 71–77.

Neufeld, G., Steiner, S., Korner, M., and Schramm, M. (1983) Trapping of the beta-adrenergic receptor in the hormone induced state. *Proc. Natl. Acad. Sci. USA* **80,** 6441–6446.

Northup, J. K., Sternweis, P. C., Smigel, M. D., Schleifer, L. S., Ross, E. M., and Gilman, A. G. (1980) Purification of the regulatory component of adenylate cyclase. *Proc. Natl. Acad. Sci. USA* **77,** 6516–6520.

O'Donnell, S. R. and Wanstall, J. C. (1987) Functional evidence for differential regulation of beta-adrenoceptor subtypes. *Trends Pharmacol. Sci.* **8,** 265–268.

Orly, J. and Schramm, M. (1976) Coupling of catecholamine receptors from one cell with an adenylate cyclase from another cell by cell fusion. *Proc. Natl. Acad. Sci. USA* **73,** 4410–4414.

Pedersen, S. E. and Ross, E. M. (1982) Functional reconstitution of beta-adrenergic receptors and the stimulatory GTP-binding protein of adenylate cyclase. *Proc. Natl. Acad. Sci. USA* **79,** 7228–7232.

Pike, L. J. and Lefkowitz, R. J. (1978) Agonist specific alterations in receptor binding affinity associated with solubilization of turkey erythrocyte membrane beta-adrenergic receptors. *Mol. Pharmacol.* **14,** 370–375.

Pike, L. J., Limbird, L. E., and Lefkowitz, R. J. (1979) Beta-adrenoreceptors determine affinity but not intrinsic activity of adenylate cyclase stimulants. *Nature* **280,** 502–504.

Pittman, R. N. and Molinoff, P. B. (1980) Interactions of agonists and antagonists with beta-adrenergic receptors on intact L6 muscle cells. *J. Cyclic Nucleotide Res.* **6,** 421–435.

Rall, T. W., Sutherland, E. W., and Berthet, J. (1957) The relationship of epinephrine and glucagon to liver phosphorylase IV: Effect of epinephrine, and glucagon on the reactivation of phosphorylase in liver homogenates. *J. Biol. Chem.* **224,** 463–475.

Rashidbaigi, A. and Ruoho, A. E. (1982) Iodoazidobenzylpindolol, a photoaffinity probe for the beta-adrenergic receptor. *Proc. Natl. Acad. Sci. USA* **78,** 1609–1613.

Robison, G. A., Butcher, R. W., and Sutherland, E. W. (1971) *Cyclic AMP* (Academic, New York).

Rodbell, M., Birnbaumer, L., Pohl, S. L., and Krans, H. M. (1971a) The glucagon-sensitive adenyl cyclase system in plasma membranes of rat liver: An obligatory role of guanine nucleotides in glucagon action. *J. Biol. Chem.* **246,** 1877–1882.

Rodbell, M., Krans, H. M. J., Pohl, S. L., and Birnbaumer, L. (1971b) The glucagon-sensitive adenyl cyclase system in plasma membranes of rat liver IV: Effects of guanyl nucleotides on the binding of ^{125}I-glucagon. *J. Biol. Chem.* **246**, 1872–1876.

Rosenthal, H. E. (1967) Graphical method for the determination and presentation of binding parameters in a complex system. *Anal. Biochem.* **20**, 525–532.

Ross, E. M. and Gilman, A. G. (1980) Biochemical properties of hormone-sensitive adenylate cyclase. *Annu. Rev. Biochem.* **49**, 533–564.

Ross, E. M., Howlett, A. C., Ferguson, K. M., and Gilman, A. G. (1978) Reconstitution of hormone-sensitive adenylate cyclase activity with resolved components of the enzyme. *J. Biol. Chem.* **253**, 6406–6412.

Roth, J. (1973) Peptide hormone binding to receptors: A review of direct studies in vitro. *Metab. Clin. Exp.* **22**, 1059–1073.

Scatchard, G. (1949) The attractions of proteins for small molecules and ions. *Ann. NY Acad. Sci.* **51**, 660–672.

Schwarzmeier, J. D. and Gilman, A. G. (1977) Reconstitution of catecholamine-sensitive adenylate cyclase activity: Interaction of components following cell-cell and membrane-cell fusion. *J. Cyclic Nucleotide Res.* **3**, 227–238.

Shorr, R. G. L., Lefkowitz, R. J., and Caron, M. G. (1981) Purification of the beta-adrenergic receptor: Idenification of the hormone binding subunit. *J. Biol. Chem.* **256**, 5820–5826.

Shorr, R. G. L., Strohsacker, M. W., Lavin, T. N., Lefkowitz, R.J., and Caron, M. G. (1982) The beta-1 adrenergic receptor of the turkey erythrocyte: Molecular heterogeneity revealed by purification and photoaffinity labeling. *J. Biol. Chem.* **257**, 12341–12350.

Stadel, J. M. (1985) Photoaffinity labeling of beta-adrenergic receptors. *Pharmacol. Ther.* **31**, 57–77.

Stadel, J. M. and Lefkowitz, R. J. (1979) Multiple reactive sulfhydryl groups modulate the functions of adenylate cyclase-coupled beta-adrenergic receptors. *Mol. Pharmacol.* **16**, 709–718.

Stadel, J. M. and Lefkowitz, R. J. (1983) The beta-adrenergic receptor: Ligand binding illuminates the mechanism of receptor–adenylate cyclase coupling. *Curr. Top. Memb. Transp.* **18**, 45–66.

Stadel, J. M., DeLean, A., and Lefkowitz, R. J. (1980) A high-affinity agonist beta-adrenergic receptor complex is an intermediate for catecholamine stimulation of adenylate cyclase in turkey and frog erythrocyte membranes. *J. Biol. Chem.* **255**, 1436–1441.

Stadel, J. M., Shorr, R. G. L., Limbird, L. E., and Lefkowitz, R. J. (1981) Evidence that a beta-adrenergic receptor-associated guanine nucleotide regulatory protein conveys guanosine 5'-0-(3-thiotriphosphate)-dependent adenylate cyclase activity. *J. Biol. Chem.* **256**, 8718–8723.

Stadel, J. M., DeLean, A., and Lefkowtiz, R. J. (1982) Molecular mechanisms of coupling in hormone receptor-adenylate cyclase systems. *Adv. Enzymol.* **53**, 1–43.

Stadel, J. M., Strulovici, B., Nambi, P., Lavin, T. N., Briggs, M. M., Caron, M. G., and Lefkowitz, R. J. (1983) Desensitization of the beta-adrenergic receptor of frog erythrocytes: Recovery and characterization of the down-regulated receptors in sequestered vesicles. *J. Biol. Chem.* **258,** 3032–3038.

Staehelin, M. and Hertel, C. (1983) [^{3}H]CGP-12177, a beta-adrenergic ligand suitable for measuring cell surface receptors. *J. Recept. Res.* **3,** 35–43.

Sternweis, P. C. and Gilman, A. G. (1979) Reconstitution of catecholamine-sensitive adenylate cyclase. *J. Biol. Chem.* **254,** 3333–3340.

Stiles, G. L., Caron, M. G., and Lefkowitz, R. J. (1984) Beta-adrenergic receptors: Biochemical mechanisms of physiological regulation. *Pharmacol. Rev.* **64,** 661–743.

Sutherland, E. W., Rall, T. W., and Menon, T. (1963) Adenyl cyclase I. Distribution, preparation and properties. *J. Biol. Chem.* **237,** 1220–1227.

Terasaki, W. L. and Brooker, G. (1978) [^{125}I]Iodohydroxybenzylpindolol binding sites on intact rat glioma cells: Evidence for beta-adrenergic receptors of high coupling efficiency. *J. Biol. Chem.* **253,** 5418–5425.

Toews, M. L. and Perkins, J. P. (1984) Agonist-induced changes in beta-adrenergic receptors on intact cells. *J. Biol. Chem.* **259,** 2227–2235.

Toews, M. L., Harden, T. K., and Perkins, J. P. (1983) High-affinity binding of agonists to beta-adrenergic receptors on intact cells. *Proc. Natl. Acad. Sci. USA* **80,** 3553–3557.

Toews, M. L., Waldo, G. L., Harden, T. K., and Perkins, J. P. (1984) Relationship between an altered membrane form and a low-affinity form of the beta-adrenergic receptor occurring during catecholamine-induced desensitization: Evidence for receptor internalization. *J. Biol. Chem.* **259,** 11844–11850.

Vauquelin, G. and Maguire, M. E. (1980) Inactivation of beta-adrenergic receptors by *N*-ethylmaleimide in S49 lymphoma cells: Agonist induction of functional receptor heterogeneity. *Mol. Pharmacol.* **18,** 362–369.

Walsh, D. A., Perkins, J. P., and Krebs, E. G. (1968) An adenosine 3', 5' monophosphate-dependent protein kinase from rabbit skeletal muscle. *J. Biol. Chem.* **243,** 3763–3765.

Weiland, G. A. and Molinoff, P. B. (1981) Quantitative analysis of drug–receptor interactions I: Determination of kinetics and equilibrium properties. *Life Sci.* **29,** 313–330.

Weiland, G. A., Minneman, K. P., and Molinoff, P. B. (1980) Thermodynamics of agonist and antagonist interactions with mammalian beta-adrenergic receptors. *Mol. Pharmacol.* **18,** 341–347.

Wessels, M. R., Mullikin, D., and Lefkowitz, R. J. (1978) Differences between agonist and antagonist binding following beta-adrenergic receptor desensitization. *J. Biol. Chem.* **253,** 3371–3373.

Williams, L. T. and Lefkowitz, R. J. (1977) Slowly reversible binding of catecholamine to a nucleotide-sensitive state of the beta-adrenergic receptor. *J. Biol.
Chem.* **252,** 7207–7213.

Williams, L. T. and Lefkowitz, R. J. (1978) *Receptor Binding Studies in Adrenergic Pharmacology* (Raven, New York).

Williams, L. T., Mullikin, D., and Lefkowitz, R. J. (1978) Magnesium dependence
of agonist binding to adenylate cyclase-coupled hormone receptors. *J. Biol.
Chem.* **253,** 2984–2989.

CHAPTER 2

Structure–Function Relationships

Marc G. Caron
and Robert J. Lefkowitz

1. Introduction

Understanding, in molecular terms, the ways in which extracellular signals are transduced across the cell membrane so as to modify key intracellular metabolic processes ultimately requires that the various elements in the pathway be isolated and their structures determined. Only then can the structural basis for such signaling systems be understood. Of the many receptors that are coupled through guanine nucleotide regulatory proteins to various cellular effectors (enzymes, ion channels), perhaps the most throughly studied has been the beta-adrenergic receptor. Moreover, much has been learned over the past five years. The receptors have been purified to homogeneity by affinity chromatography, have been reconstituted into phospholipid vesicles together with G-proteins, and the effector enzyme adenylyl cyclase and the various receptor functions probed in such reconstituted systems. Limited protein sequence information, obtained from peptides derived from the purified receptors, has permitted the design and use of oligonucleotide probes that have led to the successful cloning of the gene and cDNA for the receptor, thus, permitting

The β–Adrenergic Receptors Ed.: J. P. Perkins © 1991 The Humana Press Inc.

deduction of the complete amino acid sequence. This in turn has opened the way for the cloning of the genes for all the other adrenergic receptors, and for a variety of other receptors coupled to guanine nucleotide regulatory proteins. Moreover, extensive mutagenesis studies of the receptor have led to an emerging understanding of how the structure of the receptor determines the key functions of ligand binding, effector activation, and regulation of receptor function.

Whereas much has been learned, our understanding of the structure of these receptors is just in its infancy. Given the remarkable conservation of structure and functional properties of receptors coupled to guanine nucleotide regulatory proteins, the information obtained for the beta-adrenergic receptor becomes even more important. In this article, we briefly review recent advances dealing with the purification, reconstitution, cloning, sequencing, and mutagenesis of the beta-adrenergic receptors. At several points, the potential generality of the findings and their applicability to the various other members of this receptor class are emphasized.

1.1. Purification of the Beta-Adrenergic Receptor

The basis for the recent dramatic progress in understanding the structure and function of G-protein-coupled receptors through the cloning of their genes and/or cDNAs, initially for the beta-2 adrenergic receptor and sequently for several other members of this family, rest largely on the ability to purify these proteins in sufficient amounts to allow their biochemical characterization. G-protein coupled receptors, of which the beta-2 adrenergic receptor represents a prototype, are integral membrane proteins and are present in cells in vanishingly low concentrations. In order to characterize these proteins, one needs first to solubilize them from the plasma membrane in an active state (i.e., in a form that can still interact with specific ligands) and purify them to the extent of 30,000- 100,000-fold from their membrane environment. For the beta-adrenergic receptor, this goal was accomplished by the development of several key methods. First, as early as 1976, the beta-2 adrenergic receptor was solubilized from frog erythrocyte membranes, using the plant glycoside digitonin as the detergent (Caron and Lefkowitz, 1976). Among a variety of detergents examined, digitonin was the only one that allowed the solubilization of the receptor with retention of its pharmacological properties (i.e., recognize and bind specific ligands). Because of its properties as a mild detergent, digitonin has been used repeatedly for the solubilization and characterization of several G-protein coupled receptors.

The second successful approach for the characterization of the beta-2 adrenergic receptor was the development of an affinity chromatography procedure (Vauquelin et al., 1977; Caron et al., 1979). To this end, advantage was taken of the fact that the beta-2 adrenergic receptor interacts specifically with various antagonists with high affinity *(K_d* in the range of n*M*–p*M)*. The antagonist, alprenolol, was covalently attached via its allylic function to a chromatography support (Sepharose 4B-6B) through a long side arm (Vauquelin et al., 1977; Caron et al., 1979). The solubilized beta-adrenergic receptor was selectively retained on the affinity matrix, and after sufficient washing of the chromatography matrix, the receptor could be eluted by the addition of a competing ligand (Caron et al., 1979). This step provided a ~1000-fold purification of the receptor with a recovery of ~50% (Benovic et al., 1984). A typical profile is shown in Fig.1A. After this step, the receptor could be further purified to apparent homogeneity by molecular sieve chromatography on an HPLC matrix (Fig. 1B).

The last technical advance that allowed the early characterization of the beta-adrenergic receptor was the development of photoaffinity probes specific for the receptor (Lavin et al., 1981; Rashidbaigi and Ruoho, 1981). Using this procedure, it was possible to visualize and identify the ligand binding subunit of the receptor, even in crude membrane preparations. Two probes were developed and characterized. [^{125}I] *p*-Azidobenzylcarazolol, a derivative of the potent antagonist, carazolol, was synthesized and characterized as a useful probe (Lavin et al., 1981,1982). [^{125}I] *p*-Azidobenzylpindolol a derivative of the antagonist, pindolol, was also used as a specific photoaffinity ligand (Rashidbaigi and Ruoho, 1981). More recently, a diazarine derivative of the potent antagonist, cyanopindolol, has become available and is widely used (Burgermeister et al., 1983). These probes revealed that the ligand binding subunit of the beta-adrenergic receptor resided on a polypeptide of mol wt 64,000 in most tissues

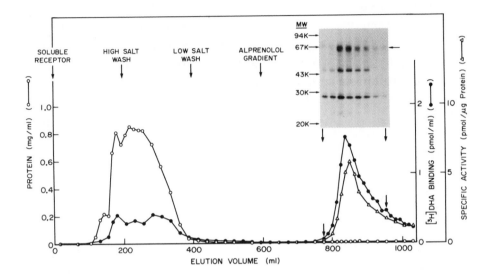

Fig. 1A. Sepharose–alprenolol chromatography of digitonin-solubilized hamster lung beta-adrenergic receptor activity. Approximately 340 pmol of digitonin-solubilized hamster lung receptor (200 mL) was applied to a 200 mL column of Sepharose–alprenolol previously equilibrated with 100 mM NaCl, 10 mM Tris-HCl, 0.05% digitonin, 2 mM EDTA, pH 7.2. The column was then washed at 4°C with 300 mL of 500 mM NaCl, 50 mM Tris-HCl, pH 7.2, 2 mM MgCl$_2$, 0.5% digitonin, followed by 300 mL of low-salt low-detergent buffer (100 mM NaCl, 10 mM Tris-HCl, pH 7.2, 2 mM MgCl$_2$, 0.05% digitonin). Elution was carried out with a gradient of 0–40 µM (±)alprenolol in the above low-salt, low-detergent buffer. Individual fractions (12.9 mL) were collected and assayed for receptor by [^{3}H]dihydroalprenolol binding (●), and for protein by the Amidoschwarz assay (0). These results were used to calculate specific activities for the eluted receptor (Δ). Additionally, 300 µL aliquots of several alprenolol eluted fractions were iodinated by the chloramine-T method and electrophoresed on a 10% SDS polyacrylamide gel. The autoradiogram of the dried gel is shown in the inset at the upper right. The arrow indicates where hamster lung beta-2 adrenergic receptor covalently labeled with the photo-affinity probe [^{125}I] pABC migrated in this experiment. The mol wt standards (MW) are shown × 1000. The alprenolol eluted receptor activity in the experiment shown (160 pmol) represents a 47% recovery of the applied digitonin-solubilized receptor activity. These data are taken from Benovic et al. (1984).

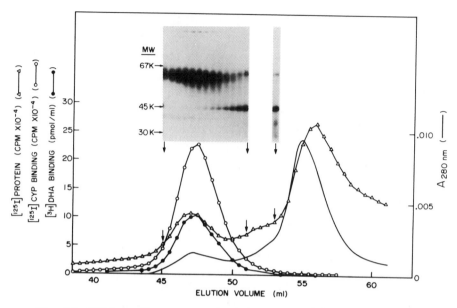

Fig. 1B. High performance steric exclusion chromatography of affinity-purified hamster lung receptor activity. Receptor containing fractions after Sepharose–alprenolol chromatography were concentrated by ultrafiltration, using an Amicon concentration cell and YM-30 membrane. The concentrated receptor was chromatographed on two TSK-4000 and one TSK-3000 steric exclusion columns, tandem-linked (total vol = 72 mL) with a buffer consisting of 0.1% digitonin, 100 mM Tris-SO$_4$, pH 7.5, at a flow rate of 1 mL/min with 0.5 mL fractions being collected 36 min after sample injection. Receptor activity was located by [^{3}H]dihydroalprenolol binding (●), or by chromatography of an aliquot of the affinity column eluate that had been incubated with [^{125}I]CYP prior to chromatography (o). The absorbance at 280 nm (-) is also shown in this profile. Additionally, an aliquot of the affinity column eluate concentrate was iodinated prior to chromatography. After chromatography, fractions were counted (Δ), and 10 μL aliquots were electrophoresed on an 8% SDS polyacrylamide gel. The resulting autoradiogram (top inset) is shown for the corresponding fractions. The mol wt standards (MW) are shown × 1000. Data are taken from Benovic et al. (1984).

(Nambi et al., 1984), whereas the beta-1 adrenergic receptor of the turkey erythrocyte showed two bands of lower mol wt (40 and 45 kDa) (Shorr et al., 1982; Rashidbaigi and Ruoho, 1982; Burgermeister et al., 1982). It was found that in most tissues in which bands of lower mol wt were observed, their presence could be significantly decreased by controlling proteolysis during the preparation of plasma membranes (Benovic et al., 1983). Eventually, it was shown that the same polypeptide that was purified by affinity and molecular sieve chromatography was the same as the one that was identified by photoaffinity labeling (Shorr et al., 1982a; Benovic et al., 1984).

The approaches described here for the characterization of the beta-adrenergic receptor have been widely useful in the characterization of several other G-protein coupled receptors. In a similar fashion the turkey and mammalian beta-1 adrenergic receptors (Shorr et al., 1982; Homcy et al., 1983; Bahouth and Malbon, 1987), the hamster alpha-1 adrenergic receptor (Lomasney et al., 1986), the human platelet alpha-2 adrenergic receptor (Regan et al., 1986), the porcine brain muscarinic receptor (Haga and Haga, 1985; Peralta et al., 1987), as well as the D_1 and D_2 dopamine receptor (Gingrich et al., 1988; Senogles et al., 1988), have all been biochemically characterized using specific affinity chromatography procedures and specific affinity and photoaffinity probes.

1.2. Reconstitution of the Biological Function of the Beta-2 Adrenergic Receptor

Once a purified receptor protein was available, it was important to document whether that protein and that protein alone was able to confer beta-adrenergic responsiveness to a cell or a system. Although it was evident from the photoaffinity labeling experiments that the purified peptide (mol wt 64,000) contained at least the ligand binding domain of the receptor, other receptor subunits could potentially be involved. Two approaches were used to this end. Purified beta-2 adrenergic receptor incorporated into phospholipid vesicles were fused with cells devoid of beta-adrenergic receptor, and shown to confer onto these cells beta-2 adrenergic stimulation of adenylyl cyclase (Cerione et al., 1983) (Fig. 2). The other approach was to reconstitute in phospholipid vesicles, using isolated components, the various protein interactions known to take place in the membrane. Thus, it was shown that purified beta-adrenergic receptor was capable of interacting pro-

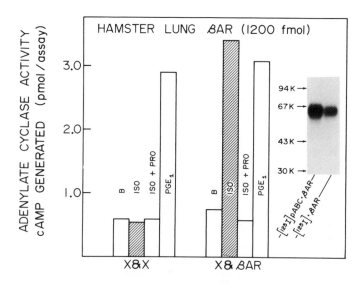

Fig. 2. Reconstitution of beta-2 adrenergic receptor responsiveness into Xenopus laevis erythrocytes. Purified hamster lung beta-2 adrenergic receptors were reconstituted in phospholipid vesicles and fused with whole erythrocytes by the propylene glycol procedure of Eimerl et al. (1980). Adenylyl cyclase activity and its responsiveness to various agents was assessed in plasma membranes. The data show that insertion of the pure beta-2 adrenergic receptor into these membranes confers responsiveness to the beta-agonist isoproterenol that can be specifically blocked by the antagonist propranolol. The inset shows an autoradiogram of an SDS-PAGE of the purified receptor preparation, revealed by radioiodination or photoaffinity labeling. Results are taken from Cerione et al. (1983).

ductively with the isolated protein G_s (Brandt et al., 1983; Cerione et al., 1984a) as evident by the ability of the receptor to mediate an agonist stimulation of the GTPase activity in, or [^{35}S]GTPγS binding to the G-protein. Ultimately, it was possible, using isolated or resolved components, to show that purified beta-adrenergic receptor could mediate agonist stimulation of adenylyl cyclase in a complete reconstituted system (Cerione et al., 1984b; May et al., 1985; Feder et al., 1986). These studies documented that the purified receptor and it alone was responsible for conferring to the other elements of the signal transduction beta-adrenergic responsiveness.

1.3. Elucidation of the Primary Structure of the Beta-Adrenergic Receptor

Until early 1986, no sequence information on any member of the G-protein coupled receptor class was available save for the structure of rhodopsin, the visual light pigment that, in a sense, represents a prototypic G-protein coupled receptor. Our laboratory succeeded in purifying several nanomoles of hamster lung beta-2 adrenergic receptor, using the procedures described above (Benovic et al., 1984). The receptor was fragmented by treatment with cyanogen bromide, and ultimately several peptides were isolated by reverse phase HPLC. Five of these were successfully sequenced, yielding stretches ranging from nine amino acid residues to 34 residues. A hamster genomic library was successfully screened with oligonucleotides based on the longest of these sequences, and several overlapping clones were obtained. Surprisingly, these genomic clones encoded, in a single continuous open reading frame uninterrupted by introns, a 418 amino acid protein (Dixon et al., 1986). The deduced sequence contained all five of the previously sequenced peptides. Moreover, since antibodies directed at one of the sequenced peptide were able to immunoprecipitate ligand-bound receptor, there was reason for great confidence that the protein that had been cloned in fact represented the hamster beta-2 adrenergic receptor. Interestingly, subsequent work from several laboratories, including our own, has demonstrated that the genes for several G-protein coupled receptors, in particular for several of the adrenergic receptors, lack introns in their coding blocks (Lefkowitz and Caron, 1988). This suggests that these genes may have arisen as processed genes from other related genes. However, it is also true that several receptors of this family do contain introns within their coding block (Bunzow et al., 1988; Cotecchia et al., 1988).

Several remarkable features emerged from a consideration of the deduced amino acid sequence of the beta-2 adrenergic receptor. First, was that hydrophobicity plots indicated the presence of seven stretches of very hydrophobic residues, each generally about 25 amino acids in length and thus, potentially being long enough to represent transmembrane spanning domains. Second, limited but significant amino acid homology was detected with bovine rhodopsin. Rhodopsin, like its prokaryote homolog bacteriorhodopsin, shares this seven membrane spanning domain arrangement (Applebury and Hargrave, 1986). The sequence homology between beta-2 adrenergic receptor and rhodopsin is about 18–20% overall, but is as high as 30% within the various membrane spanning domains. Figure 3 depicts the topography and amino acid sequences of the beta-2 adrenergic receptor and rhodopsin. It has now become apparent that all members of the G protein-coupled receptor family (about 25 cloned so far) share this highly conserved structural arrangement (O'Dowd et al., 1989b). Invariably, amino acid sequence similarity is greatest in the putative membrane spanning domains.

Other features of note from the deduced sequence are the presence of two consensus sites for *N*-linked glycosylation, toward the amino terminus, two consensus sites for phosphorylation by the cAMP-dependent protein kinase, both located on what are presumed to be cytoplasmic domains, and a very serine and threonine rich carboxyl terminus.

Shortly after the initial report (Dixon et al., 1986) on the cloning of the beta-2 adrenergic receptor, Yarden et al. (1986) reported the cloning of a cDNA for the "beta-1-like" adrenergic receptor derived from turkey erythrocytes. The successful cloning of this beta-1 adrenergic receptor was also based on the availability of partial amino acid sequence from the purified protein (Yarden et al., 1986). Although somewhat larger than the mammalian beta-2 adrenergic receptor (483 amino acid residues), its deduced sequence indicated all of the same structural features described above for the beta-2 adrenergic receptor, including the seven presumed membrane spanning domains, and a long cytoplasmic carboxyl terminal tail. Subsequent cloning of the human beta-1 adrenergic receptor by Frielle et al. (1987) further confirmed these features. Moreover, the human beta-1 adrenergic receptor was significantly closer in sequence to the turkey beta-receptor than to the hamster beta-2 adrenergic receptor (70% overall sequence identity for human beta-1 vs turkey beta-1; 50% identity for human beta-1 vs human beta-2). These data provide strong evidence that the turkey erythrocyte beta-adrenergic receptor, the pharmacological properties of which have long been viewed as atypical for a beta-1 adrenergic receptor, is in fact, a close homolog of the mammalian beta-1 adrenergic receptor.

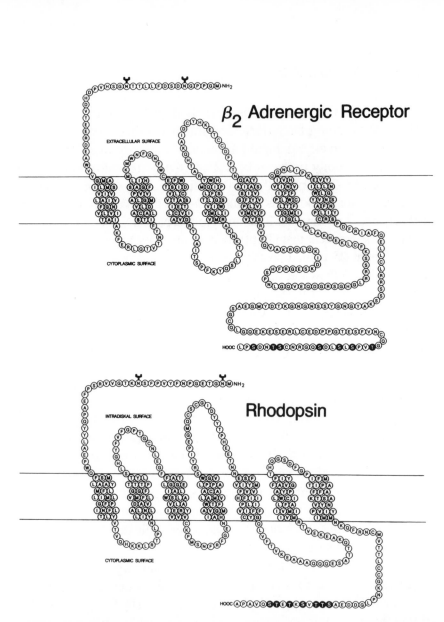

Fig. 3. Primary amino acid sequence of the hamster beta-2 adrenergic receptor and the "light receptor" rhodopsin. The sequences are arranged with respect to the plane of the plasma membrane (horizontal lines) as they have been postulated to exist in the bilayer. The extracellular or intradistal and cytoplasmic surfaces are indicated. Y shape symbols indicate sites of *N*-glycosylation toward the amino terminus of each protein. Amino acid residues, shown in dark circles, indicate segments on the beta-2 adrenergic receptor and rhodopsin carboxyl terminal that are particularly rich in serine/threonine residues. Taken from Benovic et al. (1986).

In general, within the adrenergic receptor family, approx 40–50% of the amino acid residues are identical. When comparing these receptors with more distantly related receptors such as, for example, the muscarinic cholinergic receptors, this figure drops to 30–35%; whereas, for the even more distantly related rhodopsin, the identity may drop significantly below that level. On the other hand, for subtypes within a closely related group (subsubtypes) such as beta-1 vs beta-2, or the alpha-2A and alpha-2B adrenergic receptor subtypes (Regan et al., 1988), 70–75% of the amino acids will be identical. As noted above, the sequences of the hydrophilic segments that connect the membrane spanning domains appear to be much less conserved. These criteria might become useful predictors for identifying new members of the large G-protein receptor family, as new sequences become available by recombinant DNA approaches (Libert et al., 1989).

1.4. Membrane Topography of the Beta-2 Adrenergic Receptor

Given the presence of seven hydrophobic domains in the receptor, deduced from the hydrophobicity plots, and given current understanding of the structure of bacteriorhodopsin and rhodopsin, both of which appear to contain seven membrane spanning domains (Applebury and Hargrave, 1986), it was quite reasonable to propose, as we did in 1986, that the beta-2 adrenergic and related receptors in fact span the membrane seven times. It was further proposed that the glycosylated amino acid terminus was extracellular, and that the carboxyl terminal tail was intracellular. Although to date no hard physical data, as it exists for rhodopsin (reviewed in Applebury and Hargrave, 1986), have been developed to substantiate this model (the problems being largely the technical ones of obtaining enough receptor for such studies, *see below*), several studies have presented indirect but compelling evidence for the arrangement, schematically indicated in Fig. 3. Dohlman et al. (1987) performed a study in which limited proteolysis of beta-2 adrenergic receptors was used to examine the organization of the receptor. Trypsinization of beta-2 adrenergic receptor reconstituted in phospholipid vesicles yielded two insoluble integral membrane domains of approx 38,000 and 26,000 kDa, respectively.

Identical results were obtained using intact cells, indicating that the cleavage site of the receptor is accessible at the extracellular surface of the plasma membrane. The amino terminal domain of 38,000 kDa was found to contain the site of incorporation of a photoaffinity probe, as well as the sites of glycosylation, whereas the carboxyl terminal 26,000 kDa fragment contained all the sites of in vitro phosphorylation by the cAMP-dependent protein kinase and the receptor specific beta-adrenergic receptor kinase. The various features delineated, including the length of the carboxylpeptidase Y sensitive carboxyl terminus, the extracellular location of the trypsin sensitive site, the location of the sites of phosphorylation and glycosylation, all constrain the receptor to a rhodopsin-like structure with multiple membrane spanning segments (Dohlman et al., 1987).

Moreover, recently, Wang et al. (1989) have obtained still more direct evidence for the proposed seven transmembrane arrangement of the receptor, utilizing antipeptide antibodies directed at various regions of the hamster beta-2 adrenergic receptor. Antibodies were prepared against 11 peptides corresponding to each of the hydrophilic sequences of the receptor (*see* Fig. 3). Chinese hamster ovary cells stably transfected with the beta-2 adrenergic receptor, and expressing quite a high number of sites (about 2 million/cell) were then used for *in situ* localization of the various sequences used as antigens. Indirect immunofluorescence of intact and permeabilized cells with the various sites directed antipeptide antibodies was used to assign the general topography of each of the hydrophilic sequences. In all cases, the results supported the seven membrane spanning domain arrangement and extracellular and intracellular assignment of sequences as indicated above.

1.5. Ligand Binding to the Receptor

Intensive efforts in several laboratories over the past several years have begun to delineate the major structural features of the receptors that are responsible for carrying out the two primary functions of ligand binding on the one hand, and activation of G proteins on the other. The first study in this regard was published by Dixon et al. (1987b). One reasonable speculation had been that the extracellular hydrophilic domains might be importantly involved in determining ligand binding properties, because of their obvious accessibility at the cell surface. However, Dixon et al. (1987a,b) found that deletion of much of these sequences led to absolutely no alterations in the ligand binding properties of the receptors. In contrast, any deletion that involved membrane spanning domains led to almost complete loss of binding, decreased expression, or both. This latter point is also made in a study by Kobilka et al. (1987).

The importance of the membrane spanning domains in determining the ligand binding properties of the receptors was also established by Kobilka et al. (1987) and Frielle et al. (1988), by the use of chimeric receptors. In an initial study, a series of chimeric receptor genes was constructed from portions of the human beta-2 and alpha-2 adrenergic receptors. Whereas both of these receptors bind catecholamines, the specificity of the binding for both agonists and antagonists is distinct. It was found that interchange of the various membrane spanning domains led to progressive alterations in the properties of agonist binding, confirming the importance of these membrane spanning segments in determining the properties of the ligand binding site. Interestingly, up to five of the seven transmembrane spans of the beta-adrenergic receptor (TM-I to V) could be replaced by alpha-2 receptor segments without any discernible changes in the antagonist binding properties of the beta-2 adrenergic receptor. This indicated that the specific structural determinants of agonist and antagonist binding, although overlapping, are not identical. In fact, the sixth and seventh transmembrane spans for the beta-2 receptor appear to be particularly important in determining the specificity for binding antagonists. Frielle et al. (1988), in comparing properties of various chimeras constructed from the human beta-1 and beta-2 adrenergic receptor, found that multiple membrane spanning domains contributed to the ligand binding properties; in agreement with Kobilka et al. (1987). It was also found in that study that transmembrane spanning domain 4 was particularly important for determination of agonist binding specificity with beta-1 or beta-2 subtype, whereas antagonist binding determinants were scattered in several other membrane spanning domains.

Several studies, using site directed mutants in which various amino acid residues have been altered, have strongly contributed to the notion that the ligand binding site resides within the transmembrane domains. For example, Strader et al. (1987b, 1988) and Fraser (1989) have demonstrated that Asp 113 serves as a counterion to the protonated amino group of beta-adrenergic ligands. Substitution of Asp 113 led to dramatic decreases in both agonist and antagonist binding. Consequently, when Asp 113 was changed to Glu 113, 100-fold higher concentrations of agonist were required for maximal adenylyl cyclase activation. On the other hand, when Asp 113 was changed to Asn 113 10,000-fold higher concentrations of agonist were necessary for this activation. Interestingly, all of the known receptors that have been sequenced in this family, that bind cationic amines such as the alpha-adrenergic receptors, the beta-adrenergic receptors, and the muscarinic cholinergic receptors all contain an Asp at the position equivalent to 113 in the beta-2 adrenergic receptor. In addition, an Asp residue that is highly conserved at position 79 also appears to be involved in agonist, but not antagonist, binding. Thus, conversion of this aspartate to an alanine leads to a tenfold reduction in agonist affinity with no change in antagonist affinity (Strader et al., 1988).

Recently, Strader et al. (1989) have identified by site-directed mutagenesis, two serine residues, serine 204 and 207 in the hamster beta-2 adrenergic receptor that are critical for the binding of agonists and activation of the transduction process. These two serine residues are located in the fifth transmembrane domain of the beta-2 adrenergic receptor and so far, have been found to be conserved in every cloned receptor that binds a catecholamine. Removal of the hydroxyl side chain from either residue attenuates the activity of classical full agonists to a similar extent as the one that is obtained when the hydroxyl groups on agonist ligands are modified or removed. With agonist ligands, this gives rise to classical partial agonists such as, soterenol and zinterol (MJ-9184-1) (Mukherjee et al., 1976). These data strongly suggest that hydrogen bonding between the hydroxyl side chain of these serines, and the catechol hydroxyls, may play a crucial role in agonist binding (Strader et al., 1989).

Although still quite preliminary, current models suggest that the ligand binding site of the beta-2 adrenergic receptor, and presumably, other members of this family of receptors, is formed by cooperation between several of the proposed transmembrane helices. Several of these helices would be envisaged as forming a pocket within the membrane into which various ligands fit. The exact way in which this is accomplished however is not currently known.

Another approach that has been taken to determine the sites in the receptor that might be involved in binding ligands has been that of photoaffinity or affinity labeling of receptors, followed by fragmentation by chemical or enzymatic means, and then sequencing the labeled peptides. In such a study carried out by Dohlman et al. (1988), using a bromacetylated affinity reagent and the hamster beta-2 adrenergic receptor, a single labeled peptide was isolated that, by sequencing, was shown to correspond to residues 83–96 located in the outer part of the second membrane spanning domain of the beta-2 adrenergic receptor. A similar study was carried out using the turkey erythrocyte beta-1 adrenergic receptor by Wong et al. (1988). Two different affinity labeling reagents were utilized. In that case, two different sites were labeled. One appeared to be located somewhere between membrane spanning domains 2–4, whereas the other was clearly located in membrane spanning domain 7 in a region close to that where retinal is covalently attached to opsin. Interestingly, when similar studies were carried out utilizing the M1 muscarinic receptor (Curtis et al., 1989), or the alpha-2 adrenergic receptor (Matsui et al., 1989), labeling of the third and fourth transmembrane domains, respectively, were found. Figure 4 indicates in a schematic fashion the various transmembrane helices that have been labeled in these studies, using different receptors. These studies clearly confirm that the membrane spanning domains are invariably the regions of the molecules in this receptor class that interact with ligand. However, it seems equally clear that the sites for binding the various ligands vary from receptor to receptor, that multiple sites are likely involved, and that current work has relied almost exclusively on the use of covalently inserting *antagonist*, rather than *agonist* ligands.

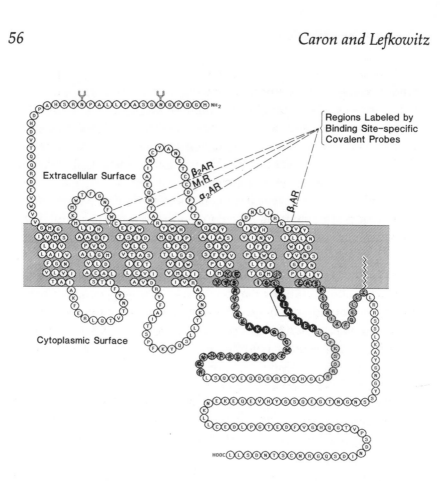

Fig. 4. Primary amino acid model of the human beta-2 adrenergic receptor highlighting functional regions of the receptor. The sequence is organized according to the model shown in Fig. 3. The brackets indicate regions of the transmembrane domains that have been covalently labeled in the beta-1, beta-2, and alpha-2 adrenergic receptors, as well as the M_1 muscarinic acetylcholine receptor, by various affinity and photoaffinity probes. These studies are described in the text. The light and dark shaded areas of sequence in the amino and carboxyl terminal part of the third cytoplasmic loop and the amino terminal of the carboxyl terminus indicate regions that have been implicated by mutagenesis studies as being involved in the interaction of the receptor with the G-protein. The wavy line at cys[341] of the beta-2 adrenergic receptor indicates the site at which palmitoylation of the receptor is thought to occur (O'Dowd et al., 1989a).

Moreover, recently, several groups provided evidence that several extracellular disulfide bonded cysteines in the beta-2 adrenergic receptor may be involved in ligand binding. The presence of disulfide bonds at or near the ligand binding site had been indicated by the fact that DTT, which reduces disulfides, has a marked inhibitory effect on beta-adrenergic receptor binding (Vauquelin et al., 1979; Dohlman et al., 1990). Kinetic studies have suggested that the effect of DTT appears to be "competitive," and moreover, that it can be specifically protected by beta-adrenergic receptor ligands, both agonists and antagonists (Clark et al., 1983). Moreover, disulfide bond reduction is required for limited proteolysis of affinity ligand-labeled beta-2 adrenergic receptor. Kinetic studies further suggest that at least two pairs of disulfide bonded cysteines may be essential for normal ligand binding. Site directed mutagenesis studies have indicated four cysteines that are critical for normal ligand binding, and for the proper expression of functional beta-adrenergic receptors at the cell surface (Chung et al., 1988; Dohlman et al., 1990). All four of these cysteines, numbers 106, 184, 190, and 191, are in the extracellular hydrophilic loops that connect the transmembrane segments. These findings suggest that in addition to the well documented involvement of the membrane spanning domains, the hydrophilic extracellular domains also may function to form or stabilize an active conformation of the ligand binding site.

1.6. Receptor Domains Involved in G-Protein Coupling

The structural domains of the beta-adrenergic receptor that are involved in the important function of coupling to G-protein and thereby effector activation, have been studied using a variety of approaches. These include limited proteolysis, site directed mutagenesis, and construction of chimeric receptors. Gratifyingly, all the experimental approaches converge on the same conclusions. Early studies by Rubenstein et al. (1987) documented that limited proteolysis of the turkey erythrocyte beta-adrenergic receptor, sufficient to remove a major portion of the third cytoplasmic loop, and carboxyl terminal tail, still did not impair coupling function as assessed by the ability to activate adenylyl cyclase. These studies suggested that much of the carboxyl terminal tail and the central part of the third cytoplasmic loop (*see* Fig. 4) were not involved in coupling to G_s.

Chimeric alpha-2–beta-2 adrenergic receptors into which sequences from the beta-2 adrenergic receptor had been inserted served to provide striking evidence for the localization of the sites of coupling to G_s. For example, in an initial study (Kobilka et al., 1988), it was shown that when the entire third cytoplasmic loop, as well as sequences from the fifth and sixth transmembrane spanning domains, as well as the second extracellular loop of the human beta-2 adrenergic receptor (segment extending from residue Tyr 174 to Val 295), was inserted into the alpha-2 adrenergic receptor, the ability of this chimera to interact with the adenylyl cyclase system was dramatically switched from inhibitory (alpha-2) to stimulatory (beta-2). Whereas the level of stimulation was only about 30–35% of that achieved with the wild type beta-2 adrenergic receptor, these early results clearly suggested a key role for this region of the beta-adrenergic receptor in activating G_s. As would be expected, the pharmacological specificity for activation of adenylyl cyclase through the chimeric receptor closely matches that of an alpha-2 rather than a beta-2 adrenergic receptor, since most of the transmembrane spanning segments were derived from the alpha-2 adrenergic receptor. Subsequent studies have indicated that substitution of the third cytoplasmic loop of the beta-2 adrenergic receptor alone without the flanking membrane segments is sufficient to confer adenylyl cyclase stimulatory activity (unpublished observations).

The exact regions of the third cytoplasmic loop that are required to convey the ability to interact with G_s have been further probed by site directed mutagenesis, substitutions, and deletions, by several laboratories. O'Dowd and Hnatowich et al. (1988) pinpointed residues at the amino and carboxyl terminal boundaries of the third cytoplasmic loop of the human beta-2 adrenergic receptor as being those responsible for interaction with G_s. It also appeared that residues in the most proximal portion of the cytoplasmic carboxyl terminal tail were involved in forming a binding surface for interaction of the receptor with G_s. Interestingly, these regions of the beta-2 adrenergic receptor (those regions of the third cytoplasmic loop and carboxyl terminal tail in closest apposition to the plasma membrane) appear to be the most highly conserved cytoplasmic regions of the receptor among the different adenylyl cyclase coupled receptors that have thus far been cloned.

Similar conclusions have been reached by Dixon, Strader, and colleagues based on deletion mutants of the third intracellular loop (Dixon et al., 1987a; Strader et al., 1987a). Interestingly, the various regions of the third cytoplasmic loop that appear to be involved in G_s coupling would be predicted to form almost perfect amphipathic helices that may be important for the interaction with G_s. In this context, it has been noted that mastoparan, a small peptide that assumes an amphipathic helical conformation when present in phospholipid micelles, is able to activate directly G-proteins. It is possible that this peptide may mimic the action of agonist occupied receptors. Based on substitutions of various alpha-2 adrenergic receptor sequences into the third loop of the beta-2 adrenergic receptor, O'Dowd et al. (1988) have speculated that the role of the amino terminus of the third loop is to bind common G-protein elements, whereas the actual specificity of which G-protein is bound may reside in the carboxyl terminus. Most of the evidence accumulated to date would suggest that the first and second cytoplasmic loops appear to play much less important and/or specific roles in coupling to G-proteins.

Although much less work has been performed with other receptors, the available information appears to confirm the importance of sequences within the amino and/or carboxyl terminus of the third cytoplasmic loop as being important for G-protein interaction. Thus, Kubo et al. (1988) showed that switching the third cytoplasmic loop between M_1 and M_2 muscarinic receptors was able to switch completely the physiological effect of the two receptors. These receptors respectively activate phospholipase C or inhibit adenylyl cyclase. Subsequently, Wess et al. (1989) were able to localize to a stretch of only 17 residues in the amino terminus of the M_1 muscarinic receptor, a structure that could completely switch the specificity of coupling of the M_2 receptor. Finally, Cotecchia et al. (1990) have shown that when the third cytoplasmic loop of the alpha-1 adrenergic receptor is inserted into the beta-2 adrenergic receptor, the resulting receptor activates phospholipase C (as with the alpha-1 receptor) but with a beta-2 adrenergic receptor agonist specificity. Thus, all of these findings are quite consistent, and with a few exceptions, seem to be applicable to several of the G-protein coupled receptors. Regions of the beta-2 adrenergic receptor that appear to be most involved in its coupling to G_s are indicated by the shading in Fig. 4.

To date, no adrenergic receptor chimera that is able to activate adenylyl cyclase to the same extent as wild type beta-2 adrenergic receptor has been reported. The possibility remains open that other cytoplasmic domains of the beta-2 adrenergic receptor may be contributing in significant ways to the activation process. There is also essentially no information available as to how the binding of ligands to the transmembrane binding pocket is able to produce conformational changes, presumably, in the third cytoplasmic loop, that lead to productive interaction with G_s.

1.7. Structural Basis of Receptor Regulation

In addition to ligand binding and G-protein coupling, the beta-adrenergic and related receptors undergo a variety of regulatory processes that rapidly control their function. Most notable among these is the process of desensitization by which continuous stimulation of a receptor leads to attenuation of receptor function. This complicated and important subject is dealt with by Perkins et al. (1990) in Chapter 3. Here, we provide only a brief summary of what is known about the structural basis of these processes.

Three processes have been implicated as potentially playing roles in the desensitization of beta-adrenergic receptors. These are referred to as uncoupling, sequestration, and downregulation. Receptor uncoupling refers to processes that, without changing either the number or subcellular distribution of the receptors, lead to a decreased efficiency in their capacity to activate G_s, thus, diminishing the ultimate biological response. Covalent modification of the receptors by phosphorylation reactions appears to be the major mechanism operating to rapidly uncouple the receptors from G_s. Whereas several protein kinases have been implicated as playing roles in these processes, the most thoroughly studied have been the cyclic AMP-dependent protein kinase, and a new cAMP-independent kinase termed βARK (for beta-adrenergic receptor kinase). The cyclic AMP-dependent protein kinase is known to phosphorylate proteins at well delineated consensus sequences. There are two consensus sequences for phosphorylation by protein kinase A on the beta-2 adrenergic receptor. As shown schematically in Fig. 5, one of these is in the distal part of the third cytoplasmic loop, and the other in the proximal part of the carboxyl terminal tail. Studies with synthetic peptides have indicated that the site in the third cytoplasmic loop is much the preferred substrate for protein kinase A (Blake et al., 1987; Bouvier et al., 1989). A variety of studies have, in fact, documented that the receptor is phosphorylated by protein kinase A during some form of desensitization (heterologous desensitization, *see* Chapter 3 for details), and that such phosphorylation functionally impairs the ability of the receptor to couple to G_s. It is interesting that both protein kinase A consensus phosphorylation sites are located adjacent to key regions of the receptor molecules that are involved in coupling to G_s as described above. Thus, it is not difficult to imagine that the charge distribution of these regions might be significantly altered by phosphorylation events, thus leading to an impairment in coupling.

The beta-adrenergic receptor kinase, in contrast, appears to phosphorylate the receptor primarily on a very serine and threonine rich domain, located at the very C terminus of the receptor. These regions are remote from those thought to be involved in coupling to G_s. However, in analogy with the reactions involved in regulation of rhodopsin function by rhodopsin kinase, it appears that an additional molecule, analogous to retinal 48K protein, is required for beta-adrenergic receptor kinase-mediated beta-adrenergic receptor desensitization. Similar to the arrestin-like molecule in the retina, it binds to beta-adrenergic receptor kinase-phosphorylated receptors, thereby somehow uncoupling them from G_s. The roles of protein kinase A and beta-adrenergic receptor kinase in rapid beta-adrenergic receptor uncoupling have been documented both by site-directed mutagenesis techniques, as well as the use of highly specific enzyme inhibitors.

Receptor sequestration refers to an agonist stimulated process in which the receptors are rapidly internalized and sequestered away from

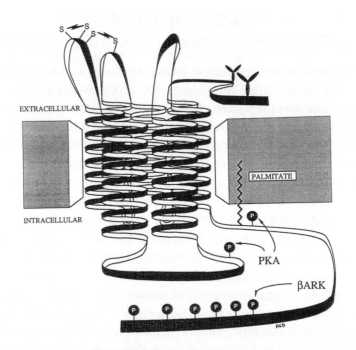

Fig. 5. Schematic transmembrane model of the beta-2 adrenergic receptor depicting various posttranslational modification of the receptor. On the
extracellular surface are shown two disulfide bonds presumably implicated
in maintaining the integrity of the ligand binding site, and the two sites of
glycosylation. On the intracellular surface are shown the sites of palmitoylation and phosphorylation by two different kinases, cyclic AMP-dependent
protein kinase A (PKA), and beta-adrenergic receptor kinase (βARK).

the cell surface. Little is known about the intracellular compartments
to which they travel, but it is clear that when agonist is removed, the
receptors can recycle to the plasma membrane and then function normally. Little is understood about the details of the process, or about
its actual role in physiological desensitization reactions. Moreover,
the structural basis of this process is not known. It seems clear that
phosphorylation of the receptors by either protein kinase A or beta-
adrenergic receptor kinase is not the trigger for the sequestration event.

    Downregulation of the receptors refers to a more slowly evolv-
ing process (1–24 h) that leads to the removal of the receptors from the

plasma membrane and their loss within the cell by unknown mechanisms. Again, essentially nothing is known about the structural basis of receptor downregulation, but recent evidence by Campbell et al. (1990) suggests that physical coupling of receptors to G_s may be a necessary prerequisite for the downregulation process. Clearly, much work will be necessary to sort out the structural basis for these receptor regulatory processes.

1.8. Covalent Modifications of the Beta-2 Adrenergic Receptor

1.8.1. Phosphorylation

As described above, receptor phosphorylation by various protein kinases appears to play an important role in regulating receptor function in desensitization.

1.8.2. Glycosylation

With few exceptions, G-protein-coupled receptors appear to be glycosylated on their amino termini. The functional significance of this glycosylation is not clear. Earlier studies had shown that enzymatic removal of carbohydrate substituents from beta-2 adrenergic receptors did not interfere with either ligand binding, or adenylyl cyclase activation (Benovic et al., 1987). More recently, mutagenesis of the glycosylation sites that completely prevents glycosylation, has also been shown to have no effect on the ability of the nonglycosylated receptors to either bind ligands or activate adenylyl cyclase (Dixon et al., 1987a). Thus, the functional role, if any, of the glycosylation remains to be determined. Speculations have centered on a function concerning intracellular trafficking of the receptors, and possibly stable maintenance of the receptors at the cell surface.

1.8.3. Disulfide Bond Formation

The hamster beta-2 adrenergic receptor contains 15 cysteine residues, the overwhelming majority of which appears to be involved in disulfide bond formation (Dohlman et al., 1990). As described above, a number of these disulfides may be importantly involved in helping to form, support, or stabilize an active conformation of the receptor that is required, both for normal expression in the plasma membrane, as well as for ligand binding function.

1.8.4. Palmitoylation

Recent reports (Ovchinnikov et al., 1988) that the visual pigment rhodopsin is covalently modified by fatty acylation, specifically, palmitoylation, at the carboxyl terminal tail, prompted us to examine the question of palmitoylation of the beta-2 adrenergic receptor. Rhodopsin has been shown to be palmitoylated at two vicinal cysteines, 322 and 323 (Ovchinnikov et al., 1988). One of these is conserved in an equivalent position in the carboxyl terminal tail of the human beta-2 adrenergic receptor (Cys 341), as well as in the carboxyl terminal tails of most members of the G-protein-coupled receptor family. We found that the beta-2 adrenergic receptor was, in fact, palmitoylated, and that mutation of cysteine 341 to glycine resulted in a nonpalmitoylated form of the receptor that exhibited a drastically reduced ability to mediate isoproterenol stimulation of adenylyl cyclase. The functional impairment of this mutated receptor was also reflected in its markedly reduced ability to form a guanine nucleotide sensitive high affinity state for binding agonist (O'Dowd et al., 1989b). These results suggest that a fourth intracellular loop of the beta-2 adrenergic receptor may exist by virtue of palmitoylation of Cys 341. As schematically indicated in Fig. 5, it may serve to anchor the amino terminal portion of the carboxyl terminal tail of the receptor in the plasma membrane. This putative fourth intracellular loop might be critical for formation of the functional G_s binding site on the receptor.

1.9. The Beta-3 Adrenergic Receptor, a Novel Beta-Adrenergic Receptor Subtype

The classification of beta-adrenergic receptors into beta-1 and beta-2 adrenergic subtypes has been in place for almost 25 y. However, from time to time, additional beta-adrenergic receptor subtypes have been proposed to be implicated in the control of various metabolic processes. One that has been controversial is a so-called beta-3 adrenergic receptor, similar in pharmacology to the beta-1 adrenergic receptor, and that has been suggested to regulate the lipolytic effect of catecholamines in human adipose tissue. Recently, Emorine et al. (1989) have isolated a cDNA that encodes a third beta-adrenergic receptor subtype from the human genome. Amino acid sequence of the encoded receptor is about 50% identical to that of either the human beta-1 or beta-2 adrenergic receptor, respectively. As expected, it contains all the classical landmarks of a G-protein-coupled receptor, including the seven membrane spanning domain arrangement. Expression of the cDNA in eukaroytic cells leads to expression of a receptor that binds beta-adrenergic receptor ligands with the atypical properties expected of the so-called "beta-3 adrenergic receptor." Interestingly, novel beta-adrenergic receptor agonists having high thermogenic, antiobesity, and antidiabetic activities in animal models are among the most potent stimulators of the beta-3 adrenergic receptor expressed in these cells.

The cloning of the gene for a beta-3 adrenergic receptor underscores the potential power of molecular cloning techniques to reveal previously unexpected molecular heterogeneity of receptor structure. As further examples, one might also cite the successful cloning of five different muscarinic cholinergic receptor cDNAs (Bonner, 1989), several different alpha-2 adrenergic receptor cDNAs, as well as several different alpha-1 adrenergic receptor cDNAs. Only a few years ago, pharmacological characterizations clearly identified only two muscarinic receptors, one alpha-1 and one alpha-2 adrenergic receptor.

2. Future Perspectives

The successful cloning of the gene for the beta-2 adrenergic receptor has opened an explosive new era in the evolution of our understanding of the structure of hormone/neurotransmitter receptors. Whereas extraordinary amounts of information have been accumulated by various laboratories in the span of just a few years, one senses that these investigations are just at a beginning. Several new frontiers might be anticipated at this time. First, the discovery of unexpected molecular heterogeneity of these receptors by molecular cloning techniques opens the way to the development of potentially much more selective subtype specific drugs. Thus, the cDNAs for various novel receptor subtypes can be expressed in eukaryotic cells. These cells, now expressing a unique subtype of receptor, can be used as screening reagents to develop new and ever more selective pharmacological agents. Moreover, Northern blotting and related techniques can be used to determine the tissue distribution of the receptors as a prelude to trying to determine their physiological roles. The history of adrenergic pharmacology in the past has been that, every time new subtypes of receptors were discovered by pharmacological techniques, new subtype selective drugs were subsequently discovered. It is hoped and anticipated that the same will be true in the era of discovery of these receptors by molecular genetics.

A second frontier concerns the development of rational approaches to drug design, based on an ultimate understanding of the three-dimensional structure of beta-adrenergic and related receptors. At present, this goal is not feasible. The major limitation is the availability of sufficient quantities of receptor protein in order to perform the necessary physical measurement, specifically X-ray crystallography. Given the successful cloning of the genes and the rapid evolution of expression technologies, it can be hoped that over the next few years techniques will be developed for very high level expression of these receptors in appropriate systems, such as eukaryote or prokaryote cells. These systems may make available, for the first time, quantities of receptor protein that will be required for crystalization and X-ray analysis. Even with the availability of such quantities of receptor protein, however, this goal will be a daunting one, given previous difficulties in obtaining atomic resolution data of membrane-bound proteins. Although the task at hand will clearly be difficult and challenging, the ultimate gain in the understanding of the fundamental structure of these molecules and the possibilities that this will open for the development of new and potentially more useful therapeutic agents certainly makes it a worthwhile and important undertaking.

References

Applebury, M. L. and Hargrave, P. A. (1986) Molecular biology of the visual pigments. *Vision Res.* **26,** 1881–1895.

Bahouth, S. W. and Malbon, C. C. (1987) Human beta-adrenergic receptors: Simultaneous purification of beta-1 and beta-2 adrenergic receptor peptides. *Biochem. J.* **248,** 557–566.

Benovic, J. L., Stiles, G. L., Lefkowitz, R. J., and Caron, M. G. (1983) Photoaffinity labeling of mammalian beta-adrenergic receptors: Metal-dependent proteolysis explains apparent heterogeneity. *Biochem. Biophys. Res. Commun.* **110,** 504–511.

Benovic, J. L., Shorr, R. G. L., Caron, M. G., and Lefkowitz, R. J. (1984) The mammalian beta-2 adrenergic receptor: Purification and characterization. *Biochemistry* **23,** 4510–4518.

Benovic, J. L., Mayor, F., Jr., Somers, R. L., Caron M. G., and Lefkowitz, R. J. (1986) Light dependent phosphorylation of rhodopsin by the beta-adrenergic receptor kinase. *Nature* **322,** 869–872.

Benovic, J. L., Staniszewski, C., Cerione, R. A., Codina, J., Lefkowitz, R. J., and Caron, M. G. (1987) The mammalian beta-adrenergic receptor: Structural and functional characterization of the carbohydrate moiety. *J. Receptor Res.* **7,** 257–281.

Blake, A. D., Mumford, R. A., Strout, H. V., Slater, E. E., and Strader, C. D. (1987) Synthetic segments of the mammalian beta-adrenergic receptor are preferentially recognized by cAMP-dependent protein kinase and protein kinase C. *Biochem. Biophys. Res. Commun.* **147,** 168–173.

Bonner, T. I. (1989) New subtypes of muscarinic acetylcholine receptors. *Trends in Pharmacol. Sci. (Suppl: Subtypes Muscarinic Receptor IV)* **10,** 11–15.

Bouvier, M., Collins, S., O'Dowd, B. F., Campbell, P. T., DeBlasi, A., Kobilka, B. K., MacGregor, C., Irons, G. P., Caron, M. G., and Lefkowitz, R. J. (1989) Two distinct pathways for cAMP mediated downregulation of the beta-2 adrenergic receptor: Phosphorylation of the receptor and regulation of its mRNA level. *J. Biol. Chem.* **264,** 16786–16792.

Brandt, D. R., Asano, T., Pederson, S. E., and Ross, E. M. (1983) Reconstitution of catecholamine-stimulated guanosine triphosphate activity. *Biochemistry* **22,** 4357–4362.

Bunzow, J. R., Van Tol, H. V. M., Grandy, D. K., Albert, P., Salon, J., Christier, C. A., Machida, C. A., Neve, K. A., and Civelli, O. (1988) Cloning and expression of a rat D_2 dopamine receptor cDNA. *Nature* **336,** 783–787.

Burgermeister, W., Hekman, M., and Helmreich, E. J. M. (1982) Photoaffinity labeling of the beta-adrenergic receptor with azide derivatives of iodocyanopindolol. *J. Biol. Chem.* **257,** 5306–5311.

Burgermeister, W., Nassal, M., Wieland, T., and Helmreich, E. (1983) A carbine-generating photoaffinity probe for beta-adrenergic receptors. *Biochem. Biophys. Acta* **729,** 219–228.

Campbell, P. T., Hnatowich, M., Hausdorff, W. P., O'Dowd, B. F., Caron, M. G., and Lefkowitz, R. J. (1990) Mutations of the human beta-2 adrenergic receptor that impair coupling to G_s interfere with receptor downregulation but not sequestration (Submitted).

Caron, M. G. and Lefkowitz, R. J. (1976) Solubilization and characterization of the beta-adrenergic receptor binding sites of frog erythrocytes. *J. Biol. Chem.* **251,** 2374–2384.

Caron, M. G., Srinivasan, Y., Pitha, J., Kociolek, K., and Lefkowitz, R. J. (1979) Affinity chromatography of the beta-adrenergic receptor. *J. Biol. Chem.* **254,** 2923–2927.

Cerione, R. A., Codina, J., Benovic, J. L., Lefkowitz, R. J., Birnbaumer, L., and Caron, M. G. (1984a) The mammalian beta-2 adrenergic receptor: Reconstitution of functional interactions between pure receptor and pure stimulatory nucleotide binding protein of the adenylate cyclase system. *Biochemistry* **23,** 4519–4525.

Cerione, R. A., Sibley, D. R., Codina, J., Benovic, J. L., Winslow, J., Neer, E. J., Birnbaumer, L., Caron, M. G., and Lefkowitz, R. J. (1984b) Reconstitution of a hormone-sensitive adenylate cyclase system. *J. Biol. Chem.* **259,** 9979–9982.

Cerione, R. A., Strulovici, B., Benovic, J. L., Lefkowitz, R. J., and Caron, M. G. (1983) Pure beta-adrenergic receptor: The single polypeptide confers catecholamine responsiveness to adenylate cyclase. *Nature* **306,** 562–566.

Chung, F. Z., Wang, C-D, Potter, P. C., Venter, J. C., and Fraser, C. M. (1988) Site directed mutagenesis and continuous expression of human -adrenergic receptors: Identification of a conserved aspartate residue involved in agonist binding and receptor activation. *J. Biol. Chem.* **263,** 4052–4055.

Clark, R. B., Green, D. A., Rashidbaigi, A., and Ruoho, A. (1983) Effect of dithiothreitol on the beta-adrenergic receptor of S49 wild type and cyc- lymphoma cells: decreased affinity of the ligand-receptor interaction. *J. Cyclic Nucleotide Res.* **9,** 203–220.

Cotecchia, S., Schwinn, D. A., Randall, R. R., Lefkowitz, R. J., Caron, M. G., and Kobilka, B. K. (1988) Molecular cloning and expression of the cDNA for the hamster alpha-1 adrenergic receptor. *Proc. Natl. Acad. Sci. USA* **85,** 7159–7163.

Cotecchia, S., Exum, S., Caron, M. G., and Lefkowitz, R. J. (1990) Regions of the alpha-1 adrenergic receptor involved in coupling to phosphatidylinositol hydrolysis and enhanced sensitivity of biological function. *Proc. Natl. Acad. Sci. USA* **87,** in press.

Curtis, C. A. M., Wheatley, M., Barrsal, S., Birdsall, N. J. M., Eveleigh, P., Padder, E. K., Poyner, D., and Hulme, E. C. (1989) Propylbenzylylcholine mustard labels an acidic residue in transmembrane helix 3 of the muscarinic receptor. *J. Biol. Chem.* **264,** 489–495.

Dixon, R. A. F., Kobilka, B. K., Strader, D. J., Benovic, J. L., Dohlman, H. G., Frielle, T., Bolanowski, M. A., Bennett, C. D., Rands, E., Diehl, R. E., Mumford, R. A., Slater, E. E., Sigal, I. S., Caron, M. G., Lefkowitz, R. J., and Strader, C. (1986) Cloning of the gene and cDNA for mammalian beta-adrenergic receptor and homology with rhodopsin. *Nature* **321,** 75–79.

Dixon, R. A. F., Sigal, I. S., Candelore, M. R., Register, R. B., Scattergood, W., Rands, E., and Strader, C. D. (1987a) Structural features required for ligand binding to the beta-adrenergic receptor. *EMBO J.* **6,** 3269–3275.

Dixon, R. A. F., Sigal, I. S., Rands, E., Register, R. B., Candelore, M. R., Blake, A. D., and Strader, C. D. (1987b) Ligand binding to the beta-adrenergic receptor involves its rhodopsin-like core. *Nature* **326,** 73–77.

Dohlman, H. G., Bouvier, M., Benovic, J. L., Caron, G., and Lefkowitz, R. J. (1987) The multiple membrane spanning topography of the beta-2 adrenergic receptor. *J. Biol. Chem.* **262,** 14282–14288.

Dohlman, H. G., Caron, M. G., Strader, C. D., Amlaiky N., and Lefkowitz, R. J. (1988) Identification and sequence of a binding site peptide of the beta-2 adrenergic receptor. *Biochemistry* **27,** 1813–1817.

Dohlman, H. G., Caron, M. G., DeBlasi, A., Frielle, T., and Lefkowitz, R. J. (1990) A role of extracellular disulfide bonded cysteines in the ligand binding function of the beta-2 adrenergic receptor. *Biochemistry,* in press.

Eimerl, S., Neufield, G., Korner, M., and Schramm, N. (1980) Functional implantation of a solubilized beta-adrenergic receptor in the membrane of a cell. *Proc. Natl. Acad. Sci. USA* **77,** 760–764.

Emorine, L. J., Marullo, S., Briend-Sutren, M.-M., Patey, G., Tate, K., Delavier-Klutchko, C., and Strosberg, A. D. (1989) Molecular characterization of the human beta-3 adrenergic receptor. *Science* **245,** 1118–1121.

Feder, D., Ian, M. J., Klein, H. W., Hekman, M., Holzhofer, A., Dees, C., Levitzki, A., Helmreich, E. J. M., and Pfeuffer, T. (1986) Reconstitution of beta-1 adrenoceptor-dependent adenylate cyclaes from purified components. *EMBO J.* **5,** 1509–1514.

Fraser, C. (1989) Site-directed mutagenesis of beta-adrenergic receptors. Identification of conserved cysteine residues that independently affect ligand binding and receptor activation. *J. Biol. Chem.* **264,** 9266–9270.

Frielle, T., Collins, S., Daniel, K. W., Caron, M. G., Lefkowitz, R. J., and Kobilka, B. K. (1987) Cloning of the cDNA for the human beta-1 adrenergic receptor. *Proc. Natl. Acad. Sci. USA* **84,** 7920–7924.

Frielle, T., Daniel, K. W., Caron, M. G., and Lefkowitz, R. J. (1988) Structural basis of beta-2 adrenergic receptors. *Proc. Natl. Acad. Sci. USA* **85,** 9494–9498.

Gingrich, J. A., Amlaiky, N., Senogles, S. E., Chang W. K., McQuade, R. D., Berger, J. G., and Caron, M. G. (1988) Affinity chromatography of the D_1 dopamine receptor from rat corpus striatum. *Biochemistry* **27,** 3907–3812.

Haga, K. and Haga, T. (1985) Purification of the muscarinic acetylcholine receptor from porcine brain. *J. Biol. Chem.* **260,** 7927–7935.

Homcy, C. J., Rockson, S. G., Countaway, J., and Egan D. A. (1983) Purification and characterization of the mammalian beta-2 adrenergic receptor. *Biochemistry* **22,** 660–668.

Kobilka, B. K., Matsui, H., Kobilka, T. S., Yang-Feng, T. L., Francke, U., Caron, M. G., Lefkowitz, R. J., and Regan, J. W. (1987) Cloning, sequencing, and expression of the gene coding for the human platelet alpha-2 adrenergic receptor. *Science* **238,** 650–656.

Kobilka, B. K., Kobilka, T. S., Regan, J. W., Caron, .G., and Lefkowitz, R. J. (1988) Chimeric alpha-2–beta-2 adrenergic receptors: Delineation of domains involved in effector coupling and ligand binding specificity. *Science* **240,** 1310–1316.

Kubo, T., Bujo, H., Akiba, I., Nakai, J., Michina, M., and Numa, S. (1988) Localization of a region of the muscarinic acetylcholine receptor involved in selective effector coupling. *FEBS Lett.* **241,** 119–125.

Lavin, T. N., Heald, S. L., Jeffs, P. W., Shorr, R. G. L., Lefkowitz, R. J., and Caron, M. G. (1981) Photoaffinity labeling of the beta-adrenergic receptor. *J. Biol. Chem.* **256,** 11944–11950.

Lavin, T. N., Nambi, P., Heald, S. L., Jeffs, P. W., Lefkowitz, R. J., and Caron, M. G. (1982) ^{125}I-labeled p-azidobenzylcarazolol, a photoaffinity label for the beta-adrenergic receptor. *J. Biol. Chem.* **257,** 12332–12340.

Lefkowitz, R. J. and Caron, M. G. (1988) The adrenergic receptors: Models for the study of receptors coupled to guanine nucleotide regulatory proteins. *J. Biol. Chem.* **263,** 4993–4996.

Libert, F., Parmentier, M., Lefort, A., Dinsart, C., VanSande, J., Maenbaut, C., Simons, M-J., Dumont, J. E., and Vassar, G. (1989) Selective amplification and cloning of four new members of the G-protein-coupled receptor family. *Science* **244,** 569–572.

Lomasney, J. W., Leeb-Lundberg, L. M. F., Cotecchia, S., Regan, J. W., DeBernardis, J. F., Caron, M. G., and Lefkowitz, R. J. (1986) Mammalian alpha-1 adrenergic receptor. Purification and characterization of the native receptor ligand binding subunit. *J. Biol. Chem.* **261,** 7710–7716.

Matsui, H., Lefkowitz, R. J., Caron, M. G. and Regan, J. W. (1989) Localization of the fourth membrane spanning domain as a ligand binding site in the human platelet alpha-2 adrenergic receptor. *Biochemistry* **28,** 4125–4130.

May, D. C., Ross, E. M., Gilman, A. G., and Smigel, M. D. (1985) Reconstitution of catecholamine-stimulated adenylate cyclase activity using three purified proteins. *J. Biol. Chem.* **260,** 15829–15833.

Mukherjee, C., Caron, M. G., Mullikin, D., and Lefkowitz, R. J. (1976) Structure-activity relations of adenylate cyclase coupled beta-adrenergic receptors: Determination by direct binding studies. *Mol. Pharmacol.* **12,** 16–31.

Nambi, P., Sibley, D. R., Caron, M. G., and Lefkowitz R. J. (1984) Photoaffinity labeling of beta-adrenergic receptors in mammalian tissues. *Biochem. Pharmacol.* **33,** 3813–3822.

O'Dowd, B. F., Hnatowich, M., Regan, J. W., Leader, W. M., Caron, M. G., and Lefkowitz, R. J. (1988) Site-directed mutagenesis of the cytoplasmic domains of the human beta-2 adrenergic receptor. *J. Biol. Chem.* **263,** 15985–15992.

O'Dowd, B. F., Hnatowich, M., Caron, M. G., Lefkowitz, R. J., and Bouvier, M. (1989a) Palmitoylation of the human beta-2 adrenergic receptor. *J. Biol. Chem.* **264,** 7564–7569.

O'Dowd, B. F., Lefkowitz, R. J., and Caron, M. G. (1989b) Structure of the adrenergic and related receptors. *Ann. Rev. Neurosci.* **12,** 67–83.

Ovchinnikov, Y. A., Abdulaev, N. G., and Bogachuk, A. S. (1988) Two adjacent cysteine residues in the C-terminal cytoplasmic fragment of bovine rhodopsin are palmitoylated. *FEBS Lett.* **230,** 1–5.

Peralta, E. G., Winslow, J. W., Peterson, G. L., Smith, D. H., Ashkenazi, A., Ramachandran, J., Schimerlik, M. I., and Capon, D. J. (1987) Primary structure and biochemical properties of an M_2 muscarinic receptor. *Science* **236,** 600–605.

Perkins, J. P., Hausdorff, W. P., and Lefkowitz, R. J. (1990) Desensitization of the beta-adrenergic receptor. Chapter 3, this book.

Rashidbaigi, A. and Ruoho, A. E. (1981) Iodoazidobenzylpindolol, a photoaffinity probe for the beta-adrenergic receptor. *Proc. Natl. Acad. Sci. USA* **78,** 1609–1613.

Rashidbaigi, A. and Ruoho, A. E. (1982) Photoaffinity labeling of beta-adrenergic receptors: Identification of the beta-receptor binding site(s) from turkey, pigeon, and frog erythrocyte. *Biochem. Biophys. Res. Commun.* **106,** 139–148.

Regan, J. W., Nakata, H., DeMarinis, R. M., Caron, M. G., and Lefkowitz, R. J. (1986) Purification and characterization of the human platelet alpha-2 adrenergic receptor. *J. Biol. Chem.* **261,** 3894–3900.

Regan, J. W., Kobilka, T. S., Yang-Feng, T. L., Caron, M. G., Lefkowitz, R. J., and Kobilka, B. K. (1988) Cloning and expression of a human kidney cDNA for a novel alpha-2 adrenergic receptor. *Proc. Natl. Acad. Sci. USA* **85,** 6301–6305.

Rubenstein, R. C., Wong, S. K.-F., and Ross, E. M. (1987) The hydrophobic tryptic core of the beta-adrenergic receptor retains G_s regulatory activity in response to agonists and thiols. *J. Biol. Chem.* **262,** 16655–16662.

Senogles, S. E., Amlaiky, N., Falardeau, P., and Caron, M. G. (1988) Purification and characterization of the D_2-dopamine receptor from bovine anterior pituitary. *J. Biol. Chem.* **263,** 18996–19002.

Shorr, R. G. L., Heald, S. L., Jeffs, P. W., Lavin, T. N., Strohsacker, M. W., Lefkowitz, R. J., and Caron, M. G. (1982a) The beta-adrenergic receptor: Rapid purification and covalent labeling by photoaffinity crosslinking. *Proc. Natl. Acad. Sci. USA* **79,** 2778–2782.

Shorr, R. G. L., Strohsacker, M. W., Lavin, T. N., Lefkowitz, R. J., and Caron, M. G. (1982b) The beta-1 adrenergic receptor of the turkey erythrocyte. *J. Biol. Chem.* **257,** 12341–12350.

Strader, C. D., Dixon, R. A. F., Cheung, A. H., Candelore, M. R., Blake, A. D., and Sigal, I. S. (1987a) Mutations that uncouple the beta-adrenergic receptor from G_s and increase agonist affinity. *J. Biol. Chem.* **262,** 16439–16443.

Strader, C. D., Sigal, I. S., Register, R. B., Candelore, M. R., Rands, E., and Dixon, R. A. F. (1987b) Identification of residues required for ligand binding to the beta-adrenergic receptor. *Proc. Natl. Acad. Sci. USA* **84**, 4384–4388.

Strader, C. D., Sigal, I. S., Candelore, M. R., Rands, E., Hill, W. S., and Dixon, R. A. F. (1988) Conserved aspartic acid residues 79 and 113 of the beta-adrenergic receptor have different roles in receptor function. *J. Biol. Chem.* **263**, 10267–10271.

Strader, C. D., Candelore, M. R., Hill, W. S., Dixon, R. A. F., and Sigal, I. S. (1989) A single amino acid substitution in the beta-adrenergic receptor promotes partial agonist activity from antagonists. *J. Biol. Chem.* **264**, 16470–16477.

Vauquelin, G., Geynet, P., Hanoune, J., and Strosberg, D. (1977) Isolation of adenylate cyclase free of beta-adrenergic receptor from turkey erythrocyte membranes by affinity chromatography. *Proc. Natl. Acad. Sci. USA* **74**, 3710–3714.

Vauquelin, G., Bottari, S., Kenarek, L., and Strosberg, A. D. (1979) Evidence for essential disulfide bonds in beta-1 adrenergic receptors of turkey erythrocyte membranes. *J. Biol. Chem.* **254**, 4462–4469.

Wang, H., Lipfert, L., Malbon, C. C., and Bahouth, S. (1989) Site-directed anti-peptide antibodies define the topography of the beta-adrenergic receptor. *J. Biol. Chem.* **264**, 14424–14431.

Wess, J., Brann, M. R., and Bonner, I. I. (1989) Identification of a small intracellular region of the muscarinic M_3 receptor as a determinant of selective coupling to PI turnover. *FEBS Lett.* **258**, 133–136.

Wong, S. K.-F, Slaughter, C., Ruoho, A. E., and Ross, E. M. (1988) The catecholamine binding site of the beta-adrenergic receptor is formed by juxtaposed membrane-spanning domains. *J. Biol. Chem.* **263**, 7925–7928.

Yarden, Y., Rodriguez, H., Wong, S. K.-F., Brandt, D. R., May, D. C., Burnier, J., Harkins, R. N., Chen, E. Y., Ramachandran, J., Ullrich, A., and Ross, E. M. (1986) The avian beta-adrenergic receptor: Primary structure and membrane topology. *Proc. Natl. Acad. Sci. USA* **83**, 6795–6799.

CHAPTER 3

Mechanisms of Ligand-Induced Desensitization of Beta-Adrenergic Receptors

John P. Perkins,
William P. Hausdorff,
and Robert J. Lefkowitz

1. Introduction

Attenuation of responsiveness to extracellular signal molecules (neurotransmitters, hormones, growth factors, and so on) is a cellular regulatory mechanism commonly observed in organisms from microbes to mammals. In the slime mold Dictyostelium, desensitization to an extracellular signal, cyclic AMP, is programmed to facilitate a periodic synchronous behavior that fosters aggregation within the cellular population (Devreotes and Zigmond, 1988). In the mammalian nervous and endocrine systems, desensitization to the effects of neurotransmitters and hormones may be a mechanism for maintenance of target cell function within normal limits (Perkins et al., 1982; Harden, 1983; Sibley and Lefkowitz, 1985; Benovic et al., 1988).

The β-Adrenergic Receptors Ed.: J. P. Perkins © 1991 The Humana Press Inc.

This chapter will focus on ligand-induced reductions in the functional capacities of the beta-adrenergic receptors. The beta-adrenergic receptor is the best characterized of the numerous cell surface receptors involved in activation of the enzyme adenylyl cyclase. The gene for this receptor has been cloned, and the primary sequences of the protein from human, hamster, rat, turkey (*see* Dohlman et al., 1987a, and references therein), and mouse (Allen et al., 1988) are known. Based on comparison of the primary structures of a variety of cell surface receptors, the beta-adrenergic receptors appear to be members of a family of proteins that interact with the signal-transducing G-proteins. Persistent stimulation of relevant target cells with any of the receptor-mediated activators of adenylyl cyclase results ultimately in a loss in responsiveness. Although desensitization occurs in general for this class of receptor, the mechanistic details are known in some depth only for the beta-adrenergic receptors linked adenylyl cyclase system.

The physiological and pharmacological consequences of excessive exposure of humans to catecholamines are reviewed in detail in other chapters of this volume and will not be discussed here. Instead, we will focus on recent progress elucidating the molecular basis of the phenomenon of catecholamine-induced desensitization of beta-adrenergic receptor function. These insights have come about in large part as a result of new understanding of the molecular properties of beta-adrenergic receptors, the guanine nucleotide-binding proteins (G_s, G_i) and adenylyl cyclase. The properties of the adenylyl cyclase system as they relate to activation by catecholamines and other effectors are reviewed in detail in another chapter of this volume. However, it is important to reiterate the nature of the interaction of beta-adrenergic receptors and G_s in order to understand the concept of "uncoupling," as it will be used to describe one component of the desensitization process.

Figure 1 is a schematic representation of some of the reactions thought to be involved in both agonist-induced activation of adenylyl cyclase, and agonist-induced desensitization and loss of beta-adrenergic receptors. The receptor is depicted as existing in two forms, one of which $((R^*)_{pm})$ is capable of functional interaction with G_s-GDP. In the absence of an agonist beta-adrenergic receptors would exist pre-

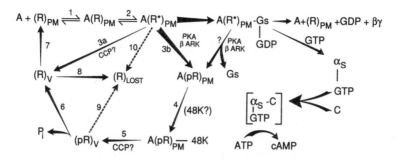

Fig. 1. The scheme depicts possible reactions subserving agonist-induced phosphorylation, internalization and downregulation of the beta-adrenergic receptors. Reactions: 1 and 2, *see* text; 3a, endocytosis of beta-adrenergic receptors independent of phosphorylation; 3b, beta-adrenergic receptor phosphorylation by protein kinase A and/or beta-adrenergic receptor kinase; 4, putative interaction of phosphorylated beta-adrenergic receptors with a 48k-like protein leading to a desensitized state; 5, endocytosis; 6, dephosphorylation of internalized beta-adrenergic receptors; 7, recycling of beta-adrenergic receptors to the plasma membrane; 8–10, possible reaction paths leading to loss of radioligand binding capacity. Also shown is the accepted reaction scheme for activation of G_s and adenylyl cyclase.

dominantly in the inactive form, $(R)_{pm}$. Agonists preferentially bind to the active form $(R^*)_{pm}$ favoring its proportionate increase within the equilibrium. A physical "coupling" between $A(R^*)_{pm}$ and G_s-GDP is proposed as the initial step in the activation of adenylyl cyclase. Subsequent steps in the activation pathway are discussed in more detail elsewhere in this volume. The receptor forms $A(R^*)_{pm}$ and $A(R^*)G_s$-GDP are also the substrates for an inactivation reaction sequence. In the current working model it is proposed that these forms of beta-adrenergic receptors are phosphorylated by a specific receptor kinase, termed beta-adrenergic receptor kinase (Benovic et al., 1988), as well as by the cyclic AMP-dependent protein kinase A (Clark et al., 1988; Hausdorff et al., 1989; Lohse et al., 1990). As a consequence of beta-adrenergic receptor kinase phosphorylation, $A(pR)_{pm}$ has been postulated to interact with a cytosolic protein analogous to arrestin (48K), a retinal protein involved in the turnoff of the light-activated signal transduced by rhodopsin (Benovic et al., 1987b). As in the retinal system, the combination of beta-adrenergic receptor phosphorylation, and association with the 48K-like protein is believed to render the receptor unable to couple with G_s-GDP, i.e., an "uncoupling" reaction has occurred.

Traditionally, desensitization has been divided into two phenomenological categories, "homologous" and "heterologous." Although recent evidence (discussed in Section 5.3.) suggests that these two categories may not be as distinct in actuality as previously thought, the terms have historically proven useful in providing a conceptual framework for the design and interpretation of experiments. The term "homologous" has been used to designate that form of desensitization in which the subsequent response to the desensitizing signal molecule is attenuated, but the response to other signal molecules is retained (Su et al., 1976; Lefkowitz et al., 1980). This situation is most readily explained as resulting from selective alterations in the receptor for the desensitizing signal. Conversely, "heterologous" desensitization is a term used to indicate that exposure to a single agonist results in loss of response to agonists for all receptors that mediate activation of adenylyl cyclase. Evidence to date indicates that heterologous desensitization is mediated by cyclic AMP, and involves receptor phosphorylation by protein kinase A (Harden, 1983; Sibley and Lefkowitz, 1985; Sibley et al., 1984; Clark et al., 1988).

A reduction in beta-adrenergic receptors-mediated adenylyl cyclase activity can be detected after only a few minutes of exposure of cells to catecholamines (Waldo et al., 1983). On a similarly rapid time scale (s-min) $A(R^*)_{pm}$, whether or not it is phosphorylated, also undergoes rapid internalization. Although yet to be proven as the case, the internalization reaction is discussed throughout this chapter in terms of the well-defined pathway (Fig. 2) for internalization of a variety of cell surface proteins that bind extracellular ligands (low density lipoprotein, asialoglycoprotein, transferrin, epidermal growth factor, insulin, and so on). This pathway has been shown to involve internalization of receptors via invagination of clathrin-coated pits on the cell surface to ultimately form vesicular structures designated "endosomes" (Goldstein et al., 1979; Mellman et al., 1986). The fate

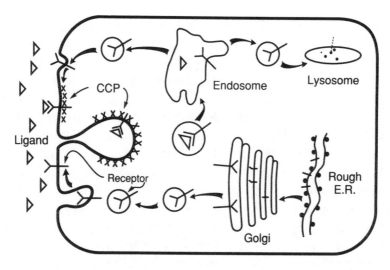

Fig. 2. The model depicts the pathways for synthesis, internalization, recycling, and degradation of integral plasma membrane proteins (receptors) that enter cells via clathrin coated pits.

of such internalized receptors is either to recycle to the cell surface or, after sorting and transfer from endosomes to lysosomes, to be degraded. Nevertheless, because the uncoupling, internalization, and recycling reactions are rapid, brief exposure of cells to a catecholamine leads only to a partial desensitization that is completely reversible. Extended exposure to an agonist can result in a more chronic state of reduced responsiveness for which beta-adrenergic receptor loss (degradation) is possibly responsible.

Depending on the degree of receptor reserve or "spareness"* of beta-adrenergic receptor, these reactions can be manifested as losses of "sensitivity" and/or maximal responsiveness to catecholamines. For example, in cells with a large receptor reserve desensitization may consist of a shift to the right in the dose-response curve for adenylyl cyclase with no alteration of the maximal response. Conversely, in the absence of spare receptors, the maximal response may also be reduced.

*Spare receptors are defined as receptors in excess of the number required for complete activation of a measured effect.

The receptor-specific consequences of agonist/receptor interactions described above are complemented in some cells by other less well-defined reactions that cause functional deficits in components of the adenylyl cyclase system other than the receptor. Such responses are probably best explained by agonist-induced alterations in G_s, G_i, or the adenylyl cyclase moiety itself, although the molecular bases of these changes in intact cells are unknown. Cyclic AMP-dependent phosphorylation of G_s has not been demonstrated. Furthermore, consensus sequence for the substrate site of cyclic AMP-dependent protein kinase A is not found in the amino acid sequence of G_s. On the other hand, adenylyl cyclase is phosphorylated by protein kinase A in vitro, with resultant decreases in enzyme activity (Yoshimasa et al., 1988). However, the occurrence of this reaction in agonist-stimulated cells has not been demonstrated.

2. Methodology

The consequences of exposure of cells to catecholamines are typically assessed using a two-step protocol. In the first step, intact cells are exposed to a catecholamine in order to induce a state of desensitization. The second step involves stopping the first reaction and assessing its consequences in terms of the reactions shown in Fig. 1. In general, the assessment step involves the use of antagonist radioligands to determine changes in the amount and/or cellular distribution of beta-adrenergic receptors, and various assays to detect changes in the physical state and functional capacity of remaining beta-adrenergic receptors to interact with G_s or activate adenylyl cyclase.

2.1. Radioligand Binding

The availability of highly selective, high affinity ^{125}I-radioligands ([^{125}I]iodopindolol, [^{125}I]iodocyanopindolol), that also are lipid soluble provides a straightforward means for counting beta-adrenergic receptors; such ligands apparently gain access to even internalized beta-adrenergic receptors at 37°C. Typically, exposure of cells to catecholamines does not alter the apparent K_D for binding of antagonist radioligands. Thus, reduction in such radioligand binding is usually interpreted as reflecting agonist-induced loss of beta-adrenergic receptors. In contrast, [^{3}H]CGP-12177, a hydrophilic beta-adrenergic receptors antagonist, apparently allows selective measurement of cell surface receptors. Thus, under appropriate conditions [^{3}H]CGP-12177 exhibits minimal binding to internalized beta-adrenergic receptors in whole cell assays (Hertel and Staehelin, 1983) or to isolated cytosolic vesicles containing beta-adrenergic receptors (Hertel et al., 1983). Interestingly, [^{125}I]iodopindolol also exhibits selectivity for cell surface beta-adrenergic receptors if assays with intact cells are carried out at 4°C (Linden et al., 1984; Toews et al., 1986). Thus, the physical translocation of beta-adrenergic receptors during exposure to agonists can be quantitated in either intact cells, or in cell lysates by the appropriate use of ligands with selective membrane permeability.

A final point on radioligand binding protocol relates to the assumption that the traditionally used lipophilic radioligands actually access the total beta-adrenergic receptor population. Strader et al. (1984) have shown that exposure of frog erythrocytes to isoproterenol results in the formation of beta-adrenergic receptors that do not bind [^{125}I]iodocyanopindolol in membranes prepared from the treated cells; however, such beta-adrenergic receptors can bind this ligand after solubilization with detergent. Whereas this is the only published report of such a form of the beta-adrenergic receptor it is also the only report of a quantitative assessment of beta-adrenergic receptor content before and after detergent solubilization (*see also* Sibley et al., 1985). Doss et al. (1981) also reported the agonist-induced accumulation of beta-adrenergic receptors in a form that did not bind [^{125}I]iodohydroxypindolol or [^{125}I]iodopindolol. In this study, such beta-adrenergic receptors slowly returned to an assayable form upon removal of the agonist, even in the presence of cycloheximide. Thus, it appears that beta-adrenergic receptors can exist in a form(s) internalized within cells such that binding of even lipophilic ligands cannot take place owing to change in either the molecular properties or the accessibility of the receptor.

2.2. Assessment of Beta-Adrenergic Receptor Function

Assays aimed at assessing changes in function fall into one of four categories:

1. Whole cell assays of the effect of the agonist on cyclic AMP accumulation;
2. Assays of the effect of agonist on cyclic AMP formation using cell lysates or membranes;
3. Assessment of the extent of high affinity (GTP-sensitive) agonist binding to beta-adrenergic receptors in washed membranes; and
4. Assessment of the capacity of beta-adrenergic receptors to activate G_s/adenylyl cyclase reconstituted into cells or phospholipid vesicles.

Unfortunately, each approach has significant limitations.

2.2.1. Whole Cell Assays

Whole cell assays of cyclic AMP accumulation reflect the combined effects of changes in beta-adrenergic receptors, G-protein, adenylyl cyclase, and phosphodiesterase activities. Whereas such assays provide a holistic perspective, they generally cannot provide information about any single reaction proposed in the model. Nonetheless, useful correlative information has been derived from such studies when the experiments were designed to take advantage of the rapid kinetics of the early reactions in the desensitization process. Properly designed experiments involve measurement of the initial rates of cyclic AMP accumulation in order to reduce the impact of changes in phosphodiesterase activity and the induction of desensitization during the assay (Toews and Perkins, 1984).

2.2.2. Assay of Adenylyl Cyclase Activity in Lysates and Membrane Preparations

Well-constructed assays of lysate or membrane activities eliminate concerns about phosphodiesterase activity and intra-assay desensitization. However, a reduction in catecholamine-stimulated adenylyl cyclase activity cannot be attributed simply to any of the specific reactions of the model, even when effects on components other than the beta-adrenergic receptors can be discounted. The problem of interpretation has two parts. The first is related to the mixture of beta-adrenergic receptor forms thought to exist, and the resultant difficulty in ascribing overall changes in beta-adrenergic receptor/adenylyl cyclase activity to changes in any particular form. This problem relates in part to the fact that beta-adrenergic receptors also can exist in different physical environments. Since three of the putative forms of beta-adrenergic receptors $((pR)_{pm}$, $(pR)_v$, and $(R)_v)$ can bind antagonist radioligands but cannot activate adenylyl cyclase, all would assay as "uncoupled" receptors. However, since not all uncoupled forms of the receptor are phosphorylated, attempts to correlate in a quantitative fashion the extent of beta-adrenergic receptor phosphorylation with loss of beta-adrenergic receptor/adenylyl cyclase activity could generate any of a number of relationships depending on the rates of the various reactions (e.g., *see* Strasser et al. (1986)).

In considering the problem of heterogeneity of receptor forms, one must realize that if the model is correct, not only will exposure to a catecholamine change the mix of receptor forms, but the ratios of the various forms will further change in time. In theory, reduction of the content of $(R^*)_{pm}$ by phosphorylation, internalization, or loss will not be distinguished by an assay that simply measures catecholamine-stimulated adenylyl cyclase activity. One recent approach that minimizes, but does not eliminate, the contributions of internalization processes to the loss of functional response, involves the selective perturbation of the desensitization and phosphorylation processes without hindering internalization. This has been accomplished by receptor mutations (Bouvier et al., 1988; Hausdorff et al., 1989) or inhibition of protein kinases (Lohse et al., 1990). In addition, procedures are available that lead to selective blockade of beta-adrenergic receptor internalization (Waldo et al., 1983; Hertel et al., 1985). Such selective perturbations help to delineate the contributions of the various forms of beta-adrenergic receptors to the desensitization process.

When cell lysates are used to assess changes in beta-adrenergic receptor/adenylyl cyclase activity, it is a relatively comfortable assumption that none of the receptor species has been selectively lost during sample preparation. However, when membrane subfractions are used, the potential for differential recovery of the different physical forms of beta-adrenergic receptors is great. For example low speed centrifugation of cell lysates may (Stadel et al., 1983a; Waldo et al.,

1983), or may not (Chuang and Costa, 1979; Mukherjee and Lefkowitz, 1977) sediment the agonist-induced vesicular forms of beta-adrenergic receptors $((pR)_v + (R)_v)$. Thus, determination of the relation between beta-adrenergic receptor degradation and reduction in beta-adrenergic receptor/adenylyl cyclase activity might be erroneously interpreted in the latter example.

The second potential difficulty of interpretation relates to the concept of receptor reserve.[*] Thus, in cells with spare receptors, the expected (and realized) consequences of loss of functional beta-adrenergic receptors by any mechanism is different than in cells without a receptor reserve.

2.2.3. High Affinity Agonist Binding

The coupling status of beta-adrenergic receptors also can be assessed by measuring the extent of high affinity (GTP sensitive) agonist binding to the receptor (Ross et al., 1977). The extent of such high affinity binding has generally been thought to reflect the formation of the complex $A(R^*).G_s$ (Fig. 1) (DeLean et al., 1980). The working model indicates the existence of functionally deficient receptors, $(pR)_{pm}$, and $(pR)_{pm}$—(48K), in the plasma membrane. Thus, to the

---

[*]In cells containing a large beta-adrenergic receptor reserve, even a substantial reduction in the number of functional beta-adrenergic receptors may not result in a reduction in adenylyl cyclase response to maximal concentrations of an agonist. When C6 rat glioma cells are exposed to isoproterenol for 20–30 min, about 40% of the cell surface beta-adrenergic receptors are translocated to a light vesicle fraction; nonetheless, the capacity of maximal concentrations of isoproterenol to stimulate cyclic AMP formation is not reduced (Toews and Perkins, 1984). However, as anticipated, the Kact for stimulation of adenylyl cyclase is shifted to a threefold higher concentration of isoproterenol. The beta-adrenergic receptor reserve in C6 cells is a well-established circumstance (Terasaki et al., 1979; Drummond et al., 1977), and in these and other cells with a beta-adrenergic receptor reserve, desensitization can be perceived as an apparent reduction in the sensitivity to isoproterenol. In contrast, in identical experiments carried out in a cell line with no beta-adrenergic receptor reserve (1321Nl human astrocytoma cells), the resulting uncoupling and internalization of beta-adrenergic receptors was associated with a proportionate reduction in the maximal rate of isoproterenol stimulated cyclic AMP formation (Su et al., 1980; Waldo et al., 1983).

extent that such species accumulate, agonist binding experiments carried out with purified plasma membranes may indicate the proportionate existence of beta-adrenergic receptors with low affinity for agonists. Results to date indicate that such forms of the receptor do not accumulate significantly in the plasma membrane of either 1321Nl or C6 cells (Harden et al., 1980; Frederich et al., 1983), whereas they do in S49 cells (Clark et al., 1985) and frog erythrocytes (Sibley et al., 1986). Interpretation of such experiments is clouded if the plasma membrane preparation is not free of the internalized form of beta-adrenergic receptors, since such receptors should exhibit only low affinity, GTP-insensitive binding for agonists (*see below*). In addition, it is important to note that functionally impaired receptors might not necessarily lose their capacity to promote high affinity agonist binding (Hausdorff et al., 1990).

2.2.4. Reconstituted Systems

An unambiguous assessment of the functionality of beta-adrenergic receptors *per se* requires direct measurement of the interaction of purified beta-adrenergic receptors with purified G_s or G_s/adenylyl cyclase. Toward this end, attempts have been made to reconstitute these components in phospholipid vesicles of defined composition. Qualitatively, such experiments have been successful when purified native forms of beta-adrenergic receptors, G_s, and adenylyl cyclase have been combined (Cerione et al., 1984; May et al., 1985); although the normal magnitude of agonist activation of adenylyl cyclase is yet to be achieved. Reconstitution methods also have been applied to the determination of beta-adrenergic receptor functionality in desensitized cells, with mixed results. These experiments will be discussed below. Here, we will point out only that these experiments have yielded important information as to molecular mechanisms of desensitization, but are very demanding in design, and difficult to control.

2.3. Assessment of Beta-Adrenergic Receptor Phosphorylation

Covalent changes in beta-adrenergic receptors derived from cells exposed to desensitizing concentrations of agonists have been monitored in two ways. Photoaffinity labeling of membrane fractions with ^{125}I-iodoazidocyanopindolol followed by polyacrylamide gel electrophoresis has revealed altered mobility of beta-adrenergic receptors following cellular pretreatment by catecholamines or by cyclic AMP analogs (Stadel et al., 1983b,1987). These changes in mobility have been presumed to reflect the phosphorylation of beta-adrenergic receptors. However, photoaffinity labeling cannot discriminate phosphorylation from other covalent modifications of the receptor, such as glycosylation and palmitoylation (O'Dowd et al., 1989). A more direct approach to monitoring phosphorylation of beta-adrenergic receptors in intact cells involves exposure of cells to $^{32}PO_4$ to label the cellular ATP pool, followed by purification by affinity chromatography. The phosphorylation status of the purified receptor is then examined by PAGE and autoradiography. This approach has traditionally been of great use in establishing the functional relevance and significance of phosphorylation (Benovic et al., 1988). It also has yielded important information regarding the stoichiometry of such phosphorylation events (Sibley et al., 1984). Quantitation also represents the most significant drawback of the technique, since at present it is impossible to determine whether one cellular pool of beta-adrenergic receptors is extensively phosphorylated and others not at all, or whether all beta-adrenergic receptors are phosphorylated to the same extent. It is thus difficult to make *quantitative* correlations between beta-adrenergic receptor phosphorylation and functional uncoupling from G_s.

3. Correlations Between Agonist-Stimulation of Adenylyl Cyclase and Induction of Desensitization

3.1. Agonist Efficacies for Activation of Adenylyl Cyclase and Induction of Desensitization

Early studies by Mukherjee, Caron, and Lefkowitz (1975) demonstratesss that beta-adrenergic receptor loss from frog erythrocytes was strictly an agonist-induced phenomenon, and that loss of agonist-stimulated adenylyl cyclase activity correlated well with loss of [^{3}H]dihydroalprenolol binding sites.*

Su et al. (1980) compared the effects of two partial agonists, zinterol (I.A. = 0.55) and soterenol (I.A. = 0.35), with those of the full agonist isoproterenol (I.A. = 1.00) in terms of inducing a reduction in isoproterenol-stimulated adenylyl cyclase activity. The conclusion from these studies was that, each compound exhibited its characteristic fractional efficacy for both activation and desensitization. Pittman et al. (1983) found an excellent correlation between fractional efficacy and the rate of beta-adrenergic receptor loss from L6 muscle cells in surface cultures. They examined 14 compounds for which efficacy (relative to isoproterenol = 1.0) varied from 0.03 to 1.08. Although it is a prediction of the model, the rate of beta-adrenergic receptor loss has not yet been shown to be proportional to the amount of internalized beta-adrenergic receptors. Thus, whereas the data base is small, it appears that partial agonists, defined in terms of efficacy for stimulation of adenylyl cyclase, exhibit similar partial efficacies for induction of desensitization and the rate of beta-adrenergic receptor loss.

Recent evidence, however, has challenged these relationships. Neve et al. (1985) have shown that, certain atypical agonists do not cause desensitization or internalization, but nonetheless elicit a marked loss of beta-adrenergic receptors. An explanation for these results in terms of the model is difficult, and requires the invention of favorable rate constants for the individual reactions. The suggestion from these workers of the existence of more than one pathway of agonist-induced beta-adrenergic receptor loss, seems reasonable at this point, and has been supported by several recent molecular analyses of beta-adrenergic receptor loss mechanisms (*see below*). We will return to this point in the section on beta-adrenergic receptor loss reactions.

---

*The membrane preparation procedure used in these early studies actually measured loss of surface receptors and thus within today's understanding, reflected beta-adrenergic receptor internalization, not beta-adrenergic receptor loss.

3.2. Agonist Potencies
for Activation of Adenylyl Cyclase
and Induction of Desensitization

In attempting to derive mechanistic insight from experiments that compare the potencies of agonists in activating cyclic AMP synthesis and in causing desensitization, several factors must be analyzed. For example, in order to establish the dose-response relationship for "desensitization," it is imperative that the effects of agonists on the initial events of the process be distinguished from effects caused by beta-adrenergic receptor loss. Agonist-induced receptor loss is a time-dependent, essentially irreversible reaction that probably also contributes to reduction in beta-adrenergic receptor/adenylyl cyclase activity following prolonged exposure to agonist. Su et al. (1979,1980) demonstrated the time dependency of the EC_{50} for isoproterenol-induced desensitization of human astrocytoma cells. The EC_{50} value for a 30 min exposure was 30 nM. At 30 min, the desensitization reaction had reached an apparent steady state (45% of control beta-adrenergic receptor/adenylyl cyclase activity); at this time, 80–90% of beta-adrenergic receptors remained detectable by radioligand binding. This EC_{50} value was identical to the value for the EC_{50} for activation of cyclic AMP synthesis in intact cells, consistent with the existence of a common intermediate in the pathways for activation and desensitization (Fig. 1). In contrast, after 24 h of exposure to isoproterenol, both beta-adrenergic receptor/adenylyl cyclase and beta-adrenergic receptors content had been reduced to 10% of control values. Now, however, the EC_{50} for this isoproterenol-induced desensitization was only 0.8 nM. This low value probably reflects the contribution to desensitization of the irreversible loss of beta-adrenergic receptors. In a series of experiments using the same cell line and short-term exposure to isoproterenol (5–30 min), the EC_{50} values were similar (10–30 nM) for activation of adenylyl cyclase, desensitization of beta-adrenergic receptor/adenylyl cyclase and beta-adrenergic receptor internalization (Su et al., 1980; Toews and Perkins, 1984). Thus, both theory and observation indicate that failure to take into account the effect of beta-adrenergic receptor loss will result in erroneous estimates of the EC_{50} for agonist-induced desensitization *per se.*

Another factor to consider in interpreting apparent EC_{50} values is that the effect of an agonist can be amplified at each step in a sequential process. For example, if desensitization of beta-adrenergic receptor/ adenylyl cyclase were the result of phosphorylation of beta-adrenergic receptors by protein kinase A, the EC_{50} might reflect the EC_{50} for activation of protein kinase A. To the extent that there was a receptor excess in terms of activation of adenylyl cyclase, and to the extent that more cyclic AMP could be made than was required to fully activate protein kinase A, the EC_{50} for desensitization by this mechanism could be markedly less than the EC_{50} for activation of adenylyl cyclase, and certainly less than the K_D for receptor occupancy. Thus, coherent analysis of the meaning of differences in EC_{50} values for various agonist-induced reactions requires knowledge of the extent to which agonist binding is amplified throughout the process examined. The proposed existence of multiple phosphorylation mechanisms in receptor desensitization (Hausdorff et al., 1989; Lohse et al., 1990), each with distinct EC_{50} values, further complicates meaningful interpretations of EC_{50} values for desensitization.

4. The Internalization and Loss of Beta-Adrenergic Receptors— Potential Mechanisms of Desensitization

4.1. The Evidence for Agonist-Induced Endocytosis of Beta-Adrenergic Receptors

An agonist-induced change in the physical location of beta-adrenergic receptors has been shown in bullfrog erythrocytes (Chuang and Costa, 1979), human astrocytoma cells (Harden et al., 1980), grass frog erythrocytes (Stadel et al., 1983a), S49 mouse lymphoma cells (Clark et al., 1985), C6 rat glioma cells (Frederich et al., 1983), and several other cell types (Kassis et al., 1986; Kassis and Sullivan, 1986), as well as in rat lung tissue (Strasser et al., 1984). The properties of beta-adrenergic receptors in this physically altered environment are most parsimoniously explained if such receptors exist in cytosolic vesicles with the ligand binding site oriented toward the inside of the vesicle. However, there is some controversy about the exact nature of the physical form of these receptors, and there are different views about their subcellular localization (Strader et al., 1984; Mahan et al., 1985b).

4.1.1. Agonist-Induced Formation of Soluble Beta-Adrenergic Receptors

Chuang and Costa (1979) provided the first evidence that catecholamines induce internalization of beta-adrenergic receptors. In these experiments, bull frog (Rana catesbianna) erythrocytes were exposed to isoproterenol. The cells were washed, then lysed, and a $30,000g$ × 30 min supernatant fraction (the cytosol) was separated from a pellet fraction (the membranes). The cytosol fraction accounted for 24% of the 30% reduction in total membrane beta-adrenergic receptors observed, i.e., 7% of the initial membrane beta-adrenergic receptors. The beta-adrenergic receptors in the cytosol fraction were reduced by only 10–15% upon centrifugation at $100,000g$ for 4 h; thus, the cytosol beta-adrenergic receptors were construed to be "soluble." This agonist-induced formation of cytosolic beta-adrenergic receptors was prevented at 0°C, was blocked completely by concanavalin A, and partially by dinitrophenol and cordycepin (Chuang et al., 1980). In a complementary set of experiments, Chuang (1982) reported that the lysosomotropic agent, chloroquine, also reduced the appearance of cytosolic receptors, but increased the accumulation of beta-adrenergic receptors in a lysosome-enriched cell fraction. These workers have interpreted their results in terms of a model similar to that shown in Fig. 2. Their unique idea is that soluble beta-adrenergic receptors are produced as a transient, post-lysosomal degradation product. Such soluble forms of beta-adrenergic receptors have not been described in other studies published to date.

4.1.2. Agonist-Induced Formation of a Nonplasma Membrane Vesicular Form of Beta-Adrenergic Receptors

If beta-adrenergic receptors undergo endocytosis it should be possible to isolate vesicles derived from endosomes that contain such receptors. Furthermore, beta-adrenergic receptors in such vesicles should be oriented so that the ligand binding site faces the lumen of the vesicle. Experiments from several laboratories have, in fact, demonstrated agonist-induced formation of a vesicular form of beta-adrenergic receptors that does not sediment in association with plasma membrane markers upon density gradient centrifugation. The beta-adrenergic receptors in these vesicle preparations are not readily accessible to hydrophilic ligands but can be detected with lipophilic radioligands (Hertel et al., 1983a,b; Toews et al., 1984). In one study, treatment of the vesicles with the pore-forming antibiotic, alimethacin, markedly increased beta-adrenergic receptor accessibility to the hydrophilic radioligand [^{3}H]CGP-12177 (Hertel et al., 1983b). Iodopindolol binds rapidly to the total beta-adrenergic receptors population of native and desensitized cells at 23–37°C; however, at 4°C, it appears not to bind to beta-adrenergic receptors in light vesicles (Toews et al., 1986). These results, although providing only indirect evidence, are consistent with the existence of beta-adrenergic receptors with binding sites exclusively facing the inner surface of semipermeable membrane vesicles.

4.1.3. Kinetics of Beta-Adrenergic Receptor Internalization

Receptor mediated endocytosis is a constitutive process that occurs with similar kinetics in most mammalian cells studied (Mellman et al., 1986). The actual rate of receptor endocytosis is determined not only by the constitutive rate of the internalization process, but by the tendency of the receptor to be found in clathrin-coated pits. In this regard, receptors can be placed in one of two general categories; those that accumulate in clathrin-coated pits without bound ligand, and those that require bound ligand to accumulate in clathrin-coated pits. If in fact beta-adrenergic receptors are internalized via this well-described pathway, they would fall into the latter category. Receptors also can be categorized in terms of their fate once internalized, i.e., those that efficiently recycle to the cell surface (e.g., receptors for low density lipoprotein, asialoglycoprotein, and transferrin), and those that are preferentially catabolized (e.g., receptors for EGF and insulin). In this regard, beta-adrenergic receptors would appear to be hybrids in that, they recycle to the cell surface efficiently for 45–60 min, but are eventually lost.

The kinetics of beta-adrenergic receptor internalization have not often been examined in detail. However, it is clear that the $t_{1/2}$ values for beta-adrenergic receptor loss from the plasma membrane are similar for a variety of mammalian cells growing as surface-attached cultures. Usually, a transient steady state is reached within 10–20 min ($t_{1/2}$ = 2–4 min), in which about 50–70% of beta-adrenergic receptors are on the cell surface, and 30–50% are internalized.

The kinetics of beta-adrenergic receptor translocations have been examined using direct methods only in human astrocytoma cells. In this cell line, exposure to isoproterenol results in the appearance of beta-adrenergic receptors in a light vesicle fraction. The reaction exhibits a lag of about 45–60 s, then proceeds with a $t_{1/2}$ of 2–3 min to a steady state at which 40–50% of beta-adrenergic receptors are internalized (Waldo et al., 1983). Hertel et al. (1983a) obtained similar results using C6 rat glioma cells in which beta-adrenergic receptor internalization was measured indirectly, using the impermeant radioligand [^{3}H]CGP-12177. Toews and Perkins (1984) used an intact cell competition binding assay (Toews et al., 1983) to establish a similar kinetic pattern for the induction by agonists of a form of the beta-adrenergic receptor that is inaccessible to hydrophilic ligands, presumably one or more of the receptor forms designated $(R)_v$ in Fig. 1. Other laboratories have made similar observations (Insel et al., 1983; Pittman and Molinoff, 1980). Thus, both direct and indirect methods of analysis of beta-adrenergic receptor internalization kinetics provide similar estimates of $t_{1/2}$ values, and such values are similar to $t_{1/2}$ values for the internalization of receptors that utilize the clathrin-coated pit pathway for endocytosis (Mellman et al., 1986).

4.1.4. Kinetics of Beta-Adrenergic Receptor Externalization

In a series of experiments not yet published, Kurz and Perkins have attempted to determine the rate of externalization of beta-adrenergic receptors in 1321N1 cells. The most straightforward protocol involved measurement of the reappearance of beta-adrenergic receptors on the cell surface as assessed by the return of isoproterenol-sensitive adenylyl cyclase activity. The $t_{1/2}$ for recovery was about 6 min. However, such measurements rely not only on the return of beta-adrenergic receptors to the plasma membrane, but also on the kinetics of recoupling reactions of unknown nature and order. When return of beta-adrenergic receptors to the cell surface was measured as return of [^{3}H]CGP-12177 binding sites on C6 cells, a $t_{1/2}$ of 5 min was obtained (Hertel and Staehelin, 1983). Direct assessment of beta-adrenergic receptor redistribution from light vesicle fractions to plasma membrane fractions of sucrose density gradients also has been carried out (Kurz and Perkins, unpublished). In one protocol, internalized beta-adrenergic receptors were selectively labeled using [^{125}I]iodopindolol in the presence of excess isoproterenol (*see* Toews et al., 1984 for methodology). The labeled receptors were observed to redistribute to the plasma membrane with a $t_{1/2}$ of 3–4 min. The presence or absence of isoproterenol or propranolol (10 μM) had no effect on the $t_{1/2}$. Thus, recycling of beta-adrenergic receptor occurs at a rate similar to that of cell surface receptors known to enter cells via clathrin-coated pits (Mellman et al., 1986).

4.1.5. *Effect of Inhibition of Endocytosis on Beta-Adrenergic Receptor Internalization*

Perkins and coworkers have compared, in the same cell line, (1321Nl astrocytoma) factors influencing internalization of beta-adrenergic receptors and receptor-mediated endocytosis of EGF. Both processes were inhibited by concanavalin A (Wakshull et al., 1985), phenylarsine oxide (Hertel et al., 1985), and reduction in cellular ATP content (Hertel et al., 1986). Reduction of the temperature to 4°C inhibits endocytosis (Mellman et al., 1986) and beta-adrenergic receptor internalization also is blocked (Toews et al., 1986). At 20–22°C, endocytosis occurs at a rate about 50% of that at 37°C, but degradation of internalized EGF and insulin is markedly reduced (Mellman et al., 1986). Beta-adrenergic receptors also are not lost on extended exposure of astrocytoma cells to isoproterenol at 22°C, but are readily internalized (Waldo et al., in preparation).

In addition, high osmolarity (0.45M sucrose) (Daukas and Zigmond, 1985), reduced intracellular K$^+$ (Larkin et al., 1983), and reduced intracellular pH (Sandvig et al., 1987) have been shown to block endocytosis via clathrin-coated pits. All of these conditions have been shown to interfere with the proper formation and/or function of clathrin-coated pits (Heuser and Anderson, 1989; Heuser, 1989). The effects of these conditions on the internalization of beta-adrenergic receptors and transferrin in astrocytoma cells and rat osteosarcoma cells have been compared (Liao and Perkins, unpublished; Jyh-Fei Liao, Ph.D. dissertation, Yale University, 1990). The effects of these conditions on transferrin internalization ranged from modest to extreme inhibition, and varied between the two cell lines. The same pattern of effects was observed with beta-adrenergic receptors as with transferrin. Thus, in addition to similar kinetic properties, the internalization of beta-adrenergic receptors, EGF, and transferrin cannot be distinguished by a variety of agents and/or conditions that reversibly block endocytosis via clathrin-coated pits.

4.2. Agonist-Induced Loss of Beta-Adrenergic Receptors

Receptor loss is typically defined in terms of the reduction in total binding sites for lipophilic antagonist radioligands, measured at 25–37°C. Internalization of beta-adrenergic receptors is not thought to be sufficient to explain loss of binding to lipophilic ligands as discussed above. However, as indicated in the scheme for receptor endocytosis (Fig. 2), internalized receptors may be directed to lysosomes, and therein degraded. If indeed agonist-induced beta-adrenergic receptor loss occurs by this pathway, certain characteristics of the process should be demonstrable (Mellman et al., 1986):

1. The loss reaction should exhibit a lag of at least 15–30 min;
2. The degradation reactions should be blocked by reducing the temperature to 20–22°C, whereas the internalization reaction should only be modestly slowed;
3. Certain lipophilic amines should inhibit lysosomal degradation of beta-adrenergic receptors, but not affect endocytosis; and
4. Beta-adrenergic receptor loss by the lysosomal pathway should lead to degradation of the primary structure of the protein; therefore, recovery of beta-adrenergic receptors would require synthesis of new receptors.

Unfortunately, few reports have appeared in which these predictions have been explicitly tested.

A major limitation in the quantitative analyses of the loss of beta-adrenergic receptors is that loss appears to occur by more than one mechanism. Doss et al. (1981) reported that in preconfluent cultures of astrocytoma cells exposure to isoproterenol for 24 h led to a 90% loss of beta-adrenergic receptors detectable by [^{125}I]iodohydroxybenzyl-pindolol. Lost receptors were quantitatively recovered in the absence of isoproterenol, even in the presence of cycloheximide. Conversely, when the identical experiment was carried out with postconfluent cultures, recovery of lost receptors was completely blocked by cycloheximide. Homburger et al. (1984) demonstrated that recovery of down regulated beta-adrenergic receptors in C6 glioma cells was not blocked by cycloheximide. Such observations clearly indicate that beta-adrenergic receptor loss cannot in every circumstance be equated to beta-adrenergic receptor degradation. Other evidence hints at alternate pathways for beta-adrenergic receptor loss. For example, Neve et al. (1985) have shown that certain partial agonists can induce beta-adrenergic receptor loss, even though they are incapable of inducing beta-adrenergic receptor internalization. Strader et al. (1984) reported that isoproterenol induced a loss of [^{125}I]iodohydroxyben-zylpindolol binding sites in frog erythrocytes, but such lost sites could be recovered upon solubilization of the beta-adrenergic receptors with detergent. Liao and Perkins (unpublished results; Jyh-Fei Liao, Ph.D. dissertation Yale University, 1990) have observed a rapid ($t_{1/2} < 5$ min) agonist-induced loss of binding sites for [^{125}I]iodocyanopindolol in ROS 17/2.8 cells. The extent of the loss was 15–25% of total beta-adrenergic receptors. On further incubation of cells in the absence of isoproterenol, such lost receptors were rapidly recovered ($t_{1/2} < 30$ min). Beta-adrenergic receptors also were lost from ROS cells by another process that exhibited a 60 min lag, proceeded with a $t_{1/2}$ of 3 h and resulted in greater than 90% loss of [^{125}I]iodocyanopindolol binding. Recovery of such lost beta-adrenergic receptors occurred with a $t_{1/2}$ of 12 h by a process that was completely blocked by cycloheximide. The latter loss reaction may be related to receptor internalization, since blockade of beta-adrenergic receptor internalization prevented this component of beta-adrenergic receptor loss (Liao and Perkins, unpublished).

Given that multiple pathways exist for agonist-induced loss of radioligand binding capacity, what can be said about agonist-induced beta-adrenergic receptor degradation *per se*? Waldo et al. (1984) utilized the heavy isotope, density shift method (Gardener and Fambrough, 1979) to directly measure beta-adrenergic receptor synthesis during recovery from beta-adrenergic receptor downregulation in post-confluent astrocytoma cells. In this instance, recovery of [^{125}I] radioligand binding was accounted for by newly synthesized beta-adrenergic receptors. Thus, in this case, receptor downregulation apparently occurred by an irreversible process, possibly lysosomal degradation. In the same cell line, it has been shown that agonist induced beta-adrenergic receptor loss occurs after a lag of 45–60 min with a $t_{1/2}$ of about 3 h (Su et al., 1979,1980). It also has been shown that, whereas isoproterenol-induced beta-adrenergic receptor internalization occurs at 22°C, beta-adrenergic receptor loss does not occur over a 12 h exposure to the agonist at 22°C (Waldo et al., in preparation). In related studies, Waldo et al. have shown that, whereas methylamine has no effect on agonist-induced beta-adrenergic receptor internalization, it markedly inhibits beta-adrenergic receptor loss. Finally, another condition that blocks beta-adrenergic receptor internalization (hypertonic sucrose), also prevents beta-adrenergic receptor loss.

A recent report by Wang et al. (1989) demonstrates that, whereas exposure of A431 cells to isoproterenol for 24 h leads to 90% loss of detectable [^{125}I]iodocyanopindolol binding sites, essentially no beta-adrenergic receptors are lost based on an immunofluorescence assay of beta-adrenergic receptors on intact or permeabilized cells. The antibodies used are directed against surface determinants of the beta-adrenergic receptors. To date, this technique has been applied to three different cell lines A431, DDT$_1$-MF2, and ROS 17/2.8 cells with similar results (personal communication). In the latter cell type, recovery of beta-adrenergic receptors after exposure to isoproterenol is blocked by cycloheximide. This finding (Wang et al., 1989) is not predicted if beta-adrenergic receptors are lost by way of degradation in lysosomes, and if shown to be generally the case, would require a complete reassessment of the mechanism of agonist-induced beta-adrenergic receptor loss.

4.3. Molecular Analyses
of Beta-Adrenergic Receptor Internalization

Several different approaches have been taken in recent years to determine the molecular entities responsible for agonist-induced internalization of beta-adrenergic receptors. In a study by Insel's group (Mahan et al., 1985a), the variants of the S49 mouse lymphoma system, all of which show some defect in the adenylyl cyclase signal transduction pathway, were examined for their capacities to undergo agonist-induced internalization of beta-adrenergic receptors. This and another study (Clark et al., 1985) clearly demonstrated that neither G_s nor protein kinase A were required for agonist-induced internalization of beta-adrenergic receptors.

Consistent with these results, mutated forms of the human and hamster beta-adrenergic receptors transfected into CHW fibroblasts and mouse L cells, respectively, in which the two consensus sequences for protein kinase A phosphorylation were disrupted or removed, underwent normal internalization in response to agonist (Strader et al., 1987a; Hausdorff et al., 1989).

Phosphorylation of the receptor by beta-adrenergic receptor kinase (*see below*) also has been postulated to be responsible for agonist-induced beta-adrenergic receptor internalization (Sibley et al., 1986; Clark et al., 1988). The involvement of this kinase has recently been tested in two ways. First, putative beta-adrenergic receptor kinase phosphorylation sites in the serine and threonine-rich carboxyl terminal segment of the receptor were deleted or replaced by glycine or alanine residues, with no effect on either the kinetics or extent of agonist-induced receptor internalization (Strader et al., 1987a; Bouvier et al., 1988; Hausdorff et al., 1989). Second, inhibition of beta-adrenergic receptor kinase activity in a permeabilized cell preparation, using low concentrations of heparin, also failed to significantly affect the internalization process (Lohse et al., 1990). A current working model is that receptor conformational changes caused by agonist occupancy alone are sufficient to trigger receptor internalization.

Recent studies by Strader and coworkers have demonstrated that deletion of either the *N*-terminal portion or a large section in the middle of the third intracellular loop of the hamster beta-adrenergic receptor markedly impaired internalization (Strader et al., 1987b; Cheung et al., 1989). They concluded that the regions of beta-adrenergic receptors involved in functional coupling to G_s were also essential for mediating internalization. Unfortunately, because this group relied solely on measurements of [^{3}H]CGP-12177 binding to assess changes in receptor distribution, in several cases it is not clear to what extent they were measuring beta-adrenergic receptor loss or internalization, or both. This distinction is important because another study indicates that the molecular determinants of the receptor required for beta-adrenergic receptor loss are not identical to those required for internalization (Campbell and Lefkowitz, in preparation). These authors constructed a series of mutated beta-adrenergic receptors in which regions of the human beta-adrenergic receptor implicated in coupling to G_s were replaced by the corresponding sections of the human alpha-2A adrenergic receptor. They observed that agonist-induced internalization followed the wild type pattern in all of the mutants, even those that were almost completely unable to activate adenylyl cyclase or promote high affinity agonist binding.

Cheung et al. (1989) reported that increasingly large truncations of the carboxyl terminus of beta-adrenergic receptors ultimately lead to reductions in internalization. A similar observation was made in what appears to be a homologous receptor system in yeast. The mating factor alpha acts by binding to a receptor with remarkable structural similarity to the beta-adrenergic receptor (Blumer et al., 1988). The interaction of alpha-factor with its receptor leads to internalization of the complex—at least in part via clathrin-coated pits (Payne et al., 1988). Genetically engineered truncations of the *C*-terminus lead to a progressive decline in internalization with progressive loss from the *C*-terminus (Reneke et al., 1988). Internalization is eliminated if the portion of the *C*-terminus near to its putative insertion into the yeast plasma membrane is eliminated. This analogy not only provides support for the contention that structures like the beta-adrenergic receptor are amenable to endocytosis via clathrin-coated pits, but leads to the prediction that amino acids in the analogous region of the *C*-terminus of the beta-adrenergic receptors are required for internalization.

4.4. *Molecular Analyses of Beta-Adrenergic Receptor Loss*

In contrast to the apparent lack of involvement of signal transducing proteins in beta-adrenergic receptor internalization, several studies have implicated G_s in certain forms of beta-adrenergic receptor loss. In the study by Mahan et al. (1985a), the different S49 mouse lymphoma variants showed distinct patterns of beta-adrenergic receptor loss in response to agonists. They tested the cyc⁻ cell type, that lacks the alpha subunit of G_s (α_s), unc, in which α_s is present, but is unable to couple to receptors, H21a, where α_s couples to receptors, but is unable to activate the cyclase moiety, and the kin⁻ variant, that lacks the catalytic subunit of protein kinase A. Both the kin⁻ and H21a cell types exhibited an essentially wild type pattern of receptor downregulation. In contrast, the cyc⁻ and unc cell types showed markedly impaired patterns of beta-adrenergic receptor loss. The authors' interpretation was that coupling of beta-adrenergic receptors to G_s itself was somehow required for the full downregulation induced by agonists, but that activation of adenylyl cyclase and protein kinase A were not. Earlier studies by Su et al. (1980), using the HC-1 hepatoma cell line that is devoid of adenylyl cyclase activity but contains beta-adrenergic receptors and G_s also revealed that agonist-induced beta-adrenergic receptor loss did not require adenylyl cyclase or the production of cyclic AMP. These results are consistent with the observations of Molinoff's group (Reynolds and Molinoff, 1986) that atypical agonists that do not detectably activate adenylyl cyclase are still able to induce some, though less, downregulation of beta-adrenergic receptors as compared to full agonists. Also consistent with this interpretation are the findings of Campbell and Lefkowitz (in preparation) who found that the abilities of mutated receptors to undergo downregulation showed a better correlation with their respective abilities to promote formation of the high affinity binding of agonists (owing to interaction with G_s) than with their capacities to stimulate adenylyl cyclase.

Somewhat different conclusions concerning the involvement of protein kinase A in downregulation were drawn from a very recent study utilizing two mutated forms of the human beta-adrenergic receptor altered by site-specific mutagenesis, and expressed in CHW cells (Bouvier et al., 1989). Disruption of the putative protein kinase A site in the third intracellular loop was sufficient to abolish beta-adrenergic receptor phosphorylation induced by dibutyryl cyclic AMP, as well as to markedly attenuate the initial rate and ultimate extent of beta-adrenergic receptor loss caused by the exogenously applied cyclic nucleotide. The cyclic AMP analogs also were found to elicit a decline over the long term in beta-adrenergic receptor mRNA levels, a finding also reported for isoproterenol (Hadcock and Malbon, 1988; Collins et al., 1989). The authors thus concluded that cyclic AMP (and presumably protein kinase A) played at least two different roles in promoting beta-adrenergic receptors downregulation, in a relatively rapid effect via receptor phosphorylation, and in a longer term action in decreasing the steady state level of beta-adrenergic receptor mRNA, possibly via a posttranscriptional mechanism (Collins et al., 1989).

However, as noted by Bouvier et al., a potentially important difference exists between their study and the others cited. Only cyclic AMP analogs and forskolin were examined regarding their capacities to induce downregulation in the Bouvier study, and in the transfected CHW cells these stimuli elicited a markedly slower and less extensive loss of BAR compared to that induced by isoproterenol (Bouvier et al., 1989). Thus, the conclusion reached with the S49 mutants (Mahan et al., 1985a), that activation of adenylyl cyclase, cyclic AMP production, and protein kinase A were not required for *agonist*-induced downregulation of beta-adrenergic receptors, does not preclude their involvement in the slower, heterologous forms of downregulation, i.e., those forms induced by agents other than beta-adrenergic agonists.

An additional complicating factor is that the molecular constructs encoding the wild type and mutant human beta-adrenergic receptors sequences used by Bouvier et al. lacked the 5' untranslated region in the genomic sequence that contains a putative cyclic AMP response element. This element is involved in a rapid, agonist-induced, and apparently cyclic AMP-mediated increase in beta-adrenergic receptors gene transcription in DDT_1MF_2 cells (Collins et al., 1989). Thus far, it is unclear whether this transient rise in mRNA levels is manifested as a transient increase in beta-adrenergic receptors expression levels, or as a slowing of the rate of beta-adrenergic receptors disappearance, and whether a similar phenomenon occurs in the S49 cells.

Just as there may be differences in the requirement for protein kinase A in agonist-induced vs heterologous beta-adrenergic receptor loss, a study by Malbon and colleagues (Hadcock et al., 1989a) suggests that long term exposure to agonists (>6 h) lowered beta-adrenergic receptor mRNA levels by a second mechanism in addition to one involving increases in protein kinase A activity. They reported that, whereas isoproterenol was predictably unable to down regulate the mRNA in most of the S49 mutants, it elicited a partial lowering of mRNA levels in the H21a variant. The authors concluded from these studies that, receptor-G_s coupling alone somehow induced a lowering of beta-adrenergic receptor mRNA levels, perhaps via another effector system linked to G_s activation, with only basal levels of protein kinase A activity required for this effect (Hadcock et al., 1989a). The same group also demonstrated that the isoproterenol effects on beta-adrenergic receptor mRNA levels are caused not by effects on transcription, but on the stability of the mRNA, decreasing its half life from ~12 to ~5 h (Hadcock et al., 1989b).

The results of several studies thus agree strikingly with one another regarding the fundamental importance of beta-adrenergic receptor-G_s coupling in agonist-induced downregulation. The studies differ, however, in the apparent requirement of basal levels of protein kinase A for decreases in beta-adrenergic receptor mRNA levels (Hadcock et al., 1989b), but not for beta-adrenergic receptor loss (Mahan et al., 1985a). Unfortunately, accompanying agonist-induced changes in beta-adrenergic receptor levels were not measured in the various mutants in the mRNA study. It also remains to be determined to what extent losses in beta-adrenergic receptor following long-term exposure to agonist are attributable to the declines in mRNA levels.

4.5. The Functional Status
of Internalized Beta-Adrenergic Receptor

The apparent molecular size (Waldo et al., 1983; Stadel et al., 1983a) and the antagonist binding properties of beta-adrenergic receptor isolated as light vesicle fractions are the same as native beta-adrenergic receptor, with the exception mentioned previously regarding ligand accessibility. However, when $(R)_v$ are examined in regard to their agonist binding properties, distinctions are apparent (Harden et al., 1980; Stadel et al., 1983a). $(R)_v$ exhibit only low affinity for agonists, and GTP has no effect on agonist binding. The absence of GTP-sensitive, high affinity binding of agonists has been interpreted as indicating an inability of $(R)_v$ to form complexes with the guanine nucleotide binding protein G_s (DeLean et al., 1980). Studies with the lipophilic agonist zinterol led DeBlasi et al. (1985) to also conclude that internalized receptors are nonfunctional with regard to adenylyl cyclase stimulation.

The possibility that the light vesicles containing $(R)_v$ might not contain G_s has been explored (Waldo et al., 1983; Stadel et al., 1983a). The results show clearly that catecholamines do not induce a translocation of G_s (or G_i or adenylyl cyclase) from the plasma membrane to the light vesicle fraction, although in one study (Waldo et al., 1983), the formation of vesicles containing a 1:1 ratio of beta-adrenergic receptors to G_s could not be excluded.

Even if one supposes that G_s is not present in the vesicles containing $(R)_v$, the question of the functional state of receptors in both light vesicles and at the plasma membrane remains. This question has been addressed using a variety of membrane fusion protocols to assess the capacity of donor R_v to reconstitute beta-adrenergic receptor functions in acceptor membranes or cells. Strulovici et al. (1983) tested the functionality of beta-adrenergic receptor isolated from a light vesicle fraction obtained from frog erythrocytes that had been exposed to catecholamines. When such vesicles were fused to cells (*Xenopus laevis* erythrocytes) that possess G_s and adenylyl cyclase, but lack beta-adrenergic receptors, the $(R)_v$ reconstituted isoproterenol-stimulated adenylyl cyclase activity. However, the relative capacities of equal numbers of beta-adrenergic receptors from desensitized or control cells to reconstitute beta-adrenergic/adenylyl cyclase activity were not compared. Thus, it remains possible that $(R)_v$ retain only partial functional capacity.

Kassis and Fishman (1984) and Clark et al. (1985) have utilized membrane–membrane fusion protocols, in the presence of detergents, to examine the functional capacity of beta-adrenergic receptors in desensitized cells. Clark et al. isolated light vesicle and heavy vesicle fractions of beta-adrenergic receptors from control and agonist-treated cyc⁻ S49 cells. The capacity of cholate extracts of wild-type S49 membranes to reconstitute beta-adrenergic receptor/adenylyl cyclase in the vesicle preparations was comparable. However, the interpretation of these results is complicated since the light vesicle fractions contained roughly one-half the total NaF/adenylyl cyclase activity, and one-third the forskolin/adenylyl cyclase activity. It seems unlikely that a purified cytosolic vesicle fraction would contain such high proportions of adenylyl cyclase activity, based on the work of others (Waldo et al., 1983; Stadel et al., 1983a). Thus, it is not clear that the light vesicle preparation utilized in the studies of Clark et al. (1985) can be compared with those of astrocytoma cells or frog erythrocytes, in terms of the level of exclusion of plasma membrane components.

Fishman and his colleagues have utilized a somewhat different approach to the question of $(R)_v$ functionality. They initially isolated a crude membrane preparation from control and agonist-exposed cells that were fused with acceptor membranes from cells devoid of beta-adrenergic receptors (Kassis and Fishman, 1984). Beta-adrenergic receptors donated by desensitized cells exhibited only 40–60% of the capability of an equal number of beta-adrenergic receptors from native cells to stimulate adenylyl cyclase. A potential problem with this study is that a mixture of membranes was used as donor of beta-adrenergic receptors; thus, beta-adrenergic receptors in plasma membranes, as well as in endosomes would be available for transfer. Since these two populations of receptor may well exist in opposite orientation in the membrane, the ratio of right-side-out and right-side-in beta-adrenergic receptors might change without a change in the total assayable beta-adrenergic receptors transferred. This is possible, since a lipophilic radioligand was used to assess beta-adrenergic receptors content. In a subsequent paper (Kassis et al., 1986), these workers attempted to address this criticism by fusing membranes from control and agonist-treated cells that had additionally been treated with phenylarsine oxide or concanavalin A. These latter two agents prevented beta-adrenergic receptor internalization without blocking the uncoupling process. Thus, even though beta-adrenergic receptors remained in the plasma membrane available to hydrophilic ligands, isoproterenol could not fully activate adenylyl cyclase. Fusion of such membranes from desensitized cells with beta-adrenergic receptor-depleted acceptor membranes revealed beta-adrenergic receptors with reduced functionality relative to controls.

Strulovici et al. (1984) took advantage of the noninternalizing turkey erythrocyte system to demonstrate directly that beta-adrenergic receptors purified from a plasma membrane fraction of desensitized erythrocytes were markedly impaired in their functionality. These and the experiments of Kassis et al. (1986) established that plasma membranes from desensitized cells contain beta-adrenergic receptors with functional deficits. Conversely, this work and previous work by others described in the next section (Waldo et al., 1983; Hertel et al., 1985; Feldman et al., 1986) clearly shows that, in certain cells, beta-adrenergic receptor translocation is *not* required for attainment of a "desensitized" state.

The studies mentioned have not definitively answered the related question of the functional status of beta-adrenergic receptors in the internalized form once the reaction has occurred. This question can be answered unambiguously only when beta-adrenergic receptors from purified light vesicles are fused into simplified acceptor vesicles of known chemical constituents. The problem is that all reconstitution experiments, especially those involving intact cells, must be interpreted within the caveat that the reversal of any uncoupled state of the beta-adrenergic receptors could occur during the fusion procedure. Clearly, since the overall process is reversible, any alteration that does account for the uncoupled behavior of the beta-adrenergic receptors must also be rapidly reversed in intact cells *(see above)*. Under such defined conditions, the machinery for reversal of the desensitized state should be absent. A quantitative comparison of the reconstitutive activity of beta-adrenergic receptors purified from light vesicles of desensitized cells and those purified from the plasma membrane of control cells has yet to be published. Of interest in this regard are the observations of Sibley et al. (1986), who demonstrated that in desensitized frog erythrocytes, beta-adrenergic receptors in the light vesicles are markedly less phosphorylated than their plasma membrane counterparts.

The direct demonstration of a chemical modification of the beta-adrenergic receptors and its possible role in the uncoupling and internalization of beta-adrenergic receptors are discussed later.

4.6. Do Agonist-Induced Desensitization, Internalization, and Loss Reflect an Ordered Sequence?

The various reactions associated with agonist-induced change in beta-adrenergic receptor function are proposed as a linear sequence (Fig. 1). Since a physical separation of beta-adrenergic receptors from G_s and adenylyl cyclase is a *sufficient* mechanism to explain uncoupling, it is a legitimate first hypothesis. However, as indicated above, a number of considerations suggests a preliminary step. Detailed kinetic analyses have detected changes occurring in beta-adrenergic receptor coupling capacity prior to detectable beta-adrenergic receptor internalization (Waldo et al., 1983). The lack of detection of such disparities by others (e.g., Linden et al., 1984) does not preclude the sequence proposed and may reflect a kinetic pattern that minimizes accumulation of $(pR)_{pm}$ in the plasma membrane. Further evidence for the existence of an uncoupling reaction in addition to internalization comes from studies in which internalization can be blocked without prevention of uncoupling. Waldo et al. (1983) demonstrated that prior treatment of astrocytoma cells with concanavalin A completely prevented beta-adrenergic receptor internalization but did not affect the agonist-induced loss of isoproterenol-stimulated adenylyl cyclase activity. Similar results were subsequently obtained by Kassis et al. (1986) using A431 and C6 cells. In related experiments Hertel et al. (1985), Kassis et al. (1986), and Feldman et al. (1986) demonstrated that phenylarsine oxide also prevented beta-adrenergic receptor internalization but did not prevent uncoupling. Conversely, as mentioned earlier, in recent studies various manipulations have markedly impaired desensitization without affecting internalization (Bouvier et al., 1988; Hausdorff et al., 1989; Lohse et al., 1990). Thus, if uncoupling and internalization are steps in a sequence, the uncoupling reaction occurs first. Further, internalization is not required for the desensitized (uncoupled) state to be expressed.

Even if the model (Fig. 1) is accurate in terms of the individual steps involved but there is variation among cell types in the relative rates of the various reactions, a large number of patterns could be found. Thus, if reaction 3b is rapid relative to reactions 4 and 5, phosphorylated receptors would accumulate in the plasma membrane. If reaction 4 and 5 are rapid relative to reaction 6, desensitized beta-adrenergic receptors would accumulate in light vesicles. If reaction 6 is rapid relative to reaction 4 and 5, phosphorylated beta-adrenergic receptors would not be found in light vesicles. Thus, the same degree of desensitization could be achieved with quite different distributions of the various uncoupled forms of beta-adrenergic receptors: $(pR)_{pm};(pR)_v; (R)_v$. Such an argument suggests the exercise of caution in attempting a comparison of these events in different cell types, and certainly when comparing cultured mammalian cells and erythrocytes, since the overall kinetics of beta-adrenergic receptor desensitization and translocation appear much slower in the erythrocyte.

5. Phosphorylation of Beta-Adrenergic Receptors: Probable Mechanism of Rapid Desensitization

The previous sections have described the evidence that agonist-induced internalization of beta-adrenergic receptor is probably not the major mechanism responsible for the rapid functional desensitization observed in most systems. Another potential model of desensitization invoked the formation of a beta-adrenergic receptor-G_i complex, resulting in a desensitized state of the beta-adrenergic receptor. This hypothesis was based on the demonstration by two groups that G_i was capable of interacting with beta-adrenergic receptors in reconstituted systems (Asano et al., 1984; Cerione et al., 1985). In an attempt to test this idea, Clark et al. (1986) pretreated cells with pertussis toxin. Their rationale was based on the observations that ADP-ribosylation of G_i prevents its normal interaction with certain receptors (e.g., alpha-2 adrenergic; muscarinic). Greater than 85% inhibition of G_i function in S49 cells failed to prevent agonist-induced desensitization of beta-adrenergic receptor function. Waldo, Harden, and Perkins (unpublished results) carried out the same analysis in human astrocytoma cells with the same result. Thus, evidence to date does not support the validity of this possibility.

In contrast, a fair amount of evidence accumulated over the past decade (outlined below) has indicated that uncoupling of beta-adrenergic receptors from G_s may be sufficient to yield the desensitization phenotype. This section will discuss the current level of knowledge regarding the nature and role of agonist-induced receptor phosphorylation in the desensitization process.

5.1. Components Required for Agonist-Induced Beta-Adrenergic Receptor Desensitization

Several observations support the concept that the beta-adrenergic receptor *per se* is the component of the isoproterenol-sensitive signal transduction pathway that is altered in desensitization. Initially, it simply seemed logical that the receptor would be the altered species since other aspects of G_s and adenylyl cyclase function appeared to be unaltered in membranes from catecholamine-treated cells, and responsiveness to other hormones often was apparently not changed.

Experimental evidence pointing to an alteration in beta-adrenergic receptors originally came from studies of the cyc⁻ mutant of the S49 mouse lymphoma cell line. These cells do not express the gene for $G_{s\alpha}$ (Harris et al., 1985), and thus adenylyl cyclase activity cannot be stimulated by hormones in these cells (Coffino et al., 1976). A key experiment by Green and Clark (1981) involved reconstitution of membranes from naive and catecholamine-treated cyc⁻ cells with G_s from wild-type S49 cells. The catecholamine-sensitive adenylyl cyclase activity in membranes from agonist-treated cyc⁻ cells could not be reconstituted by wild-type G_s to control levels. In essentially the converse of this experiment, it was shown that G_s extracted from wild-type S49 cells pretreated with catecholamine was equal to G_s from native cells in reconstituting naive cyc⁻ membranes. Both results indirectly indict the beta-adrenergic receptor as the altered entity.

Direct evidence that the beta-adrenergic receptor itself undergoes some kind of functional modification was greatly facilitated by the ability to purify, essentially to homogeneity, beta-adrenergic receptors from intact cells (Shorr et al., 1982; Benovic et al., 1984). By taking advantage of this technique, Strulovici and colleagues were able to demonstrate that receptors purified from desensitized turkey erythrocyte cells were functionally impaired as assessed in a reconstitution system (Strulovici et al., 1984). Similar conclusions were reached by Kassis et al. (1986) using membrane fusion techniques and several mammalian cell lines.

The covalent nature of this functional modification was demonstrated in studies that detailed the altered electrophoretic mobilities of photoaffinity labeled receptors from desensitized turkey erythrocytes (Stadel et al., 1982). Subsequent experiments in which cells were prelabeled with $^{32}PO_4$ revealed that phosphorylation of the receptor occurs following exposure of the cells to desensitizing concentrations of agonist, with approx. 2–3 mol of phosphate incorporated into each mol of beta-adrenergic receptors (Stadel et al., 1983b; Sibley et al., 1984). The kinetics and extent of agonist-induced beta-adrenergic receptor phosphorylation and desensitization also have been shown to closely correlate with each other, and to exhibit similar dose-dependences for agonist (Sibley et al., 1984,1985,1986). Furthermore, treatment of desensitized turkey erythrocyte membranes with phosphatases also has been shown to relieve desensitization (Stadel et al., 1988). Importantly, these correlations between phosphorylation and desensitization were made in systems when the predominant patterns of agonist-induced desensitization were manifested as either predominantly heterologous (turkey erythrocyte: Sibley et al., 1984; Stadel et al., 1988), or predominantly homologous (frog erythrocyte: Sibley et al., 1985,1986).

5.2. Specific Kinases Involved

Several lines of evidence suggest that phosphorylation of beta-adrenergic receptors by protein kinase A is intimately involved in agonist-induced desensitization. The potential involvement of the kinase in these processes in intact cells was strongly suggested by studies in which exposure of turkey erythrocytes to exogenously added cyclic AMP analogs led to both phosphorylation and desensitization of beta-adrenergic receptors (Sibley et al., 1984). That phosphorylation of beta-adrenergic receptors by protein kinase A has functional consequences has been directly shown with plasma membrane fractions (Nambi et al., 1985; Kunkel et al., 1989) and purified beta-adrenergic receptors reconstituted into phospholipid vesicles (Benovic et al., 1985). In each case, preincubation of the beta-adrenergic receptors preparations with the catalytic subunit of protein kinase A resulted in functional impairments in the G_s-coupling properties of the receptor.

Similarly, several studies have established a role for protein kinase A phosphorylation of beta-adrenergic receptors in agonist-induced desensitization of intact cells. Nambi et al. (1985) demonstrated in a cell-free homogenate that protein kinase inhibitor, a specific inhibitor of protein kinase A, partially inhibited both phosphorylation and desensitization of beta-adrenergic receptors induced by agonist. Comparable results were obtained by Lohse et al. (1990), using a permeabilized cell preparation. Studies with the variants of the S49 lymphoma cell system also implied a fundamental role for protein kinase A in the heterologous desensitization induced by nanomolar concentrations of agonist (Clark et al., 1988). These studies have generally been interpreted to implicate a key functional role for protein kinase A in *heterologous* desensitization. However, disruption of the two consensus sequences (RRSS) for protein kinase A phosphorylation on beta-adrenergic receptors by site-specific mutagenesis (Mutant A in Fig. 3) resulted in marked impairments in the abilities of agonists to promote beta-adrenergic receptor phosphorylation and desensitization (Hausdorff et al., 1989) induced by either nanomolar or micromolar concentrations of agonist. Similarly, in the permeabilized cell system, Lohse et al. (1990) found that protein kinase inhibitor markedly inhibited desensitization induced by micromolar concentration of agonists. These results indicate that protein kinase A may *always* be involved in desensitization induced by agonists, even during so-called homologous desensitization induced by high concentrations of agonist (*see* next section).

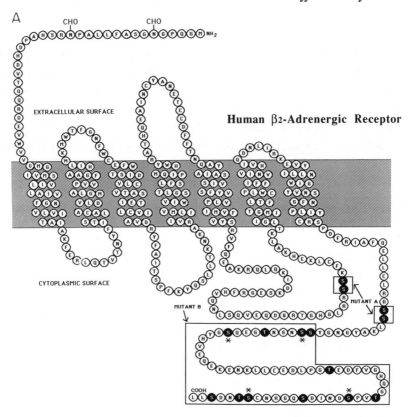

Fig. 3. Membrane topography of wild-type and mutant human beta$_2$-adrenergic receptors. Filled circles are serine or threonine residues substituted in the mutants by alanine, except for those marked with an asterisk, that were substituted for by glycine. Residues substituted for in mutant AB include those replaced in mutants A and B. Adapted from Hausdorff et al., 1989.

It is not known how phosphorylation of beta-adrenergic receptors by protein kinase A actually functionally uncouples receptors from G_s. It is worth noting, however, that the consensus sequences for protein kinase A phosphorylation are located immediately adjacent to the carboxyl terminus of the third intracellular loop, and the amino terminus of the cytoplasmic tail. Both regions have been proposed to be involved in coupling of the receptor to G_s (O'Dowd et al., 1988; Hausdorff et al., 1990; Strader et al., 1987b). Recent evidence suggests that the phosphorylation site located in the third intracellular loop has greater functional significance in the densensitization elicited by nanomolar levels of agonist (Clark et al., 1989).

Although the above studies have clearly implicated protein kinase A in the desensitization process, there is substantial evidence that another agonist-induced desensitization pathway is operative. First, the extent of desensitization elicited by cyclic AMP analogs is generally markedly less than that induced by agonists (Sibley et al., 1984; Nambi et al., 1985). In addition, several groups have demonstrated that agonist-induced desensitization occurs in the mutant S49 lymphoma cells and HC-1 cells (Shear et al., 1976; Coffino et al., 1976; Su et al., 1980; Green and Clark et al., 1981; Strasser et al., 1986), despite the absence of agonist-stimulated protein kinase A activity.

The involvement of a second kinase in the desensitization process was implicated when it was shown that, agonists promote beta-adrenergic receptor phosphorylation even in the kin-mutant of S49 lymphoma cells (Strasser et al., 1986). In addition, careful examination of the electrophoretic mobilities of photoaffinity- and ^{32}P-labeled beta-adrenergic receptors purified from cells pretreated with cyclic AMP or agonist also led Stadel et al. (1987) to conclude that, two kinases phosphorylate beta-adrenergic receptors in intact cells. These findings led to the isolation and purification of a novel kinase (Benovic et al., 1987a) based on its ability to preferentially phosphorylate only the agonist-occupied form of beta-adrenergic receptors. Starting with the amino acid sequence of peptide fragments of the purified kinase, a cDNA encoding the gene for beta-adrenergic receptor kinase has recently been cloned from a bovine brain cDNA library (Benovic et al., 1989). Based on the limited sequence homology with other known kinases, this gene has been proposed to represent the first member of a multi-gene family. Interestingly, however, inspection of the proposed amino acid sequence revealed significant homology (~33%) of the putative beta-adrenergic receptor kinase catalytic domain with the corresponding regions of the catalytic subunits of protein kinase A and with protein kinase C (Benovic et al., 1989).

Several lines of evidence support the hypothesis that this kinase mediates agonist-specific desensitization. First, phosphorylation by crude beta-adrenergic receptor kinase preparations of purified beta-adrenergic receptors reconstituted in phospholipid vesicles greatly attenuated the functional coupling of the receptor with G_s. Essentially pure preparations caused phosphorylation but evoked little desensitization. This observation, coupled with the marked restoration of desensitization when retinal arrestin (48K) was added, along with purified beta-adrenergic receptor kinase (Benovic et al., 1987b), suggests the involvement of a 48K-like molecule in analogy to the rhodopsin/rhodopsin kinase system of the retina.

The carboxyl terminal segment of beta-adrenergic receptors has been shown by enzymatic digestion analyses to be the predominant locus of phosphorylation by beta-adrenergic receptor kinase in vitro (Dohlman et al., 1987b). Mutated versions of beta-adrenergic receptors expressed in mammalian cells in which serine and threonine residues in this region (Fig. 3) were deleted or replaced by alanine and glycine residues (Mutant B), showed markedly decreased levels of agonist-induced phosphorylation and desensitization (Bouvier et al., 1988; Hausdorff et al., 1989; Liggett et al., 1989) following a short-term (<3 h) incubation of cells with micromolar concentrations of agonist. In addition, pretreatment of permeabilized A431 cells with concentrations of heparin that almost completely abolish the ability of beta-adrenergic receptor kinase to phosphorylate purified beta-adrenergic receptors in vitro similarly reduced the subsequent extent of beta-adrenergic receptor phosphorylation and desensitization induced by micromolar concentration of agonists (Lohse et al., 1989; Lohse et al., 1990). In contrast, the beta-adrenergic receptor phosphorylation (Hausdorff et al., 1989) and desensitization (Hausdorff et al., 1989; Lohse et al., 1990) induced by *nanomolar* levels of agonist are not affected by removal of the putative beta-adrenergic receptor kinase sites or treatment with heparin. One interpretation of this result is that agonist occupancy is required for beta-adrenergic receptor kinase activation in intact cells, as well as in in vitro systems (Benovic et al., 1986).

5.3. Complexity of the Desensitization Process

The results obtained from several of the studies described above suggest that the influence of protein kinase A on agonist-induced phosphorylation and desensitization of beta-adrenergic receptors may have been seriously underestimated in many studies. For example, on a conceptual level, it is difficult to imagine an instance in which pretreatment of cells with micromolar concentrations of agonist would not activate protein kinase A, since only nanomolar levels of agonist are generally required for maximal increases in cellular cyclic AMP levels (Liggett et al., 1989). Since protein kinase A has traditionally been believed to mediate "heterologous" desensitization, one would expect that a purely "homologous" pattern of desensitization would rarely, if ever, exist. There are two major reasons why the contribution of protein kinase A to desensitization may have been overlooked. First, Clark and colleagues have recently shown that lowering the concentration of free magnesium ion in adenylyl cyclase assays to <1–2 mM "unmasks" the contribution of protein kinase A phosphorylation to functional uncoupling of beta-adrenergic receptors in S49 cells induced by low concentrations of agonist (Clark et al., 1987). These results have been extended to lymphocytes (Feldman, 1989), and A431 cells (Lohse et al., 1990) that were desensitized with micromolar concentrations of agonist. Much higher concentrations of the divalent cation have traditionally been used in measurements of adenylyl cycalse activity.

Second, most studies of desensitization have utilized only micromolar and supramicromolar concentrations of agonist, both during the desensitization process and in the adenylyl cyclase assay itself. However, recent studies (Hausdorff et al., 1989; Clark et al., 1989; Lohse et al., 1990) have emphasized the importance of constructing full dose-response curves in the enzymatic assay since impairments in functional coupling putatively owing to protein kinase A phosphorylation are most obviously manifested in decreased potencies of agonist to restimulate the adenylyl cyclase response. Even when full dose-response curves are constructed, marked decreases in potency in the absence of decreases in efficacy can be overlooked (e.g., *see* Garcia-Sainz and Michel, 1987).

Another variable that has recently assumed great importance in the understanding of the molecular mechanisms underlying desensitization is the duration of cellular exposure to desensitizing concentrations of agonist. For example, although the studies with the mutated forms of beta-adrenergic receptors have been very useful in elucidating the contributions of various kinases to desensitization, it must be kept in mind that, in general, differences in patterns of desensitization among the mutants disappear following exposure of the cells to agonist for several hours (Strader et al., 1987a; Bouvier et al., 1988; Cheung et al., 1989). At present, the molecular basis for this finding is unclear. One possibility is that receptor phosphorylation is only important in short term desensitization, and that other mechanisms (such as receptor loss) play more important roles following long-term exposure to agonist.

A second possibility is that agonist occupancy triggers some other signal leading to functional uncoupling, and that phosphorylation simply hastens the process, as has been observed with the nicotinic cholinergic receptor (Huganir and Greengard, 1987). In support of this hypothesis, the data of Benovic et al. (1987b) have suggested that additional cytosolic factors, perhaps a homolog of retinal arrestin, are normally required to mediate the desensitization process.

Finally, it cannot be excluded that even after long term agonist exposure, receptor phosphorylation is indeed the major mechanism uncoupling beta-adrenergic receptors from G_s, but that in the experimentally engineered absence of phosphorylation, the internalization and loss of receptors, proceeding unimpaired, ultimately serves to uncouple the receptors.

Figure 4 (A and B) depicts a current model of desensitization induced by cellular exposure to low (nanomolar) or high (micromolar) concentrations of agonist. Protein kinase A appears to be the principal mediator of desensitization induced by low concentrations of catecholamines that are similar to those in the peripheral circulation. In contrast, both beta-adrenergic receptor kinase and protein kinase A appear to mediate desensitization resulting from exposure of cells to micromolar concentrations of agonist, levels found mainly in neural synapses (Bevan and Su, 1974). Interestingly, the highest concentrations of beta-adrenergic receptor kinase mRNA were found in tissues receiving substantial adrenergic innervation, and in the brain (Benovic et al., 1989).

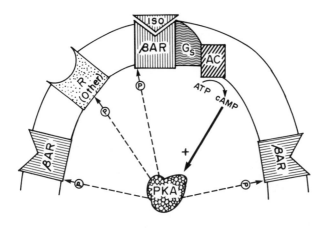

LOW [AGONIST]

Fig. 4A. Schematic model of kinases proposed to mediate agonist-induced phosphorylation and desensitization of beta-adrenergic receptors. "Low" (nanomolar) concentrations of ISO bind to only a small fraction of the cellular complement of beta-adrenergic receptors, thereby activating the stimulatory G protein (Gs), and adenylyl cyclase. Because of the existence of a receptor reserve, these concentrations of ISO nonetheless maximally stimulate cellular cyclic AMP levels (Liggett et al., 1989), and in turn activate the cyclic AMP-dependent protein kinase. Activated protein kinase A can then phosphorylate beta-adrenergic receptors, both in its agonist-occupied and -unoccupied forms, as well as other hormone receptors (R). The functional consequences of this phosphorylation consist predominantly of decreases in the potency of ISO (and that of the agonist binding to R) to restimulate adenylyl cyclase. Since beta-adrenergic receptor kinase requires agonist-occupied receptors to serve as substrates, only the few beta-adrenergic receptors that are occupied will be phosphorylated by that kinase.

6. Conclusion

In summary, several changes in the physical location and the functional status of the beta-adrenergic receptors occur on exposure of cells to agonist, and contribute to the process of desensitization of receptor function. Agonist-induced receptor phosphorylation by at least two kinases leads to rapid desensitization. Although much has

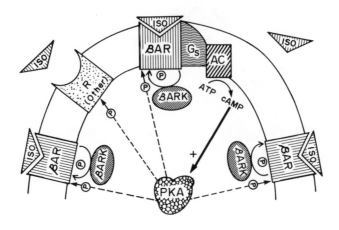

HIGH [AGONIST]

Fig. 4B. Schematic model of kinases proposed to mediate agonist-induced phosphorylation and desensitization of beta-adrenergic receptors. "High" (micromolar) concentrations of ISO bind to almost all available beta-adrenergic receptors, activating protein kinase A, and leading to phosphorylation of the receptors. In addition, beta-adrenergic receptor kinase can now phosphorylate a substantial fraction of the cellular pool of receptors leading to the profound desensitization seen under these desensitizing conditions.

been learned about these various processes in the past few years, there are many unanswered questions. Among these are the following:

1. What is the role of rapidly occurring receptor internalization in desensitization?
2. How does this process occur?
3. How do molecular components such as beta-adrenergic receptor kinase and an arrestin homolog cooperate to regulate receptor function?
4. How many such receptor kinases are there, and how broad are their regulatory actions?
5. What are the physiological contributions of protein kinase A and beta-adrenergic receptor kinase in beta-adrenergic receptors desensitization?

6. How is receptor "downregulation" mediated, and what is its contribution to long term desensitization?

The stage now appears set for answers to these and other questions that should lead to a much clearer understanding of the basic mechanisms that underlie these important regulatory processes.

References

Allen, J. M., Baetge, E. E., Abrass, I. B., and Palmiter, R. D. (1988) Isoproterenol response following transfection of the mouse β-2 adrenergic receptor gene into Y1 cells. *EMBO J.* **7**, 133–138.

Asano, T., Katada, T., Gilman, A. G., and Ross, E. M. (1984) Activation of the inhibitory GTP-binding protein of adenylate cyclase, G_i by β-adrenergic receptor in reconstituted phospholipid vesicles. *J. Biol. Chem.* **259**, 9351–9354.

Benovic, J. L., Shorr, R. G. L., Caron, M. G., and Lefkowitz, R. J. (1984) The mammalian β-2 adrenergic receptor: Purification and characterization. *Biochemistry* **23**, 4510–4518.

Benovic, J. L., Mayor, Jr., F., Staniszewski, C., Lefkowitz, R. J., and Caron, M. G. (1987a) Purification and characterization of the β-adrenergic receptor kinase. *J. Biol. Chem.* **262**, 9026–9032.

Benovic, J. L., Kuhn, H., Weyand, I., Codina, J., Caron, M. G., and Lefkowitz, R. J. (1987b) Functional desensitization of the isolated β-adrenergic receptor kinase: Potential role of an analog of the retinal protein arrestin (48-kDa protein) *Proc. Natl. Acad. Sci.* **84**, 8879–8882.

Benovic, J. L., DeBlasi, A., Stone, W. C., Caron, M. G., and Lefkowitz, R. J. (1989) β-adrenergic receptor kinase: Primary structure delineates a multigene family. *Science* **246**, 235–240.

Benovic, J. L., Bouvier, M., Caron, M. G., and Lefkowitz, R. J. (1988) Regulation of adenylyl cyclase-coupled β-adrenergic receptors. *Ann. Rev. Cell Biol.* **4**, 405–428.

Benovic, J. L., Strasser, R. H., Caron, M. G., and Lefkowitz, R. J. (1986) β-adrenergic receptor kinase: Identification of a novel protein kinase that phosphorylates the agonist-occupied form of the receptor. *Proc. Natl. Acad. Sci. USA* **83**, 2797–2801.

Benovic, J. L., Pike, L. J., Cerione, R. A., Staniszewski, C., Yoshimasa, T., Lefkowitz, R. J., and Caron, M. G. (1985) Phosphorylation of the mammalian β-adrenergic receptor by cyclic AMP-dependent protein kinase: Regulation of the rate of receptor phosphorylation and dephosphorylation by agonist occupancy and effects on coupling of the receptor to the stimulatory guanine nucleotide regulatory protein. *J. Biol. Chem.* **260**, 7094–7101.

Bevan, J. A. and Su, C. (1974) Variation of intra- and perisynaptic adrenergic transmitter concentrations with width of synaptic cleft in vascular tissue. *J. Pharm. Exp. Ther.* **190**, 30–38.

Blumer, K. J., Reneke, Johanna, J. E., and Thomer, J. (1988) The STE2 gene product is the ligand-binding component of the α-factor receptor of Saccharomyces cerevisiae. *J. Biol. Chem.* **263**, 10836–10842.

Bouvier, M., Hausdorff, W. P., DeBlasi, A., O'Dowd, B. F., Kobilka, B. K., Caron, M. G., and Lefkowitz, R. J. (1988) Removal of phosphorylation sites from the β-2 adrenergic receptor delays onset of agonist-promoted desensitization. *Nature* **333**, 370–373.

Bouvier, M., Collins, S., O'Dowd, B. F., Campbell, P. T., DeBlasi, A., Kobilka, B. K., MacGregor, C., Irons, G. P., Caron, M. G., and Lefkowitz, R. J. (1989) Two distinct pathways for cyclic AMP-mediated downregulation of the β-2 adrenergic receptor. *J. Biol. Chem.* **264**, 16786–16792.

Cerione, R. A., Codina, J., Benovic, J. L., Lefkowitz, R. J., Birnbaumer, L., and Caron, M. G. (1984) The mammalian β-adrenergic receptor: Reconstitution of functional interactions between pure receptor and pure stimulatory nucleotide binding protein of the adenylate cyclase system. *Biochemistry* **23**, 4519–4525.

Cerione, R. A., Staniszewski, C., Benovic, J. L., Lefkowitz, R. J., Caron, M. G., Gierschik, P., Somers, R., Spiegel, A. M., Codina, J., and Birnbaumer, L. (1985) Specificity of the functional interactions of the β-adrenergic receptor and rhodopsin with guanine nucleotide regulatory proteins reconstituted in phospholipid vesicles. *J. Biol. Chem.* **260**, 1493–1500.

Cheung, A. H., Sigal, I. S., Dixon, R. A. F., and Strader, C. D. (1989) Agonist-promoted sequestration of the β-2 adrenergic receptor requires regions involved in functional coupling with G_s. *Mol. Pharm.* **34**, 132–138.

Chuang, D.-M. and Costa, E. (1979) Evidence for internalization of the recognition site of β-adrenergic receptors during receptor subsensitivity induced by (-)-isoproterenol. *Proc. Natl. Acad. Sci. USA* **76**, 3024–3028.

Chuang, D.-M. (1982) Internalization of β-adrenergic receptor binding sites: Involvements of lysosomal enzymes. *Biochem. Biophys. Res. Commun.* **105**, 1466–1472.

Chuang, D.-M, Kinnier, W. J., Farber, L., and Costa, E. (1980) A biochemical study of receptor internalization during β-adrenergic receptor desensitization in frog erythrocytes. *Mol. Pharmacol.* **18**, 348–355.

Clark, R. B., Friedman, J., Dixon, R. A. F., and Strader, C. D. (1989) Identification of a specific site required for rapid heterologous desensitization of the β-adrenergic receptor by cyclic AMP-dependent protein kinase. *Mol. Pharmacol.* **36**, 343–348.

Clark, R. B., Kunkel, M. W. Friedman, J., Goka T. J., and Johnson, J. A. (1988) Activation of cyclic AMP-dependent protein kinase is required for heterologous desensitization of adenylyl cyclase in S49 wild-type lymphoma cells. *Proc. Natl. Acad. Sci. USA* **85**, 1442–1446.

Clark, R. B., Friedman, J., Johnson, J. A., and Kunkel, M. W. (1987) β-adrenergic receptor desensitization of wild-type but not cyc⁻ lymphoma cells unmasked by submillimolar Mg^{2+}. *FASEB J.* **1**, 289–297.

Clark, R. B., Friedman, J., Prashad, N., and Ruoho, A. E. (1985) Epinephrine-induced sequestration of the β-adrenergic receptor in cultured S49 WT and cyc⁻ lymphoma cells. *J. Cyc. Nuc. Prot. Phos. Res.* **10**, 97–119.

Clark, R. B., Goka, T. J., Proll, M. A., and Friedman, J. (1986) Homologous desensitization of β-adrenergic receptors in lymphoma cells is not altered by the inactivation of Ni (G_i), the inhibitory guanine nucleotide regulatory protein. *Biochem. J.* **235**, 399–405.

Coffino, P., Bourne, J. R. Friedrich, U., Hochman, J., Insel, P. A., Lemaire, I., Melmon, K. L., and Tomkins, G. M. (1976) Molecular mechanisms of cyclic AMP action: A genetic approach. *Recent Prog. Horm. Res.* **32**, 669–684.

Collins, S., Bouvier, M., Bolanowski, M. A., Caron, M. G., and Lefkowitz, R. J. (1989) cyclic AMP stimulates transcription of the β-2 adrenergic receptor gene in response to short-term agonist exposure. *Proc. Natl. Acad. Sci. USA* **86**, 4853–4857.

Daukas, G. and Zigmond, S. H. (1985) Inhibition of receptor mediated but not fluid-phase endocytosis in polymorphonuclear leukocytes. *J. Cell. Biol.* **101**, 1673–1679.

DeBlasi, A., Lipartiti, M., Motulsky, J., Insel, P. A., and Fratelli, M. (1985) Agonist-induced redistribution of β-adrenergic receptors on intact human monomuclear leukocytes: Redistributed receptors are nonfunctional. *J. Clin. Endocrinol. Metab.* **61**, 1081–1088.

DeLean, A., Stadel, J. M., and Lefkowitz, R. J. (1980) A ternary complex model explains the agonist-specific binding properties of the adenylate cylase-coupled β-adrenergic receptor. *J. Biol. Chem.* **255**, 7108–7 117.

Devreotes, P. and Zigmond, S. (1988) Chemotaxis in eukaryotic cells. *Ann. Rev. Cell Biol.* **4**, 649–686.

Dohlman, H. G., Caron, M. G., and Lefkowitz, R. J. (1987a) A family of receptors coupled to guanine nucleotide regulatory proteins. *Biochemistry* **26**, 2657–2664.

Dohlman, H. G., Bouvier, M., Benovic, J. L., Caron, M. G., and Lefkowitz, R. J. (1987b) The multiple membrane spanning topography of the β-2 adrenergic receptor. *J. Biol. Chem.* **262**, 14282–14288.

Doss, R. C., Perkins, J. P., and Harden, T. K. (1981) Recovery of β-adrenergic receptors following long term exposure of astrocytoma cells to catecholamine: Role of protein synthesis. *J. Biol. Chem.* **256**, 12281–12286.

Drummond, A. H., Baguley, B. C., and Staehelin, M. (1977) Beta-adrenergic regulation of glycogen phosphorylase activity and adenosine cyclic 3'5'-monophosphate accumulation in control and desensitized C-6 astrocytoma cells. *Mol. Pharmacol.* **13**, 1159–1169.

Feldman, R. D. (1988) β-Adrenergic desensitization reduces the sensitivity of adenylate cyclase for magnesium in permabilized lymphocytes. *Mol. Pharmacol.* 35, 304–310.

Feldman, R. D., McArdle, W., and Lai, C. (1986) Phenylarsine oxide inhibits agonist-induced changes in photoaffinity labeling but not agonist-induced desensitization of the β-adrenergic receptor. *Mol. Pharmacol.* 30, 459–462.

Frederich, R. C. Jr., Waldo, G. L., Harden, T. K., and Perkins, J. P. (1983) Characterization of agonist-induced β-adrenergic receptor-specific desensitization in C62B glioma cells. *J. Cyc. Nucl. Prot. Phos. Res.* 9, 103–118.

Garcia-Sainz, J. A. and Michel, B. (1987) Homologous β-adrenergic desensitization in isolated rat hepatocytes. *Biochem. J.* 246, 331–336.

Gardener, J. M. and Fambrough, D. M. (1979) Acetylcholine receptor degradation measured by density labeling: Effects of cholinergic ligands and evidence against recycling. *Cell* 16, 661–674.

Goldstein, J. L., Anderson, R. G. W., and Brown, M. S. (1979) Coated pits, coated vesicles and receptor-mediated endocytosis. *Nature* 279, 679–685.

Green, D. A. and Clark, R. B. (1981) Adenylate cyclase coupling proteins are not essential for agonist-specific desensitization of lymphoma cells. *J. Biol. Chem.* 256, 2105–2108.

Hadcock, J. R. and Malbon, C. C. (1988) Downregulation of β-adrenergic receptors: Agonist-induced reduction in receptor mRNA levels. *Proc. Nad. Acad. Sci. USA* 85, 5021–5025.

Hadcock, J. R., Ross, M., and Malbon, C. C. (1989a) Agonist regulation of β-adrenergic receptor mRNA. *J. Biol. Chem.* 264, 13956–13961.

Hadcock, J. R., Want, H., and Malbon, C. C. (1989b) Agonist-induced destabilization of β-adrenergic receptor mRNA. *J. Biol. Chem.* 264, 19928–19933.

Harden, T. K. (1983) Agonist-induced desensitization of the β-adrenergic receptor linked adenylate cyclase. *Pharmacol. Rev.* 35, 5–32.

Harden, T. K. Cotton, C. U., Waldo, G. L., Lutton, J. K., and Perkins, J. P. (1980) Catecholamine-induced alteration in sedimentation behavior of membrane bound β-adrenergic receptors. *Science* 210, 441–443.

Harris, B. A., Robishaw, J. D., Mumby, S. M., and Gilman, A. G. (1985) Molecular cloning of the complementary DNA for the alpha subunit of the G protein that stimulates adenylate cyclase. *Science* 229, 1274–1277.

Hausdorff, W. P., Bouvier, M., O'Dowd, B. F., Irons, G. P., Caron, M. G., and Lefkowitz, R. J. (1989) Phosphorylation sites on two domains of the β-2 adrenergic receptor are involved in distinct pathways of receptor desensitization. *J. Biol. Chem.* 264, 12657–12665.

Hausdorff, W. P., Hnatowich, M., O'Dowd, B. F., Caron, M. G., and Lefkowitz, R. J. (1990) A mutation of the β-adrenergic receptor impairs agonist activation of adenylyl cyclase without affecting high affinity agonist binding. Distinct molecular determinants of the receptor are involved in physical coupling to and functional activation of G_s. *J. Biol. Chem.* 265, 1388–1393.

Hertel, C., Coulter, S. J., and Perkins, J. P. (1985) A comparison of catecholamine-induced internalization of β-adrenergic receptors and receptor-mediated endocytosis of epidermal growth factor in human astrocytoma cells. *J. Biol. Chem.* **260**, 12547–12553.

Hertel, C., Coulter, S. J., and Perkins, J. P. (1986) The involvement of cellular ATP in receptor-mediated internalization of epidermal growth factor and hormone-induced internalization of β-adrenergic receptors. *J. Biol. Chem.* **261**, 5974–5980.

Hertel, C. and Staehelin, M. (1983) Reappearance of β-adrenergic receptors after isoproterenol treatment in intact C6-cells. *J. Cell Biol.* **97**, 1538–1543.

Hertel, C., Mueller, P., Portenier, M., and Staehelin, M. (1983a) Determination of desensitization of β-adrenergic receptors by ^{3}H-CGP-12177. *Biochem. J.* **216**, 669–674.

Hertel, C., Staehelin, M., and Perkins, J. P. (1983b) Evidence for intravesicular β-adrenergic receptors in membrane fractions from desensitized cells: Binding of the hydrophilic ligand CGP-12177 only in the presence of alamethicin. *J. Cyc. Nuc. Prot. Phos. Res.* **9**, 119–128.

Heuser, J. (1989) Effects of cytoplasmic acidification on clathrin lattice morphology. *J. Cell Biol.* **108**, 401–411.

Heuser, J. E. and Anderson, R. G. W. (1989) Hypertonic media inhibit receptor-mediated endocytosis by blocking clathrin-coated pit formation. *J. Cell Biol.* **108**, 389–400.

Homburger, V., Pantaloni, C., Lucas, M., Gozlan, H., and Bockaert, J. (1984) β-adrenergic receptor repopulation of C6 glioma cells after irreversible blockade and downregulation. *J. Cell Physiol.* **121**, 589–597.

Huganir, R. L. and Greengard, P. (1987) Regulation of receptor function by protein phosphorylation. *TIPS* **8**, 472–477.

Insel, P. A., Mahan, L. C. Motulsky, H. J., Stoolman, L. M., and Koachman, A. M. (1983) Time-dependent decreases in binding affinity of agonists for β-adrenergic receptors of intact S49 lymphoma cells: A mechanism of desensitization. *J. Biol. Chem.* **258**, 13597–13605.

Kassis, S. and Fishman, P. H. (1984) Functional alteration of the β-adrenergic receptor during desensitization of mammalian adenylate cyclase by β-agonists. *Proc. Natl. Acad. Sci. U.S.A.* **81**, 6686–6690.

Kassis, S. and Sullivan, M. (1986) Desensitization of the mammalian β-adrenergic receptor: Analysis of receptor redistribution on nonlinear sucrose gradients. *J. Cyc. Nuc. Prot. Phos. Res.* **11**, 35–46.

Kassis, S., Olasmaa, M., Sullivan, M., and Fishman, P. H. (1986) Desensitization of the β-adrenergic receptor-coupled adenylate cyclase in cultured mammalian cells: Receptor sequestration versus receptor function. *J. Biol. Chem.* **261**, 12233–12237.

Kunkel, M. W., Friedman, J., Shenolikar, S., and Clark, R. B. (1989) Cell-free heterologous desensitization of adenylyl cyclase in S49 lymphoma cell membranes mediated by cyclic AMP-dependent protein kinase. *FASEB J.* **3, 9**, 2067–2074.

Larkin, J. M., Brown, M. S., Goldstein, J. L., and Anderson, R. G. W. (1983) Depletion of intracellular potassium inhibits coated pit formation and receptor-mediated endocytosis in fibroblasts. *Cell* **33**, 273–285.

Lefkowitz, R. J., Wessels, M. R., and Stadel, J. M. (1980) Hormones, receptors and cyclic AMP: Their role in target cell refractoriness. *Curr. Top. Cell Regul.* **17**, 205–230.

Liggett, S. B., Bouvier, M., Hausdorff, W. P., O'Dowd, B. F., Caron, M. G., and Lefkowitz, R. J. (1989) Altered patterns of agonist-stimulated cyclic AMP accumulation in cells expressing mutant β-2 adrenergic receptors lacking phosphorylation sites. *Mol. Pharm.* **36**, 641–646.

Linden, J., Patel, A., Spanier, A. M. and Weglicki, W. B. (1984) Rapid agonist-induced decrease of ^{125}Ipindolol binding to β-adrenergic receptors. *J. Biol. Chem.* **259**, 15115–15122.

Lohse, M. J., Lefkowitz, R. J., Caron, M. G., and Benovic, J. L. (1989) Inhibition of β-adrenergic receptor kinase prevents rapid homologous desensitization of β-2 adrenergic receptors. *Proc. Natl. Acad. Sci. USA* **86**, 3011–3015.

Lohse, M. J., Benovic, J. L., Caron, M. G., and Lefkowitz, R. J. (1990) Multiple pathways of rapid β-2 adrenergic receptor desensitization: Delineation with specific inhibitors. *J. Biol. Chem.* **265**, 3202–3211.

Mahan, L. C., Koachman, A. M., and Insel, P. A. (1985a) Genetic analysis of β-adrenergic receptor internalization and downregulation. *Proc. Natl. Acad. Sci. USA* **82**, 129–133.

Mahan, L. C., Motulsky, J. J., and Insel, P. A. (1985b) Do agonists promote rapid internalization of β-adrenergic receptors? *Proc. Natl. Acad. Sci. USA* **82**, 6566–6570.

May, D. C., Ross, E. M., Gilman, A. G., and Smigel, M. D. (1985) Reconstitution of catecholamine-stimulated adenylate cyclase activity using three purified proteins. *J. Biol. Chem.* **260**, 15829–15833.

Mellman, I., Fuchs, R., and Helenius, A. (1986) Acidification of the endocytic and exocytic pathways. *Ann. Rev. Biochem.* **55**, 663–700.

Mukherjee, C., Caron, M. G., and Lefkowitz, R. J. (1975) Catecholamine-induced subsensitivity of adenylate cyclase associated with loss of β-adrenergic receptor binding sites. *Proc. Natl. Acad. Sci. USA* **72**, 1945–1949.

Nambi, P., Peters, J. R., Sibley, D. R., and Lefkowitz, R. J. (1985) Desensitization of the turkey erythrocyte β-adrenergic receptor in a cell-free system. *J. Biol. Chem.* **260**, 2165–2171.

Neve, K. A., Barrett, D. A., and Molinoff, P. B. (1985) Selective regulation of β1 and β-2 adrenergic receptors by atypical agonists. *J. Pharm. Exp. Ther.* **235**, 657–664.

O'Dowd, B. F., Hnatowich, M., Regan, J. W., Leader, W. M., Caron, M. G., and Lefkowitz, R. J. (1988) Site-directed mutagenesis of the cytoplasmic domains of the human β-2 adrenergic receptor. *J. Biol. Chem.* **263**, 15985–15992.

O'Dowd, B. F., Hnatowich, M., Caron, M. G., Lefkowitz, R. J., and Bouvier, B. (1989) Palmitoylation of the human β-2 adrenergic receptor. *J. Biol. Chem.* **264,** 7564–7569.

Payne, G. S., Baker, D., van Tuinen, E. and Schekman, R. (1988) Protein transport to the vacuole and receptor-mediated endocytosis by clathrin heavy chain-deficient yeast. *J. Cell. Biol.* **106,** 1453–1461.

Perkins, J. P., Harden, T. K., and Harper, J. P. (1982) in *Handbook of Experimental Pharmacology,* (Nathanson, J. A. and Kebabian, J. W., eds.) Springer-Verlag, Berlin, pp. 185–224.

Pittman, R. N. and Molinoff, P. B. (1980) Interaction of agonists and antagonists and β-adrenergic receptors on intact L6 muscle cells. *J. Cyclic Nucleotide Res.* **6,** 421–435.

Pittman, R. N. and Molinoff, P. B. (1984) Interactions of full and partial agonists with β-adrenergic receptors on intact L6 muscle cells. *Mol. Pharmacol.* **24,** 398–408.

Pittman, R. N., Reynolds, E. E., and Molinoff, P. B. (1983) Relationship between intrinsic activities of agonists in normal and desensitized tissue and agonist-induced loss of β-adrenergic receptors. *J. Pharm. Exp. Ther.* **230,** 614–618.

Reneke, J. E., Blumer, K. J., Courchesne, W. E., and Thorner, J. (1988) The carboxy-terminal segment of the yeast α-factor receptor is a regulatory domain. *Cell* **55,** 221–234.

Reynolds, E. E. and Molinoff, P. B. (1986) Downregulation of beta adrenergic receptors in S49 lymphoma cells induced by atypical agonists. *J. Biol. Chem.* **239,** 654–660.

Ross, E. M., Maguire, M. E., Sturgill, T. W., Biltonen, R. L., and Gilman, A. G. (1977) Relationship between the β-adrenergic receptor and adenylate cyclase. *J. Biol. Chem.* **252,** 5761–5775.

Ross, E. M., Howlett, A. C., Ferguson, K. M., and Gilman, A. G. (1978) Reconstitution of hormone-sensitive adenylate cyclase activity with resolved components of the enzyme. *J. Biol. Chem.* **253,** 6401–6412.

Sandvig, K., Olsnes, S., Petersen, O. W., and van Deurs, B. (1987) Acidification of the cytosol inhibits endocytosis from coated pits. *J. Cell Biol.* **105,** 679–689.

Shear, M., Insel, P. A., Meloan, K. L., and Coffino, P. (1976) Agonist-specific refractoriness induced by isoproterenol. *J. Biol. Chem.* **251,** 7572–7576.

Shorr, R. G. L., Strohsacker, M. W., Lavin, T. N., Lefkowitz, R. J., and Caron, M. G. (1982) The β-1 adrenergic receptor of the turkey erythrocyte: Molecular heterogeneity revealed by purification and photoaffinity labeling. *J. Biol. Chem.* **257,** 12341–12350.

Sibley, D. R. and Lefkowitz, R. J. (1985) Molecular mechanism of receptor desensitization using the β-adrenergic receptor-coupled adenylate cyclase system as a model. *Nature* **317,** 124–129.

Sibley, D. R., Peters, J. R., Nambi, P., Caron, M. G., and Lefkowitz, R. J. (1984) Desensitization of turkey erythrocyte adenylate cyclase. *J. Biol. Chem.* **259,** 9742–9749.

Sibley, D. R., Strasser, R. H., Benovic, J. L., Daniel, K., and Lefkowitz, R. J. (1986) Phosphorylation/dephosphorylation of the β-adrenergic receptor regulates its functional coupling to adenylate cyclase and subcellular distribution. *Proc. Natl. Acad. Sci. USA* **83,** 9408–9412.

Sibley, D. R., Strasser, R. H., Caron, M. G., and Lefkowitz, R. J. (1985) Homologous desensitization of adenylate cyclase is associated with phosphorylation of the β-adrenergic receptor. *J. Biol. Chem.* **260,** 3883–3886.

Stadel, J. M., Nambi, P., Lavin, T. N., Heald, S. L., Caron, M. G., and Lefkowitz, R. J. (1982) Catecholamine-induced desensitization of turkey erythrocyte adenylate cyclase. *J. Biol. Chem.* **257,** 9242–9245.

Stadel, J. M., Strulovici, B., Nambi, P., Lavin, T. N., Briggs, M. M., Caron, M. G., and Lefkowitz, R. J. (1983a) Desensitization of the β-adrenergic receptor of frog erythrocytes: Recovery and characterization of the downregulated receptors in internalized vesicles. *J. Biol. Chem.* **258,** 3032–3038.

Stadel, J. M., Nambi, P., Shorr, R. G. L., Sawyer, D. F., Caron, M. G., and Lefkowitz, R. J. (1983b) Catecholamine-induced desensitization of turkey erythrocyte adenylate cyclase is associated with phosphorylation of the β-adrenergic receptor. *Proc. Natl. Acad. Sci. USA* **80,** 3173–3178.

Stadel, J. M., Rebar, R., and Crooke, S. T. (1988) Alkaline phosphatase relieves desensitization of adenylate cyclase-coupled β-adrenergic receptors in avian erythrocyte membranes. *Biochem. J.* **252,** 771–776.

Stadel, J. M., Rebar, R., and Crooke, S. T. (1987) Catecholamine-induced desensitization of adenylate cyclase coupled β-adrenergic receptors in turkey erythrocytes: Evidence for a two-step mechanisms. *Biochemistry* **26,** 5861–5866.

Strader, C. D., Sibley, D. R., and Lefkowitz, R. J. (1984) Association of internalized β-adrenergic receptors with the plasma membrane: A novel mechanism for receptor downregulation. *Life Sciences* **35,** 1601–1610.

Strader, C. D., Sigal, I. S., Blake, A. D., Cheung, A. H., Register, R. B., Rands, E., Zemcik, C., Candelore, M. R., and Dixon, R. A. F. (1987a) The carboxyl terminus of the hamster β-adrenergic receptor expressed in mouse L cells is not required for receptor sequestration. *Cell* **49,** 855–863.

Strader, C. D., Dixon, R. A. F., Cheung, A. H., Candelore, M. R., Blake, A. D., and Sigal, I. S. (1987b) Mutations that uncouple the β-adrenergic receptor from Gs and increase agonist affinity. *J. Biol. Chem.* **262,** 16430–16443.

Strasser, R. H., Sibley, D. R., and Lefkowitz, R. J. (1986) A novel catecholamine activated adenosine 3', 5'-phosphate independent pathway for β-adrenergic receptor phosphorylation in wild-type and mutant S49 lymphoma cells. *Biochem.* **25,** 1371–1377.

Strasser, R. H., Stiles, G. L., and Lefkowitz, R. J. (1984) Translocation and uncoupling of the β-adrenergic receptor in rat lung after catecholamine promoted desensitization in vivo. *Endocrinology* **115,** 1392–1400.

Strulovici, B., Cerione, R. A., Kilpatrick, B. F., Caron, M. G., and Lefkowitz, R. J. (1984) Direct demonstration of impaired functionality of a purified desensitized β-adrenergic receptor in a reconstituted system. *Science* **225**, 837–840.

Strulovici, B., Stadel, J. M., and Lefkowitz, R. (1983) Functional integrity of desensitized β-adrenergic receptors: Internalized receptors reconstitute catecholamine-stimulated adenylate cyclase. *J. Biol. Chem.* **258**, 6410–6414.

Su, Y. F., Harden, T. K., and Perkins, J. P. (1979) Isoprotemol-induced desensitization of adenylate cyclase in human astrocytoma cells. *J. Biol. Chem.* **254**, 38–41.

Su, Y. F, Harden, T. K., and Perkins, J. P. (1980) Catecholamine specific desensitization of adenylate cyclase: Evidence for a multistep process. *J. Biol. Chem.* **255**, 7410–7419.

Su, Y. F., Cubeddu-Ximenez, L., and Perkins, J. P. (1976) Regulation of adenosine 3′ 5′-monophosphate content of human astrocytoma cells: Desensitization to catecholamines and prostaglandins. *J. Cyclic Nucl. Res.* **2**, 257–270.

Terasaki, W. L., Linden, J., and Brooker, G. (1979) Quantitative relationship between β-adrenergic receptor number and physiologic responses as studied with a long-lasting β-adrenergic antagonist. *Proc. Natl. Acad. Sci. USA* **76**, 6401–6405.

Toews, M. L., Harden, T. K., and Perkins, J. P. (1983) High-affinity binding of agonists to β-adrenergic receptors on intact cells. *Proc. Natl. Acad. Sci. USA* **80**, 3553–3557.

Toews, M. L. and Perkins, J. P. (1984) Agonist-induced changes in β-adrenergic receptors in intact cells. *J. Biol. Chem.* **259**, 2227–2235.

Toews, M. L., Waldo, G. L., Harden, T. K., and Perkins, J. P. (1984) Relationship between an altered membrane form and a low affinity form of the β-adrenergic receptor occurring during catecholamine induced desensitization: Evidence for receptor internalization. *J. Biol. Chem.* **259**, 11844–11850.

Toews, M. L., Waldo, G. L., Harden, T. K., and Perkins, J. P. (1986) Comparison of binding ^{125}I-Iodopindolol to control and desensitized cells at 37° and on ice. *J. Cyc. Nuc. Pro. Phos. Res.* **11**, 47–62.

Wakshull, E., Hertel, C., O'Keefe, E. J., and Perkins, J. P. (1985) Cellular redistribution of β-adrenergic receptors in a human astrocytoma cell line: A comparison with the epidermal growth factor receptor in murine fibroblasts. *J. Cell. Biol.* **29**, 127–141.

Waldo, G. L., Doss, R. C., Perkins, J. P., and Harden, T. K. (1984) Use of a density shift method to assess β-adrenergic receptor synthesis during recovery from catecholamine-induced downregulation in human astrocytoma cells. *Mol. Pharmacol.* **26**, 424–429.

Waldo, G. L., Northup, J. K., Perkins, J. P., and Harden, T. K. (1983) Characterization of an altered membrane form of the β-adrenergic receptor produced during agonist-induced desensitization. *J. Biol. Chem.* **258**, 13900–13908.

Wang, H. Y., Berrios, M., and Malbon, C. (1989) *In situ* localization of β-adrenergic
   receptors in A431 cells: Effect of chronic exposure to agonists. *Biochem. J.* **263,**
   533–538.
Yoshimasa, T., Bouvier, M., Benovic, J. L., Amlaiky, N., Lefkowitz, R. J., and Caron,
   M. G. (1988) Regulation of the adenylate cyclase signalling pathway: Potential
   role for the phosphorylation of the catalytic unit by protein kinase A and protein
   kinase C, in *Molecular Biology of Brain and Endocrine Peptidergic Systems*
   (McKerns, K. W. and Chretien, M., eds.) Plenum, New York, pp. 123–139.

CHAPTER 4

Reconstitution of the Beta-Adrenergic Receptor and Its Biochemical Functions

Elliott M. Ross

1. Introduction

The animal cell plasma membrane is a complex information-processing organelle. A typical mammalian plasma membrane may contain as many as a dozen different hormone receptors that regulate five or more GTP-binding proteins (G proteins), a diverse group of "effector" proteins (adenylate cyclase, phospholipases and phosphodiesterases, channels and carriers), and other relevant regulatory proteins (calmodulin, kinases and phosphatases, and so on). Furthermore, the regulation of each signal-transducing pathway is not independent of the others—they interact to form a highly regulated information-transducing switchboard.

The biochemist's traditional approach to such complex systems has been to take them apart, identify their protein components, and then put them back together to test how they work. This approach led to the elucidation of metabolic pathways composed of soluble enzymes,

The β-Adrenergic Receptors Ed.: J. P. Perkins © 1991 The Humana Press Inc.

but the strategies for reconstituting multiprotein systems became more important as membrane-bound or other supramolecular systems came under study. The reconstitution of membrane-bound energy-transducing and transport proteins has become paradigmatic of this approach. Racker (1986) has recently reviewed the field in a highly personal monograph that contains an excellent discussion of experimental strategies. (Racker's influence on this review and its author is obvious and significant.) The study of DNA replication has also demanded the development of complex reconstitutive assay systems that allow multiple sequential reactions to proceed faithfully on the surface of what is functionally a solid substrate (Kornberg, 1980). The membrane systems that process newly synthesized proteins and move them among organelles are now being successfully analyzed by this approach as well (Walter and Lingappa, 1986). In each of these areas, the reconstitutive approach to assaying biochemical function has been necessary because of the number of coupled reactions and enzymes involved, and because of the requirement for large-scale structure and organization. In the end, the approach has also provided valuable information on the structures themselves.

Reconstitutive assays have, over the last ten years, yielded much of the information on the function of adenylate cyclase, its related proteins and the receptors that regulate them. This review will stress the strategies, goals, and limitations of these assays more than their results, which are covered in detail in other chapters. Although many experiments cited here come from work on the beta-adrenergic receptor, these studies have generally depended on experimental approaches that were developed for transport proteins or the enzymes of oxidative phosphorylation. Other reviews on the reconstitution of the beta-adrenergic adenylate cyclase system have appeared recently and, in general, are consistent with the arguments put forth here (Racker, 1986; Levitzki, 1985; Klausner et al., 1984; Lefkowitz et al., 1985).

2. Why Use Reconstituted Assay Systems?

The strategy and techniques for developing a reconstituted assay system are dictated by the experimental questions that the investigator wants to ask of it. These goals decide the choice of assays, the use of purified or crude proteins and lipids, and the methods that are used to achieve reconstitution. In general, the reasons for reconstituting a membrane-bound receptor–effector system are few but important:

1. To enumerate and identify the components necessary to perform a given regulatory function;
2. To quantify those components in crude preparations or during purification;
3. To study their activities and interactions in a physically and chemically defined system; or
4. To study the effects of membrane lipids and bilayer structure upon them.

2.1. Identification and Enumeration of Proteins

The use of reconstitutive assays to detect the presence of multiple protein components that are required for the display of a specific function is the biochemists application of the classic genetic technique of complementation analysis. Crude reconstituted systems were used initially to detect many of the components of the beta-adrenergic regulatory pathway: cyclic AMP, protein kinases, GTP, and G_s. Initial assays were difficult in each case, but the observation that a component could be removed and its function restored by a second fraction was adequately clear to allow the missing component's identification. Relatively crude and indirect reconstitutive assays allowed both the identification of G_s as a distinct component in the pathway and its localization between the receptor and adenylate cyclase.

Details vary, but in vitro complementation analysis demands that a preparation first be abused such that an assayable activity is lost (by solubilization, fractionation, chemical treatment, or mutagenesis). If the activity can be restored by the addition of a second, *complementary* preparation that also lacks the activity of interest, a strong argument exists that each preparation contains a different component necessary for the activity in question. Such a demonstration does not define the function of either component—they may act sequentially, in tandem, one upon the other, or in other more complex patterns. Neither component defined in such an assay need directly produce the proximal effect that is measured. Other components of the system may exist also and might be found in both preparations.

In reconstitutive experiments that are designed to detect the components of a multiprotein system, the need for sensitive and specific assays is far more important than absolute quantification. Sensitivity is crucial because a reconstituted activity or pattern of regulation is rarely as good as the original. An initial reconstitution of a membrane-bound, multiprotein reaction that is 10% as efficient as the native system is generally a triumph, but it is not worth much if the only assay for the reconstituted activity is sensitive only down to the 20% level.

The need for specific assays is just as great. The activities of transmembrane signaling systems are highly sensitive to their environment. Manipulation of detergents, lipids, pH, other ions, and contaminating guanine nucleotides have caused activities to disappear, and the addition of other preparations have apparently restored them. The description of a number of bogus regulatory factors has resulted from such experiments (some are listed by Ross and Gilman, 1980). In crude systems, negative controls almost always include the demonstration that a putative protein factor behaves appropriately in response to several different treatments that might be expected to inactivate it (heat, proteolysis, and the like). Skepticism is particularly appropriate at this stage.

The actual specific activity of a reconstituted system is a less crucial issue, especially at first. If criteria of specificity are met and if the background is low, an initially inefficient reconstitution may still be adequate to argue for the resolution of necessary components. For example, reconstitution of a response to beta-adrenergic receptors may be small, but characteristic structure–activity relationships for agonists and blockers can argue strongly that it is legitimate.

2.2. Quantification

Reconstitutive assays are also applicable to the quantitative measurement of membrane-bound receptors and enzymes. If the protein has a suitably specific enzymatic or ligand-binding activity that is retained upon solubilization, then a reconstitutive assay is not required and is usually more trouble than it is worth. However, several membrane-bound proteins lose one or more of their activities upon solubilization, and their reincorporation into a suitably depleted membrane or a lipid bilayer is necessary. This is obviously the case for ion channels or transport activities, but other proteins also reversibly lose enzymatic or ligand-binding activities when solubilized. Examples are the voltage-sensitive Na^+ channel, which binds scorpion toxin only when incorporated in a bilayer (Tamkun et al., 1984; Feller et al., 1985) and G_s, which does not efficiently catalyze GTP hydrolysis in detergent solution (Brandt et al., 1983). Oddly, the digitonin-solubilized beta-adrenergic receptor cannot reversibly bind [^{125}I]-iodohydroxybenzylpindolol, but binding is restored after its reconstitution into phospholipid vesicles (Fleming and Ross, 1980). In contrast, the soluble receptor binds [^3H]dihydroalprenolol (Caron and Lefkowitz, 1976) or [^{125}I]iodocyanopindolol (Brandt and Ross, 1986) with virtually unchanged affinity.

Certain membrane-bound enzymes that remain functionally active after solubilization may display quantitatively different properties when in a phospholipid membrane. This may reflect a requirement for a specific phospholipid or a more general response to its physical environment. The many effects of lipids on the activities of membrane-bound proteins has been well reviewed (Sandermann, 1978; Gennis and Jonas, 1977). The exquisite structure–activity relationships that have been deciphered for beta-hydroxybutyrate dehydrogenase form an outstanding example of the potential specificity of such interactions (Isaacson et al., 1979; Churchill et al., 1983). Protein–protein interactions are also altered by solubilization, and, in some cases, these effects are intrinsic to the interactions of one or the other protein with a component of the bilayer. Thus, adenylate cyclase is activated more efficiently by G_s when phosphatidylcholine is present (Ross, 1982; Smigel, 1986) and protein kinase C is optimally active when its protein substrate is presented in a membrane that contains the activator phosphatidylserine (Kaibuchi et al., 1981). Frequently, if a phospholipid head group is the predominant regulatory determinant, micelles composed of lysophospholipids or of phospholipids with short acyl chains may substitute for the more common membrane-forming lipid to yield an active system that is apparently soluble (Baron and Thompson, 1975; Rydstrom et al., 1976; Cortese et al., 1982; Ross, 1982). Note that such an instance blurs the boundary between lipid and detergent (Tausk et al., 1974).

Reconstitutive assays are required when a protein has no easily assayable activity of its own, but can be measured only insofar as it alters the activity of another membrane-bound protein. This was initially the case for G_s which was identified only as an activator of adenylate cyclase (Ross and Gilman, 1977). When designing such an assay, a relatively arbitrary unit of reconstitutive activity is usually developed according to stimulation of the second protein. It is the investigator's responsibility to develop the assay to the point where the unit is an accurate measure of the amount of the relevant protein. This involves the usual demands that the observed activity be proportional to the amount of protein added but, in the case of are constitutive assay, the absolute reproducibility of an assay also depends on standardizing the amounts and concentrations of the other components of the assay system. The reported enzyme should be added in saturating amounts to insure reproducibility. For example, one can perform Gs assays that are linear with respect to the amount of G_s added but that are invalid because the amount of adenylate cyclase used is insufficient (Sternweis et al., 1981; Larner and Ross, 1981). This principle has been violated in several published studies.

Reconstitutive assays are particularly valuable for measuring the regulatory interactions of receptors and G proteins. Liposomes that contain receptors and G proteins have been used to study the mechanism of receptor-catalyzed G protein exchange (Asano and Ross, 1984; Brandt and Ross, 1986; Cerione et al., 1985b; May and Ross, 1988), the role of the beta gamma subunits of G proteins (Im et al., 1987, 1988; Cerione et al., 1986b,1987), and the effects of proteolytic cleavage on individual domains of the receptor (Rubenstein et al., 1987; Boege et al., 1987).

An important determinant of protein–protein interactions in membranes is the effective concentration that occurs when two membrane-bound proteins are co-reconstituted. Consider two proteins, each at a nominal solution concentration of $10^{-7}M$, that have been reconstituted into unilamellar vesicles of egg PC (phosphatidylcholine) at a phospholipid concentration of $10^{-4}M$. By being reconstituted, the proteins have been concentrated into a volume about 15,000-fold smaller than that of the total aquaeous solution, yielding functional local concentrations of each protein of $1.5 \times 10^{-3}M$. If one considers that both proteins are now oriented in parallel and have lost one degree of freedom of rotational diffusion, their functional concentration is estimated at over $5 \times 10^{-3}M$, increasing each of their concentrations about 5×10^4-fold and their collisional frequency over 10^9-fold. If two such proteins bound each other with low affinity, their interactions in solution would be negligible. However, after reconstitution into phospholipid vesicles, they could interact quite efficiently, even though there had been no change whatever in their structures, intrinsic affinity for each other, or activities. It is likely that many protein–protein interactions that are said to occur "only in membrane bilayers" are actually limited only by the low concentration of each protein when they are solubilized. The concept of "lateral concentration," the number of molecules/unit area of bilayer, will recur throughout this review.

Skepticism and rigorous choice of controls can make reconstitutive assays specific, useful, and quantitative. Their strengths lie in their ability to cope with crude preparations, membrane-dependent activities, and proteins that seemingly have no activities by themselves. These assays may also show a more physiologically relevant activity of an enzyme than does ligand binding. Caution in design makes them believable.

2.3. Effects of Bilayer Structure or Specific Lipids on Activities of Receptors and Enzymes

The beta-adrenergic receptor and adenylate cyclase are exposed to two solvents: two aqueous compartments and the hydrocarbon phase of the plasma membrane bilayer. Even at a simple level, such localization puts geometric and diffusional constraints on proteinprotein interaction and raises questions about the effects of the composition and physical state of the bilayer on the activities of individual proteins. Regulatory interactions between integral membrane proteins of the sort that couple receptors with G proteins, and G proteins with effectors, are an essentially untouched area in membrane biochemistry. In general, one can approach the roles of bilayer structure and of individual lipids on protein-mediated biochemical events in one of two ways, each of which has its strengths and weaknesses.

Until recently, essentially all studies of the effect of membrane composition on the adenylate cyclase system were carried out either in whole cells or in native membranes where the lipid composition was manipulated, whether metabolically, genetically, or by in vitro exchange. We reviewed such studies in 1980 (Ross and Gilman, 1980) and there have been few conceptual advances in this area. These approaches have the advantage that the membranes and the proteins involved are probably perturbed relatively little, especially in the metabolic or genetic approach. However, the interpretation of such experiments is often difficult. Cells typically display multiple compensatory changes in cholesterol and phospholipid content in response to either mutational or nutritional variations in the composition of their membranes. For example, the growth of LM fibroblast cells in media containing mono-, di-, or tri-methylated ethanolamines changes the amounts of PC and PE (phosphatidylethanolamine) in the plasma membrane and introduces novel N-methylated PEs (Engelhard et al., 1978). However, the relative amounts of other membrane lipids and the saturation of lipid acyl chains also vary in such cells. Therefore, attributing a change in the activity of a membrane-bound enzyme to a particular lipid or physical parameter is not trivial, even if a constant amount of enzyme is synthesized and integrated into the plasma membrane. The last two assumptions are themselves rarely confirmed. Because the physiologic role of proteins involved with transmembrane signaling is to regulate the responsiveness and metabolic activities of a cell, one would expect that significant alteration of the plasma membrane would cause compensatory changes in the synthesis or activity of these proteins. These uncertainties force the invstigator to monitor carefully the absolute concentrations of each protein and, optimally, to apply adequate physical tests of membrane structure to be convinced that there is a consistent and predictable variation of the observed activity according to the lipid composition of the membrane and some measurable physical parameter. The last precaution is relatively ambiguous. Parameters of membrane structure are generally linked, and it is unlikely that one could vary only one aspect of a bilayer, such as its viscous retardation of lateral diffusion, without altering another, such as thickness, retardation of rotational diffusion, or the partitioning into structural or compositional domains.

The use of reconstituted systems composed of purified lipids and proteins obviates many of these objections. As discussed below, purified receptors, G proteins, and adenylate cyclase can be reconstituted in known quantities into vesicles composed of purified natural phospholipids or synthetic phospholipids with known acylchain compositions. This allows the investigator to study activities and regulatory interactions in a medium of known composition without the problem of cytoskeleton constraints or the effect of other intrinsic proteins, and with knowledge of the precise phospholipid composition of the system. It is a homogeneous system where the lipid-to-protein ratio can be methodically manipulated, lipid composition can be quantitatively changed, and the stoichiometric ratios of individual proteins can be altered at will. Although physical controls still must be applied, there are far fewer uncertainties in such a reconstituted system.

The use of purified and reconstituted components in a completely artificial system points out clearly the caveat of the reconstituted-system strategy: It must be demonstrated clearly that effects observed in reconstituted systems actually reflect the activities and regulatory patterns that occur in a living cell. At the simplest level, it must be demonstrated that such effects are on the activity under study, rather than on the process of the reconstitution itself. One must be careful to monitor the recovery of activities after reconstitution, and it is best to use independent and mutually confirmatory assays in detailed studies. A reconstituted vesicle composed of two synthetic phospholipids is not a native plasma membrane. There are potentials both for observing nonphysiological phenomena and, obviously, for failing to observe physiological phenomena that are mediated by missing components. Given this caveat, however, many details of beta-adrenergic regulation of adenylate cyclase have been approachable in reconstituted systems that have not yielded to study in native membranes. In general, the predictive value of such systems has been excellent, and such studies have led to new insights that have been confirmed in studies of intact cells or native membranes.

3. Techniques of Reconstitution

The reconstitution of membrane-linked functions varies widely in the complexity of techniques and of the preparations used. Technical sophistication and investigative sophistication are not identical; the most appropriate technology for answering a specific question may be quite simple.

3.1. Assay

The first and most important step in developing a reconstituted system is the choice of an adequately specific and sensitive assay for the biochemical activity or regulatory behavior of interest. The first question is what function should be assayed as a measurement of reconstitution. As an extreme case, consider the mitochondrial coupling factor F_1, which is an ATPase in solution. The reconstituted protein can act as an ATP synthase proton pump, coupling factor for oxidative phosphorylation, and a "plug" of the proton leak caused by F_o (Racker, 1976). Similarly, G_s can be assayed as a regulator of the affinity with which agonist ligands bind to receptors (Ross et al., 1978; Cerione et al., 1984), an agonist-stimulated GTPase or guanine nucleotide-binding protein (Brandt et al., 1983; Brandt and Ross, 1986), an activator of adenylate cyclase (Ross and Gilman, 1977), or a factor that enhances the sensitivity of adenylate cyclase to forskolin (Darfler et al., 1982). For both enzymes, the available assays vary in their information content, convenience, sensitivity, and selectivity in the face of interfering enzymatic activities. When one is trying to establish a new reconstitutive assay, it is imperative that the most sensitive assay be used, because the initial reconstitution is unlikely to be highly efficient: It is far easier to improve a small signal than to search for a signal where none has been detected. Depending upon the goals of the study and the sort of system to be reconstituted, one should also attempt to assay the activity most proximal to the protein of interest. An activity intrinsic to a specific protein will, in general, be much less variable among experiments than will one that requires a complex interaction among several proteins in a reconstituted system. It will also be easier to recreate in an artificial in vitro system.

On a more positive note, many activities of the adenylate cyclase system are far easier to assay in reconstituted preparations containing purified or partially purified proteins than they are in membranes. This largely results from decreased interfering enzymatic activities and decreased nonspecific ligand-binding activities caused by the sheer mass of irrelevant protein in membrane preparations. The best example is that of the catecholamine-stimulated GTPase. This activity was, and still is, difficult to measure accurately in native membranes, and it is generally underestimated because of the measures that are needed to suppress interfering nucleoside triphosphatase activities (Cassel and Selinger, 1976). By contrast, catecholamine-stimulated GTPase is easily assayed in reconstituted phospholipid vesicles that contain beta-adrenergic receptor and G_s because the low background is dependent only on the purity of the $[\gamma\text{-}^{32}P]GTP$ used as substrate (Brandt et al., 1983).

4. Reconstitution by Fusion of Cells or Membranes

Both conceptually and technically, the simplest way of reconstituting the interaction of two membrane-bound proteins in an interpretable way is to fuse two cells that each express only one of the two proteins. Fusion allows membrane proteins from functionally complementary cells to interact in the membrane of the heterokaryon. Both intact cells and plasma membrane preparations can be fused with relatively high efficiency by an increasingly wide array of procedures. Reconstitution by fusion was first used to identify distinct components of the cyclase system by Schramm's group (Orly and Schramm, 1976; Schramm et al., 1977; Schulster et al., 1978; Schramm, 1979; Neufeld et al., 1980). The strategy has now been widely exploited to demonstrate the presence of multiple factors needed for responses to hormones or to verify the effect of a specific perturbation on one or another component of the adenylate cyclase system (Schwarzmeier and Gilman, 1977; Pike et al., 1979; Strulovici et al., 1983; Kassis et al., 1984).

All of the advantages attached to reconstitution by membrane fusion relate to its simplicity. It is not necessary to solubilize or purify any of the proteins of interest, and the procedure is relatively quick—heterokarya or fused membranes can be prepared in about one hour. More important, reconstitution by fusion maintains proteins of interest in their proper topological orientations, in native membrane environments (their own or that of the complementary cell type), and at a physiologically appropriate lateral concentration. Because purified proteins that have been reconstituted into phospholipid vesicles can subsequently be introduced into biological membranes by fusion, this procedure is also applicable to purified or partially purified membrane proteins (*see* Citri and Schramm, 1980; Eimerl et al., 1980).

Reconstitution by fusion yields data that are primarily qualitative, as discussed below. Consequently, this technique has been most applicable to simple complementation analysis. (It is the first step in the creation of viable hybrids between complementary cells.) Fusion is also applicable to measuring the effects of specific perturbations on proteins within a membrane. Such perturbations can be chemically induced by the investigator or may be natural cellular mechanisms, such as those related to desensitization or potentiation. The fusion of two complementary cell types was also used by Schulster et al. (1978) to demonstrate the lateral mobility of at least one of the proteins of the hormone-stimulated adenylate cyclase system in much the same way that Frye and Edidin (1970) first demonstrated the mobility of cell surface proteins that had been labeled with fluorescent dyes. Such experiments have been the only ones that have spoken directly, if only qualitatively, to the question of the mobility of the proteins involved in adenylate cyclase regulation. However, it should be noted that some or all of these proteins are associated with or influenced by the cytoskeletal architecture that underlies the plasma membrane (Rudolph et al., 1977; Insel and Koachman, 1982; Rasenick et al., 1981; Carlson et al., 1986). This architecture is necessarily rearranged or even destroyed during the formation of a heterokaryon.

4.1. Mechanisms of Membrane Fusion

The fusion of biological membranes or lipid bilayers is the process by which the two surfaces meet *en face* and reform to yield a new surface, on which molecules on either membrane that were oriented toward the contacting faces become oriented in the same direction. Molecules oriented away from the interacting faces also take on a parallel orientation in the opposite direction. Fusion of spheres (cells) yields a larger sphere (heterokaryon), with extracellular faces remaining extracellular. Fusion of laminar sheets at a single point yields a toroidal sheet. After fusion, components of each original membrane have the potential to mix with the lipid and protein components of the other.

Membrane fusion is of biological interest because it underlies secretion, endocytosis, cytokinesis, and the intracellular sorting of membrane proteins. It is experimentally useful in introducing DNA, viruses, and drugs into cells. Although the mechanism of fusion has been intensely studied over the past 20 y, no generally accepted, universal theory of membrane fusion exists. In general, fusion can be viewed as a two-step process. The membranes must first be brought into close juxtaposition, and this must be followed by local disordering of each membrane such that the bilayer structures can be broken and reformed. The initial aggregation of cells or membranes can be promoted by a divalent ion, such as Ca^{2+}, hemagglutinins; centrifugation into a tightly packed pellet; or any of several other processes. Each serves to abut the membranes and exclude the intervening layer of water. In most cases, this removal of the water layer is also intimately connected with the second step—the disordering of the phospholipid bilayer. Many of the compounds mentioned above are thus fusogens in and of themselves. However, some detergents, including lysophospholipids, are efficient fusogens, and these presumably work primarily as local disordering agents. Potential mechanisms of membrane fusion and the structures of putative intermediates have been discussed elsewhere (Cullis et al., 1986; Wilschut and Hoekstra, 1986; Duzgunes, 1985; Poste and Nicolson, 1978).

Practically, relatively few fusogens are used for the creation of heterokarya or for the formation of fused membrane fragments. One of the earliest practical fusogens to be recognized was Sendai virus, an enveloped mammalian virus whose major hemagglutinating coat protein promotes fusion between the membrane of the virus and that of the cell to be infected. Under different conditions, Sendai virus can fuse to and infect a cell, cause cell lysis, or promote the fusion of multiple cells. Sendai virus hemagglutinin binds to glycolipids on the cell surface to promote cell–cell crosslinking and subsequent fusion (Haywood, 1974a,b). Schramm and coworkers used Sendai virus in their early fusion studies, in which cells lacking either beta-adrenergic receptors or functional adenylate cyclase systems were fused to give a reconstituted beta-adrenergic stimulation of cyclase activity (Orly and Schramm, 1976; Schulster et al., 1978). With proper experimental control, fusion mediated by Sendai virus can be highly efficient and relatively gentle to the cells or membranes being fused. However, one must grow and prepare the virus, and it is intrinsically a highly variable reagent. Fusion by Sendai virus is usually undertaken by mixing the cells or membranes with virus at high concentrations under isotonic and isosmotic conditions with, in the case of cells, appropriate energy sources in the media. Fusion can be promoted by pelleting the cells, although this is not generally necessary. Interested readers are referred to the work of Schramm and colleagues and to a review by Poste and Pasternak (1978; *see also* Toister and Loyter, 1973; Peretz et al., 1974).

The most common fusogen, when working with both membranes and intact cells, is polyethylene glycol (PEG) in the 6000-dalton mol mass range. The mechanism by which PEG promotes fusion remains unclear (Duzgunes, 1985). PEG is usually used as a fusogen at about 50% (w/v). It is mixed with cells or membranes at room temperature under isotonic conditions, and fusion is usually promoted by centrifuging the cells or membranes to increase contact. The PEG is washed away, as well as possible, after fusion has occurred. It is worth noting that the efficacy of different lots of PEG and of PEG from different suppliers varies markedly, and it is likely that both contaminants and polymer chain length are important variables. Both hydrophobic and inorganic ionic contaminants have been implicated. There is also great variation in the toxicity and lytic activity of different batches of PEG, probably reflecting the quantity of toxic contaminants. Although this is not an obviously important factor in short-term experiments in which heterokarya need not survive, one should note that PEG-treated membranes have definitely been exposed to and probably contaminated with various surfactant compounds.

PEG-induced fusion has been widely used to reconstitute membrane-linked functions by fusing complementary membranes or cells (Schramm, 1979; Schwarzmeier and Gilman, 1977; Korner et al., 1982; Kassis and Fishman, 1982). Schramm and coworkers have also extended this technique to allow them to reintroduce detergent-solubilized components, such as hormone receptors and G_s, into membranes of whole cells (Eimerl et al., 1980; Citri and Schramm, 1980). The detergent-solubilized proteins were precipitated with added phospholipid to form what was probably a mixture of unilamellar and multilamellar vesicles. The precipitate was then fused with whole cells having the complementation phenotype using PEG, thereby allowing association of the added proteins with proteins in the acceptor membrane. This two-part reconstitution protocol avoids adding detergent to the acceptor cells. It has been used effectively, primarily by the laboratories of Schramm and Lefkowitz (Cerione et al., 1983), for the assay of activated G_s and of beta-adrenergic receptors in whole cells. Although it has been less used than other methods of reconstitution, it has clearly proved to be a valuable intermediate step in developing more refined systems. In general, PEG and Ca^{2+} have been the fusogens of greatest utility in this procedure, and the reader is referred to the work of Schramm for useful methods.

Electrofusion is a third technique for inducing both cell–cell and vesicle–vesicle fusion; it has gained recent popularity because of its utility in introducing recombinant DNA into cells. In this procedure, cells are aligned in an electric field, and fusion is then induced by a rapid discharge of electricity at a much higher voltage (Zimmermann and Vienken, 1982). Applications to the production of heterokarya for biochemical assay are still not abundant, and the efficiency of electrofusion may not be great enough for such direct studies. However, the technique is relatively gentle and avoids the introduction of chemical impurities into the system. Commercial electrofusion apparatuses are available.

The interpretation of reconstitution experiments that are based on cell–cell, cell–membrane, or membrane–membrane fusion is limited, generally by the difficulty in quantifying the extent of fusion and in determining whether fusion occurred predominantly among similar or complementary cells (A–A and B–B vs A–B). The fusion of whole cells can be driven to yield large numbers of polykarya, suggesting that fusion is largely efficient, but this does not speak directly to whether cells fuse only with similar cells. Fusion of heterogeneous cells or membranes can be monitored by observing the mixing of cell-surface immunologic markers, covalently attached labels (Frye and Edidin, 1970), or functionally interacting proteins (Miller et al., 1976). More quantitative measurements of membrane fusion depend on either the mixture or dilution of fluorophores that are dissolved in the bilayers or reacted with surface components before fusion (Struck et al., 1981; Morris et al., 1985; Parente and Lentz, 1986; Nir et al., 1986; Klappe et al., 1986; Wojcieszyn et al., 1983). Routinely, however, reconstitution by fusion has been applied primarily in qualitative experiments in which the appearance absence of a sizable response was the desired signal. Because such reconstitutions utilize intrinsically impure systems, background enzymatic activities are often relatively high. Thus, the creation of hormonal stimulation of a preexistent activity is the most practical signal to monitor.

5. Detergent Solubilization and Interactions Among Solubilized Components

The first step in the characterization and purification of a membrane-bound protein is its solubilization, which is most commonly under taken using detergents. Many properties of membrane proteins can be studied effectively in detergent solution, and several events important to the beta-adrenergic regulation of adenylate cyclase activity can be observed in detergent-solubilized preparations. For example, both G_s and calmodulin can efficiently stimulate the activity of adenylate cyclase in detergent solution, although this interaction is facilitated by phosphatidylcholine. Another manifestation of this interaction is the efficient solubilization of a highly stable G_s–cyclase complex in the presence of activating ligands such as Al^{3+}/F^-, GTPγS or Gpp(NH)p, and forskolin. Receptors and G proteins generally do not interact productively in detergent solution. However, soluble receptor–G protein complexes have been described for the beta-adrenergic receptor

(Limbird and Lefkowitz, 1978; Limbird et al., 1980; Neufeld et al., 1983) and receptors for glucagon (Welton et al., 1977), FSH (Abou-Issa and Reichert, 1977), alpha-2 adrenergic amines (Smith and Limbird, 1981), dopamine (Senogles et al., 1987), and opiates (Koski et al., 1981). To varying degrees, these complexes are stabilized by the addition of agonist, and by the "locking" of agonist into the complex by treatment with N-ethylmaleimide (Korner et al., 1982; Vauquelin et al., 1980). The native complex tends to dissociate in response to GTP, and numerous examples of GTP-sensitive, high-affinity states of receptors have been described.

It should be noted that almost all of these proteins are also less stable in detergent solution than in membranes, leading to nonlinear kinetics and variable activities. For measuring these activities, it is generally better to use detergents at lower concentrations than are necessary initially to solubilize the proteins. However, one must be aware that dilution of the detergent can often cause the reformation of stable, large lipid-protein aggregates that include vesicles and sheets. Such inadvertent and uncontrolled reconstitution of membrane-like structures in the absence of obvious turbidity can confuse the question of whether one is studying a soluble or insoluble system. The simplest control for such aggregation is gel-exclusion chromatography.

The use of detergents to solubilize active membrane-bound enzymes and receptors is now well-established. A limited amount of theory, combined with experience and determination, can usually overcome the idiosyncratic behavior of detergents and the lability of the native proteins to yield an active, solubilized preparation. Numerous reviews (Helenius and Simons, 1975; Helenius et al., 1979; Tanford and Reynolds, 1976; Hjelmeland and Chrambach, 1984; Kagawa, 1972) and an excellent monograph (Tanford, 1980) are available on the theory and practice of detergent solubilization, and the present chapter will only mention a few general principles.

It is necessary to define solubilization of a membrane protein as its physical removal from the membrane and its suspension in a micellar solution of detergent as a monodisperse species of macro-molecular, not organellar, dimensions. Criteria of solubility that refer only to time and rate of centrifugation are generally inadequate, because many detergent–protein complexes sediment only slowly, if at all, because of their high partial specific volumes. It is therefore possible to disperse and suspend relatively large, multicomponent chunks of membrane using detergents; witness early descriptions of "soluble" hormone-stimulated adenylate cyclase (*see* Ross and Gilman, 1980) and oxidative phosphorylation. The requirement for solubilization of monodisperse species, usually characterized by gel filtration or sucrose density-gradient centrifugation, is not just a nicety of physical biochemistry. It is a prerequisite to the optimal use of powerful chromatographic techniques to separate the protein of interest from contaminants.

Although many practical aspects of different detergents are highly idiosyncratic, certain rules and useful, quantifiable physical properties of detergents have emerged (*see* Helenius et al., 1979; Tanford and Reynolds, 1976; Hjelmeland and Chrambach, 1984 for reviews). The formation of detergent micelles, the entities that actually solubilize lipids and proteins, is central. The formation of micelles is governed by the unique physical properties of a detergent molecule that cause it to bind differentially to itself or to water (Tanford, 1980). Thus, for example, the size of a micelle, typically 4–100 molecules, is a constant for a specific detergent under a given set of conditions.

Because detergent micelles are usually composed of many molecules, their formation is highly cooperative. A discrete concentration at which micelles form, the critical micelle concentration (CMC), is therefore definable. The CMC is the most informative physical property of a detergent. Because it is the concentration of *total* detergent above which micelles exist, the CMC is also equal to the concentration of *monomeric* detergent that is in equilibrium with micelles when the total detergent concentration is above the CMC (Tanford, 1980, Chapter 14). A detergent is an efficient solubilizing agent only above its CMC, because the micelle is the species that is responsible for maintaining phospholipids in suspensions. The concentration of micellar detergent, which is equal to the total concentration of detergent minus the CMC, is key in determining capacity for solubilization. Rivnay and Metzger (1982) defined the term $\rho = (\text{Detergent} - \text{CMC})$ as the ratio of micelle to lipid, which can be optimized objectively to improve solubilization (*see also* Klausner et al., 1984).

Table 1
Frequently Used Detergents

Detergent	Charge	CMC, mM	Mol wt[d]	Micellar number	v, mL/g
Cholic acid	−	3–10[c]	409	2–6[c]	0.75[i]
Deoxycholic acid	−	0.9–2[c]	393	2-~500[c]/4-30[b]	0.778[b]
Chaps	+/−	–1.4[h]	615	10[h]	−
Cetyltrimethyl-ammonium bromide	+	–0.9[a]	274	~70[a]	0.995[b]
Digitonin	0	0.1-0.3[e]	1229	~60[a]	~0.74[g]
Triton X-100	0	0.25[a]	~620	~140[a]	0.908[b]
Lubrol 12A9 (PX)	0	~0.1[f]	~600	~100[b]	0.958[b]
Lubrol WX (Cirrasol ALN-WF)	0	−	~940	~100[b]	0.929[b]
Tween 60	0	0.025[c]	~1200	~60[c]	~0.9
Octyl glucoside	0	25[c]	292	~25	−
Lauroyl sucrose[j]	0	0.05[f]	525	−	−

These data are taken primarily from reviews by Helenius and Simons (1975), Tanford and Reynolds (1976), and Helenius et al. (1979). They are given as examples and not as reliable working constants, because many of these values are highly sensitive to ionic strength, pH, and the presence of impurities. Abbreviations and trade names include v, partial specific volume; CHAPS, 3-[(cholamidopropyl) dimethyl-ammonio]-1-propanesulfonate; Triton X-100, (*p-t*-octyl-phenoxy-polyethyleneoxide (n~9.5); Lubrol 12A9, dodecyl and tetradecyl polyethyleneoxide (n~9.5); Lubrol WX, hexadecyl and octadecyl polyethyleneoxide (n~16); Tween 60, octadecylsorbitan-polyethyleneoxide (n~20).

[a]Helenius and Simons, 1975.
[b]Tanford and Reynolds, 1976.
[c]Helenius et al., 1979.
[d]Molecular weights are of free acids or bases.
[e]Fleming and Ross, 1980.
[f]E. M. Ross, unpublished. Determined according to fluorescence enhancement of 10^{-7} M 1,6-diphenyl-1,3,5-hexatriene.
[g]Hubbard, 1953.
[h]Hjelmeland and Chrambach, 1984.
[i]Cited in Steele et al., 1978.
[j]*See* Hekman et al., 1984; Feder et al., 1986.

The CMC values of many detergents are tabulated (Mukerjee and Mysels, 1971; Helenius and Simons, 1975; Kagawa, 1972; Helenius et al., 1979; Becher, 1967; *see also* Table 1). However, because the CMC varies, it is advisable to measure the CMC of a detergent under the conditions in which it will be used. Numerous methods of determining the CMC exist, but assays based on the fluorescence-enhancement of hydrophobic fluorophores as they partition into micelles are easy and usually adequately accurate. Many fluorophores have been used for such assays (for example, *see* Schrock and Gennis, 1977; Fleming and Ross, 1980). Note that a high CMC warns of a high monomer concentration that can cooperatively bind to and denature proteins (*see* Tanford, 1980, Chapter 14). More positively, a detergent with a high CMC will be easy to remove by gel filtration or dialysis during reconstitution.

Micellar size is important in two ways. If a detergent forms large micelles, protein–micelle complexes may display chromatographic properties more characteristic of the detergent micelle than of the protein being purified. A detergent that has a small micellar size, such as cholate, is also often easier to remove than one that forms large micelles.

Both the CMC and the micellar size vary according to the composition of the medium. The CMC of ionic detergents decreases markedly with increasing ionic strength because of charge-buffering effects (Tanford, 1980). A similar effect on micelle size is observed. Consequently, elevated ionic strength (>0.1M) is generally used to increase the concentration of solubilizing micelles while decreasing the concentration of the potentially denaturing monomer. In an extreme case, Strittmatter and Neer (1980) and Ross (1981) used ~1M (NH$_4$)$_2$SO$_4$ to decrease the CMC of cholate and thus allow the solubilization and fractionation of adenylate cyclase, which is usually quite unstable in cholate solution at lower ionic strength. The pH of the solution can also alter the CMC if the detergent has an ionizable group with a pI in a usable range. This effect is, however, more abrupt (all or none) and is rarely used intentionally. The micellar size is altered by the ionic strength in a manner similar to that of the CMC. By buffering local charge, salt favors the formation of larger micelles (*see* Tanford, 1980).

This property of ionic detergents has not been widely manipulated, however.

The relative hydrophobic or hydrophilic tendencies of a detergent are clearly important. However, because detergents form micelles in aqueous media and inverted micelles in organic solvents, their relative hydropathy ("strong feeling about water," Kyte and Doolittle, 1982) is difficult to quantify meaningfully in terms of partition coefficients or molar free energy of transfer from one medium to the other. The hydropathy of nonionic detergents has been described by a constant termed the "hydrophilic–lipophilic balance" (HLB). The HLB number can be measured, or it can be estimated based on the detergent's structural formula, using empirically derived constants (The Atlas HLB System; Becher, 1967). HLB numbers for commercial detergents have been tabulated (McCutcheon's Detergents and Emulsifiers; Helenius and Simons, 1975; Kagawa, 1972). HLB numbers for useful detergents tend to fall between 12 and 14, but the predictive value of this number is probably much less than originally thought.

The choice of an ionic or nonionic detergent, as well as the secondary choice of anionic, cationic, or zwitterionic detergent, is frequently based more on experience than theory. In the case of ion-exchange chromatography, it may be useful to have a detergent that does not bind ionically to the matrix. However, one may also use a detergent that binds and imparts an adjustable reverse-phase property to an otherwise hydrophilic chromatographic matrix. DEAE chromatography in cholate-containing buffers has proven useful in the purification of G proteins (Sternweis et al., 1981). When one uses a salt gradient to elute proteins from a DEAE column in cholate-containing buffer, an increasing concentration of cholate is eluted as well, thus decreasing the hydrophobicity of the column. This is, in effect, a combination of ion exchange and hydrophobic chromatography. Clearly, preequilibration of the column with detergent before applying sample is important if such a procedure is to succeed. Although it is frequently said that ionic detergents are harsher or stronger than nonionic detergents, this is not always the case. This generalization reflects the fact that the CMC for frequently used ionic detergents are higher than those for the most common nonionic detergents. Specifically designed nonionic and zwitterionic detergents that have high CMC (octyl glucoside, lauroyl sucrose, CHAPS, and the like) are increasingly available.

Impurities in commercial detergents compose a parameter that, although hard to quantify, is of major importance. Particularly with detergents that are prepared primarily for industrial use (Tritons, Lubrols, Tweens) and those that are natural products (bile acids, digitonin), extreme variability among manufacturers and lots is routine. There are many sources of *p-tert*-octylphenoxy-polyoxyethylenes (Tritons, Nonidets, and so on), but the purchased products are almost never the same. Remedies to this problem include selecting a single vendor and manufacturer, screening lots before buying a large quantity, and repurifying the detergent. All nonionic detergents should be deionized on a mixed-bed ion exchanger before use, and bile acids should be recrystallized or purified on DEAE (Ross and Schatz, 1978) unless highly purified material is purchased. Last, polyoxyethylene detergents accumulate peroxides during storage. Assays for peroxides and methods for their removal are available (Johnson and Siddiqi, 1970; Lever, 1977; Glavind and Hartmann, 1955), but storing stock solutions only briefly, under N_2 or Ar, is preferable.

Given this information, the choice of detergent is up to the investigator. Initially, one tries a detergent based on past experience, either one's own or that of others. The detergent is then titrated over a range of concentrations that is based both on the CMC and on the concentration of membrane protein present. One hopes to see a monotonically increasing amount of the desired activity in the supernatant and a parallel fall in activity in the pellet, although changes in the activity of an enzyme or receptor caused by solubilization can be expected to distort this quantitatively perfect picture. Total activity in the mixture before centrifugation should also be monitored. The generalization that solubilizing membrane-bound enzymes is best at a relatively high concentration of protein may reflect the fact that integral membrane proteins are often stabilized by lipids, and increasing the initial concentration of membranes serves to increase the concentration of lipid in the detergent supernatant. The addition of a membrane lipid to the solubilization mixture or to subsequently purified fractions can be useful and has been crucial in some cases (Agnew and Raftery, 1979; Tamkun et al., 1984). Last, as exemplified by the problems surrounding digitonin, consistent purity, cost, and solubility are not trivial considerations.

Beyond these criteria and known physical properties lies the investigator's experience with the essentially idiosyncratic behavior of the specific membrane protein in different detergents. For example, it is not known why digitonin is such a good detergent for the solubilization of beta-adrenergic receptors, why the deservedly unpopular detergent Tween 60 is useful for the affinity chromatography of adenylate cyclase (Pfeuffer et al., 1985a,b; Smigel, 1986), why lysolecithin is so effective for the mitochondrial NADP-NAD transhydrogenase (Rydstrom et al., 1976), or why such a popular detergent as Triton X-100 has proven virtually useless for any protein involved with transmembrane signaling functions.

Some detergents that have proven particularly useful in studies of the signal transduction systems of plasma membranes are listed in Table 1. No obvious rules of thumb come to mind. It is noteworthy that in reconstitution of the adenylate cyclase system, it has frequently been necessary to mix multiple detergents that individually are advantageous for working with one or another protein (*see* May et al., 1985). This perceived problem has recently been obviated in studies from the groups of Helmreich, Levitzki, and Pfeuffer (Feder et al., 1986).

6. Reconstitution of Solubilized Proteins into Biological Membranes

The reconstitution of solubilized membrane proteins into native membranes has been an effective mode of assay for membrane-bound enzymes, particularly enzymes that are part of a larger multiprotein complex whose other components may not be known. Many of the transport proteins, G proteins, and proteins of oxidative phosphorylation were identified in this way. This strategy allows one to assay a protein whose intrinsic function may be difficult to define by placing it in membrane that is replete with the other proteins necessary for the display of its physiologic effect. Such a reconstitutive assay permits the protein's purification so that its intrinsic biochemical functions can be studied in more detail. It also provides a convenient way to monitor the proteins' activity without purification for use in studies of development or regulation in vitro. The essential feature of such reconstitutions is that, with luck, a suitably depleted acceptor membrane can be found or designed such that only a single protein or activity is missing, providing essentially an unperturbed membrane environment and all of the other components necessary for the solubilized protein to express its activity. The converse is also true: Reconstitution of a purified protein into a native membrane to restore a specific activity implies only that the purified protein is necessary for the expression of that activity, and does not exclude the possible existence of unknown but equally necessary components.

The introduction of a detergent-solubilized protein into a membrane is conceptually the reverse of the process of its solubilization. Detergent must be removed slowly, such that the protein can partition preferentially into the acceptor membrane. The process is technically more difficult, however. The soluble protein and the acceptor membrane must be mixed such that the acceptor membrane is not itself solubilized, the protein must be encouraged to integrate into the membrane or bind to it, rather than to aggregate nonspecifically in the aqueous medium, and the orientation of the protein in the reconstituted membrane should be controlled. The techniques that have been successfully used for such reconstitutions are numerous, and only the strategies will be reviewed here.

The essential step in any of these procedures involves the removal of detergent. Frequently, simply the dilution of a small volume of detergent-solubilized protein into a larger volume of acceptor membrane suspension is adequate for reconstitution of the protein without significantly perturbing the acceptor membranes. Simple dilution is adequate for such proteins as G_s (Ross et al., 1978; Sternweis and Gilman, 1979), which probably do not require extensive interaction with the hydrocarbon core of the membrane bilayer. The relevant parameters in such a reconstitution by dilution are the volume and concentration of the detergent solution in which the protein is added, the volume of the diluting suspension of membranes and the concentration of membrane phospholipid, and the CMC of the detergent. In general, one wants to maintain an initial detergent concentration slightly above the CMC in order to maintain monomeric protein, but dilution must be sufficient to reduce the volume to below the CMC so that the acceptor membranes are not solubilized. Furthermore, the final ratio of the total amount of detergent to the total amount of phospholipid in the acceptor membrane should be minimized by increasing the concentration of acceptor membrane as much as possible in order to minimize the disruption of the membrane. This modified detergent:lipid ratio is essentially the ρ of Rivnay and Metzger (1982) that was discussed in the context of solubilization. The detergent also acts as a "chaotropic" solute in the acceptor membrane, and its final concentration should be minimized. Frequently, reconstitution is best done in a relatively large volume followed by centrifugation to concentrate the reconstituted membrane and remove detergent monomer. Removal of detergent can also be facilitated by treating either the protein solution or the reconstitution mixture with an adsorptive resin, such as Dowex™ XAD (Bio-Beads; Bio-Rad Laboratories), Sephadex™ LH-20 (Pharmacia), Extractigel™ (Pierce Chemical Co.), or the like. It should be noted that adsorptive resins also adsorb phospholipid and hydrophobic proteins; their selectivity for detergents is not absolute.

The quantitative analysis of the activities expressed by reconstituted membranes of this sort is not straightforward. In general, if a "saturating" concentration of acceptor membrane is used, one can be reasonably sure that activity is proportional to the amount of protein added. This statement is, however, a tautology. The reconstitution of cyc⁻ membranes with G_s can yield adenylate cyclase activities that are consistently indicative of the absolute amount of G_s added. However, this reliability holds true only with G_s from a single source under a single, well-defined set of conditions. In practice, variation in specific activities among experiments or among laboratories can be great, depending on the intrinsic capacity of the acceptor membrane to support the reaction of interest. The basic problem is related to the usual lack of stoichiometric information available and ignorance of the specific roles of detergent and lipid.

Structural information is also lacking: it is not obvious how a presumably integral membrane can insert itself into an already formed bilayer. This question has been particularly annoying in terms of the constitution of G_s or other nonretinal G proteins into plasma membranes. When a crude solution of G_s in Lubrol™ 12A9 was first reconstituted into membranes of cyc⁻, it was noted that G_s activity was not tightly associated with the membranes, even though the added G_s could mediate agonist-stimulated adenylate cyclase activity. Only if the membranes were incubated at elevated temperature with a persistently activating ligand of G_s, such as GDP/Al³⁺/F⁻ or GTPγS, did the G_s remain bound to the membrane upon centrifugation (Ross and Gilman, 1977; Howlett et al., 1979). Subsequently, Sternweis and Gilman (1979) showed that the reassociation of cholate-solubilized G_s with cyc⁻ membranes required the presence of a substantial amount of cholate in the reconstitution mixture. This suggested that partial dissolution and reassembly of the acceptor membrane might be required in this protocol that did not call for an activating ligand. In light of the recent findings of Sternweis (1986) that it is the βγ subunits of G proteins that associate with the bilayer and that the α subunit is associated primarily by βγ, these earlier experiments should probably be reinterpreted in terms of the exchange of added $G_{s\alpha}$ with βγ already constitutively in the acceptor membrane. In another particularly well-documented case, it was shown that insertion of cytochrome b_5 into preformed phospholipid vesicles yields an unusual configuration of the protein in which the normally membrane-spanning, hydrophobic domain bends such that the two hydrophilic regions of the molecule are on the same side of the membrane (Enoch et al., 1979; Takagaki, 1983a,b). Thus, systems such as these are primarily useful for quantifying a protein according to an arbitrary specific activity rather than for the detailed biochemical study of its mechanism of activity.

7. Reconstitution into Artificial Phospholipid Vesicles

The reconstitution of purified integral membrane proteins into phospholipid vesicles has traditionally provided the experimental model system of choice for their detailed biochemical study. Using purified proteins at known concentrations in a defined phospholipid environment allows the study of regulation and mechanism as well as effects of the composition and structure of the membrane bilayer on the proteins' functions. Because ion channels that have been reconstituted into vesicles can now be studied by patch-clamp techniques, reconstitution into vesicles is no longer limited to chemical signaling systems (Tank et al., 1982; Miller, 1984). The corollary to the experimental freedom of a reconstituted system is the need to demonstrate that a manifestly artificial system is in fact behaving as a good model for the plasma membrane signaling system under study.

The basic technology of reconstituting protein–phospholipid vesicles was developed in the late 1960s, and conceptually it has changed very little. In general, one mixes the detergent-solubilized protein or proteins of interest with a dispersion of phospholipids, and allows the lipids and proteins to associate slowly as detergent is removed. The choices facing the investigator are the state of purity of protein with which to start, the detergent and lipids to be used, and the means of removal of detergent. Development of protocols for reconstitution of integral membrane proteins is still essentially empirical, and it seems that successful protocols for no two proteins are identical. Experience has yielded some good starting points and strategies, however, and success is worth the effort. A broader discussion of overall strategy has been provided recently by Racker (1986). The reconstitution of beta-adrenergic receptor–G_s–adenylate cyclase systems have been described by Cerione et al. (1984; 1985a,b; 1986a,b), Levitzki, Helmreich, and coworkers (Gal et al., 1983; Hekman et al., 1984; Feder et al., 1986; Keenan et al., 1982), Schramm and coworkers (Citri and Schramm, 1980), and my coworkers and me (Pedersen and Ross, 1982, 1985; Asano and Ross, 1984; Asano et al., 1984a,b; Brandt and Ross, 1986; Brandt et al., 1983; Rubenstein et al., 1987; Moxham et al., 1988; May et al., 1985).

It is worth remembering that the technology of in vitro reconstitution of artificial membranes has so far applied almost nothing of what is known about the mechanism of biogenesis of native cellular membranes. The determinants of post-translational insertion of integral proteins into membranes, the roles of signal recognition particle and docking protein, the requirement of a membrane potential for protein translocation, the triggering effect of cleaving a signal peptide, and the likely need to unfold a protein at least partially in order to move it across a membrane are all potentially relevant to rational reconstitution in vitro (*see* Wickner and Lodish, 1985). Understanding the structure of transbilayer proteins and the physics of protein-bilayer interaction should be similarly useful.

7.1. Purity

Most investigators have found it easiest to develop reconstitution protocols using relatively crude solubilized preparations of the proteins of interest. Stability of the proteins and their activities is usually highest, and losses from nonspecific adsorption of proteins to dialysis bags or adsorptive resins are lower. Furthermore, possibly unidentified stabilizing or facilitating components, including lipids, will not have been removed. As the protein components of interest are identified and as experience with the reconstitution accumulates, it becomes easier to purify the components of the system and retain confidence in one's procedures. This stepwise approach also allows the investigator to optimize and standardize assays for proteins that are frequently unstable during purification. In the case of the beta-adrenergic receptor, our group spent several years using crude or partially purified receptor preparations for reconstitution-based experiments (Pedersen and Ross, 1982), and Lefkowitz's group developed their protocols for reconstitution by initially fusing pure receptors with intact cells (Cerione et al., 1983).

7.2. Detergent

The choice of the detergent or detergents used in the reconstitution procedure is almost invariably dictated by the proteins: in what detergents are they purified, and in what detergents are they stable. These considerations are dealt with in the preceding section. The only criterion for detergents that is unique to the process of reconstitution is the ability to remove detergents from the lipid–protein mixture. Removal of detergents by gel filtration or dialysis is essentially dependent on using detergents of small micelle size and/or high CMC. The classic example is cholate, but CHAPS and octyl glucoside are also suitable choices. A short-chain alkyl-polyoxyethylene with a high CMC has also been described, but has not been widely used (Egan et al., 1976). For a detergent with a relatively large micelle that cannot cross the dialysis membrane, the rate of detergent removal will be directly proportional to the concentration of monomer, which is equal to the CMC. Thus, Lubrol 12A9 will dialyze approx 100 × more slowly than will cholate or octyl glucoside. If gel filtration is used to remove detergent, a similar relationship will hold, because the principal species that is retarded on the column is the monomer. A small micellar size will also facilitate removal of detergent, because small micelles can pass through dialysis tubing of intermediate pore size. Small micelles can also be separated from large lipid–protein aggregates and nascent vesicles by gel filtration on a support that has an appropriate pore size. If detergent is to be removed by adsorption to a hydrophobic matrix, there is little to recommend one detergent over another. Adsorption of detergents to hydrophobic beads seems equally applicable to ionic and anionic detergents of high and low CMCs.

A frequently overlooked function of detergent in a reconstitution is to maintain the dispersion of lipid as well as protein. This can allow lipid–protein association to occur prior to formation of a membrane or can add adequate disorder to an already-formed acceptor membrane for the insertion of an integral protein. An adequate initial concentration of detergent may be as important as a low final concentration. For example, we were unable to coreconstitute purified beta-adrenergic receptor and G_s with phospholipids until we dispersed the lipid with added deoxycholate. Furthermore, the initial concentration of detergent and the rate of its removal can also alter the physical properties of the reconstituted vesicles (*see* Rhoden and Goldin, 1979; Wang et al., 1979; Zumbuehl and Weder, 1981).

Feder et al. (1984) recently stressed the desirability of using a single detergent to suspend all proteins and lipids prior to the beginning of a reconstitution. Although this makes the design of the reconstitutive protocol conceptually simple and facilitates the chemical monitoring of detergent removal, it does not seem to be logically necessary, nor has it been generally useful in the field. The other extreme, however, where four different detergents were used for a single reconstitution (May et al., 1985), is not necessarily desirable either.

7.3. Detergent Removal

Techniques for removal of detergent from a mixture of proteins and bilayer-forming lipids include (1) dilution; (2) dialysis, gel filtration, or centrifugation; or (3) adsorption to a hydrophobic support. Each has proven useful in some instances in the reconstitution of receptor–effector systems. The choice of method will dictate the efficiency of reconstitution and the recovery of protein. It will also determine in a more or less controlled way the size of the vesicles that are formed, the heterogeneity of their size, and the occurrence of multilamellar vesicles, which are generally not useful for biochemical probes of receptor function.

Bilayer-forming phospholipids are essentially insoluble in water, and most detergents display a CMC greater than $10^{-5}M$. Consequently, dilution of a detergent–lipid–protein mixture or dilution of solubilized protein into a suspension of vesicles will frequently lead to the formation of phospholipid vesicles that contain any integral proteins that were present in the initial mixture. Reconstitution of vesicles occurs because the detergent will be monomeric below its CMC, whereas the phospholipids will still be at a concentration sufficient to cause their aggregation to form bilayers. This tendency allows reconstitution without actual removal of detergent, although significant detergent can be expected to be found as a contaminant in the vesicle bilayers that are formed. However, collecting the vesicles by centrifugation and resuspending them in a detergent-free medium allows the preparation of substantially detergent-free vesicles with relatively little manipulation. Dilution has been quite useful for the reconstitution of transport proteins and for some of the enzymes of oxidative phosphorylation (Racker et al., 1975; Helenius et al., 1981; Eytan et al., 1976). Its efficiency can often be enhanced by sonication of the diluted mixture (Racker, 1973) or by a freeze–thaw cycle (Kasahara and Hinkle, 1977). In general, it has not been used extensively with signal transducing proteins, primarily because the low concentrations of lipids used in these reconstitutions makes adequate dilution unwieldy. Extensive losses from adsorption of lipid and protein to surfaces of the container are also observed. Dilution has been successfully applied to the rhodopsin–transducin system (Tsai et al., 1984).

The primary utility of reconstitution by dilution is the rapid assay of activities that are not easily detectable in the presence of solubilizing concentrations of detergent. Thus, Witkin and Harden (1981) showed that the beta-adrenergic receptor solubilized in digitonin could be readily assayed using iodopindolol if the preparation was simply diluted prior to incubation with the ligand. Separation of free and bound ligand was then performed by filtration on glass fiber filters. Dilution followed by centrifugation also proved useful for promoting the reassociation of G_s with adenylate cyclase in phosphatidylcholine vesicles (Ross, 1982). Regardless of general utility, dilution is easy to test and may work well.

Removal of detergents by gel filtration has been extremely useful in reconstituting the beta-adrenergic receptor and its related proteins. This procedure yielded the first reconstitution of beta-adrenergic receptors into vesicles (Fleming and Ross, 1980) and has since been modified for use with different combinations of purified receptors, G proteins, and adenylate cyclase (Asano et al., 1984b; Brandt and Ross, 1986; Haga et al., 1985; Florio and Sternweis, 1985; Kelleher et al., 1983; *see also* Sweet et al., 1985). The procedure yields primarily unilamellar phospholipid vesicles of 1000–5000 A diameter (Pedersen and Ross, 1982; May et al., 1985). Recovery of activities and total protein can be quite good. Seventy-five percent of total G_s and over 50% of beta-adrenergic receptor that is applied to such a column can be recovered in the vesicle fraction. Reconstitution of phospholipid vesicles by gel filtration has usually been performed on a hydrophilic support such as Sephadex™ G-25 or G-50 (Asano et al., 1984b; Haga et al., 1985), or on a support with a larger pore size to remove soluble proteins (Sternweis, 1986; Florio and Sternweis, 1985). Vesicles that form during gel filtration elute in the void volume, presumably with any contaminating water-soluble proteins that are large enough to be excluded from the matrix. Because many detergent micelles are also large enough to be excluded from such matrices, a column that is large relative to the sample size (usually 10–20-fold) is used. A probable mechanism of reconstitution involves the constant removal of a concentration of detergent greater than or equal to the CMC. Thus, vesicles form slowly as detergent is depleted during chromatography. It is also likely that this procedure includes a component of adsorption of detergent to the nominally hydrophilic gel exclusion matrix. This has allowed reconstitution from mixtures containing detergents with low CMC and large micelles, such as Lubrol 12A9, digitonin, and lauroyl sucrose. In these cases, the required column volumes, although large, have not been of theoretically adequate size to remove detergent totally if an amount only equal to the CMC were being depleted.

Detergent dialysis, although the classic technique in the field, has been of limited utility in working with receptor–effector systems. In my experience, dialysis has led to large adsorptive losses of protein and lipid because of the low concentrations that were initially present. The abundance of the nicotinic cholinergic receptor and rhodopsin has overcome this barrier, and these proteins have been efficiently reconstituted by dialysis (Hong and Hubbell, 1973; Hong et al., 1982; Fung, 1983). The receptor for immunoglobulin E from mast cells has also been reconstituted into vesicles by dialysis (Rivnay and Metzger, 1982), but its biochemical function remains unknown and was therefore unassayed.

Preferential adsorption of detergents from detergent–lipid–protein mixtures was first utilized by Gaylor and Delwiche (1969) and Holloway (1973) to reconstitute the enzymes of microsomal mixed-function oxidases. Using improved adsorptive resins, this approach has now become a standard for working with receptor–G protein vesicles, primarily through the work of Cerione and coworkers (1984, 1985a,b, 1986a,b). Two resins are commonly used: Biobeads™ are a washed styrene-divinylbenzene copolymer (Dowex-like); Extractigel™ is a proprietary product (Pierce) that is claimed to be less permeable to macromolecules and, therefore, to offer a higher recovery of protein. Whichever is used, either the gel is added to the mixture to be reconstituted and removed by centrifugation or filtration, or the mixture may be chromatographed through a small column of the gel. Removal of detergent can be efficient (95–98%) and is relatively rapid. Several studies of reconstituted vesicles that contain beta-adrenergic receptor and G_s have indicated that the use of Extractigel™ gives a good recovery, frequently above 75% (*see* Cerione, 1984; Feder et al., 1986). In unpublished experiments, we found that Biobeads also adsorb significant amounts of phospholipid and protein, however. There is no significant literature on the systematic evaluation of these adsorptive matrices, but recent studies from multiple laboratories make them appear to be extremely attractive.

Monitoring the removal of detergent is vital, but it is frequently overlooked if the desired activity has been restored. It is almost always limited by the convenience and sensitivity of one's assay for detergents. In general, one is dependent upon the availability of radiolabeled detergent for use as a tracer. Radiolabeled bile acids are commercially available; labeled CHAPS, octyl glucoside, or other synthetic detergents can be obtained or synthesized. It should be remembered, however, that labeled impurities can be deceiving. If the tracer is, for example, only 95% pure, then retention of radioactivity in the reconstituted vesicles can cause an overestimate of residual detergent if the labeled impurity is more difficult to remove than is bulk detergent. For example, in initial attempts to monitor removal of tritium-labeled deoxycholate from a crude preparation of receptor-G_s vesicles, we found that 1–2% of the tritium remained in the vesicle fraction after gel filtration and centrifugation. However, thin layer chromatographic analysis indicated that the radioactivity was not deoxycholate, but a significantly less polar contaminant. Common assays that rely on extraction of dyes into micelles or on fluorescence enhancement of amphipathic fluorophores are seldom sensitive enough to detect the low levels of detergent that should be present after the formation of phospholipid vesicles. They are also rarely specific enough to avoid interference by phospholipids. Assays based on extraction of the detergent and chromatographic purification are specific, but are rarely sensitive enough unless one is blessed with a highly abundant protein.

It should be remembered that the removal of detergent by any technique is generally a first-order process. Thus, removing 99% of the detergent takes twice as much effort as removing 90%, and removing 99.9% takes three times as much. If one begins with a fivefold excess of detergent over lipid, a 5% detergent contamination in the vesicle bilayer will remain even after 99% of the detergent has been removed. This may be tolerable in some cases, and restoration of a desired function may result. However, it is inadequate if one wishes to draw conclusions relating to the influence of native bilayers on receptor-effector coupling. Therefore, many investigators have found that it is worthwhile to combine two methods of detergent removal (gel filtration plus centrifugation, adsorption plus centrifugation, and so forth).

7.4. Choice of Lipid for Reconstitution

Anecdotes suggest that using crude lipid mixtures gives better yields and better results in developing reconstituted assay systems than does the use of a single lipid. A frequent first choice is a crude phospholipid fraction prepared from either soybeans (asolectin) or egg yolk. Cost and availability argue for the soy or egg yolk lipids. The head group and acyl chain compositions of soy and egg phospholipids are known (see Ansell et al., 1973, Chapters by White and Galliard). The differences between them are not obviously meaningful a priori, although egg phospholipids are generally less saturated than their soy counterparts. Commercial soy or egg lipid preparations of increasing purity are available through reliable vendors, and there is extensive experience in their use. It is advisable to wash the cruder preparations with acetone and an aqueous salt solution to remove cholesterol and lysophospholipids, respectively. Schramm's group (Citri and Schramm, 1980; Kirilovsky et al., 1985) has been quite successful in the use of soy lipids for reconstituting impure preparations of β-adrenergic receptor with G_s. Cerione and coworkers (1984; 1986b; Lefkowitz et al., 1985) and the groups of Helmreich and Levitzki (Hekman et al., 1984; Feder et al., 1986) have also used asolectin for more extensive studies using purified proteins.

An alternative to these preparations is the use of a total phospholipid extract of the membranes from which the proteins to be reconstituted were initially extracted. As our laboratory began to use increasingly purified beta-adrenergic receptors in combination with purified G_s we found that a crude Folch extract of turkey erythrocyte plasma membranes, from which the receptor had been purified, provided the optimal mixture of lipids for receptor–G_s coupling and for reliable assays of these proteins (Brandt et al., 1983; Asano et al., 1984b). Both we (Fleming and Ross, 1980) and Kelleher et al. (1983) found that unknown components of egg lecithin caused high levels of nonspecific binding of beta-adrenergic ligands after reconstitution. In some cases, this nonspecific binding was decreased by the addition of propranolol in a rather confusing pattern.

The use of a single phospholipid considerably simplifies analysis of the physical structure of the reconstituted vesicles. For studies where detailed biochemical or biophysical data are desired, it is clearly preferable to work in vesicles that contain as few lipid components as possible. However, whereas several enzymes have retained activity after reconstitution into vesicles composed of a single synthetic lipid with a known acyl chain composition (Churchill et al., 1983; Enoch et al., 1979; Racker and Hinkle, 1974; *see* Sandermann, 1978), such systems have not yet been perfected for receptor–G protein–effector systems. Thus, although we have been able to reconstitute efficient coupling of the beta-adrenergic receptor, G_s, and adenylate cyclase in vesicles composed of PE plus PS (phosphatidylserine) (3:2 optimal ratio), we have been unable to eliminate the second component. The PE:PS mixture has also been reported to be optimal by Feder et al. (1986). We have also been unable to replace natural PE and PS with synthetic lipids that have a defined acyl-chain composition without diminishing coupling and recovery of activity. Furthermore, we found that cholesterol or a cholesterol-ester, in addition to the two phospholipids, serves to markedly stabilize the beta-adrenergic receptor after it has been activated by treatment with thiols (Pedersen and Ross, 1985). It may be significant that the PE:PS:cholesterol mixture closely mimics the composition of the inner monolayer of the mammalian plasma membrane, but the importance of this correlation is difficult to assess biologically.

The simplest strategy for optimizing the phospholipid composition of a reconstituted vesicle is to begin by mimicking the composition of the mixture of natural lipids that is used in the original successful reconstitution and then removing lipids that are found to be unnecessary. For example, we found that using avian erythrocyte membrane phospholipids yielded efficient reconstitution of beta-adrenergic receptor from that source. We then analyzed these lipids chromatographically and prepared a mixture of purified lipids that contained the major components of the original preparation. It was then relatively easy to eliminate components that were not required and to optimize the relative proportions of the PS and PE.

The use of synthetic phospholipids with defined acyl side chains is clearly desirable for detailed biophysical studies. In many cases, their physical properties are well understood; when these properties have not been documented, one at least has a chance of defining the structures and formal phases present within the reconstituted bilayers. In this sense, PCs with symmetric acyl chains are the lipids of choice because they have been so carefully studied. The use of dimyristoyl-PC, dipalmitoyl-PC, dioleoyl-PC or 1-palmitoyl, 2-oleoyl-PC allows one to make relatively precise interpretations of changes in activity that may be caused by altering the lipids or the temperature. Excellent examples of such studies are to be found in work on beta-hydroxybutyrate dehydrogenase (Churchill et al., 1983). Such systems remain an experimental goal to be sought after by investigators in the transmembrane-signaling field.

The question of whether the beta-adrenergic receptor or other components of the cyclase system require specific lipids for their function or stability has been approached by reconstitutive studies in several laboratories. Highly purified beta-adrenergic receptor, G_s and adenylate cyclase have been reconstituted successfully using either soy PC (Cerione et al., 1985b; Hekman et al., 1984) or a mixture of PE and PS (Asano et al., 1984a,b; Brandt and Ross, 1986; May et al., 1985; Feder et al., 1986), suggesting that specificity for lipids is not strict. In our experience, there is a broad optimum at a PE:PS ratio of 1.5. The PS may be totally replaced by PG (phosphatidylglycerol) and the PE may be replaced by PC, with little loss of activity or beta-adrenergic regulation of G_s (R.C. Rubenstein and E.M. Ross, unpublished observation). In contrast to these studies, Kirilovsky and coworkers found that when a deoxycholate extract of erythrocyte membranes was gel filtered in deoxycholate-containing buffer to deplete endogenous lipids, their ability to reconstitute beta-adrenergic regulation of a crude G_s fraction was lost unless a neutral lipid was included in the mixture of PE, PS, and PC that was used to form the vesicles (Kirilovsky et al., 1985). Cholesterol arachidonate or cholesterol oleate, α-tocopherol, *trans*-retinol, and an acetone extract of crude asolectin were all active, although cholesterol and cholesteryl-hemisuccinate were not. Neutral lipids were effective at about 8% of total lipid, but concentration dependence was not shown. It is difficult to reconcile these studies with the results obtained by others using the purified proteins, unless an endogenous neutral lipid cofractionates with receptor during 10,000-fold purification in digitonin or digitonin plus lauroyl sucrose. The purification of G_s includes gel filtration in cholate in addition to multiple other fractionation procedures carried out both in cholate and Lubrol 12A9, and lipid contamination from this source is therefore unlikely. We have not sacrificed enough pure receptor to show unambiguously that it does not contain stoichiometric amounts of a neutral lipid. However, such association would require truly high affinity association that seems inconsistent with the need for bulk amounts of tocopherol.

Levitzki's group recently published the provocative finding that a specific beta-adrenergic affinity reagent, *N*-bromoacetylpindolol, could inactivate beta-adrenergic receptor binding activity, either in native membranes or in partially purified preparations, without covalently reacting with the receptor (Bar-Sinai et al., 1986). Instead, they found that the label reacted preferentially with an organic-extractable component that may be a glycolipid. Furthermore, after treating receptors with this affinity label, beta-adrenergic binding activity was restored by the addition of a crude glycolipid fraction. This suggests that a glycolipid or other lipid having a reactive nucleophilic group may be in close association with the beta-adrenergic receptor and either contribute to the site of beta-adrenergic ligand binding or be responsible in part for maintenance of the structure of the receptor.

Can an in vitro reconstitutive protocol define structural or functional requirements of a membrane protein for specific lipids or merely define operationally which lipids yield efficient reconstitution? This question is not wholly semantic. One could imagine a functionally suboptimal combination of lipids being required for efficient reconstitution through a specific protocol. Such a worst case has not been observed in numerous studies of other membrane-bound enzyme systems, but different lipids have been found optimal for the reconstitution of a single enzyme when two different techniques were used (*see* Sandermann, 1978). It seems reasonable that after one has normalized for the lateral concentration of protein (i.e., the lipid: protein ratio), those lipids that yield high activities in reconstituted systems should be similarly optimal in vivo. However, this remains an assumption. A more profound question is whether, in a native membrane, there are specific associations between the proteins of interest and membrane lipids that might influence their activities.

7.5. The Problem of Scale

A technical problem that has so far been relatively unique to studies of receptor–effector coupling has been the need to work on a small scale. The scale results from the small amounts of receptor, G protein, and effectors that can be reasonably purified. To observe regulatory protein–protein interactions in reconstituted vesicles, each vesicle must contain at least one of each protein, and preferably more. If vesicles are only in the 100-nm diameter range and the protein concentration is low, small amounts of total lipid must be used as well, frequently less than 1 mg/mL in less than 1 mL. The reconstitution of low concentrations of lipid and protein in relatively small volumes raises persistent problems of adsorption to dialysis bags, gel filtration matrices, and detergent-adsorbing resins. Using the gel filtration protocol that is now standard in our laboratory, we have found it imperative to decrease the volume of the gel filtration column as much as possible, to coat surfaces with siliconizing compounds and/or lipids, and to maintain the highest feasible concentrations of phospholipid and protein. Even so, the loss of yield of receptor or G_s activities during reconstitution have almost always reflected the physical loss of the protein itself during reconstitution, rather than its denaturation.

8. Characterization of Receptors, Couplers, and Effectors in Reconstituted Phospholipid Vesicles

Essential characterization of a reconstituted vesicle preparation includes determining its content of specific proteins, their activities, and their ability to interact. Because reconstitution frequently alters the activities and interactions among proteins, specific activities may vary dramatically between reconstituted vesicles and native membranes. Therefore, the investigator must establish criteria for determining the reproducibility and relative activity of his preparations. We have found that the quantitative assay of ligand-binding, GTP hydrolysis, adenylate cyclase, and associated reactions are technically easier and yield data of higher quality in reconstituted vesicles that contain purified proteins than in plasma membrane preparations. This no doubt reflects freedom from contaminating activities and greater physical homogeneity of the preparation. Using these activities to estimate the amount of each protein present is not always straightforward, however.

Ligand-binding assays for the beta-adrenergic receptor using [^{125}I]iodocyanopindolol or [^{3}H]dihydroalprenolol have generally been very reliable in measuring reconstituted beta-adrenergic receptor. The addition of detergent, extra lipid, or other proteins do not perturb binding; assayed concentrations of receptor binding sites have generally corresponded to concentrations estimated from amino acid analyses. G proteins and adenylate cyclase have been less reliably quantified based on their activities after reconstitution. We found that reconstitution of G_s into vesicles decreased the number of available binding sites for [^{35}S]GTPγS, but that these sites could be reexposed upon solubilization of the vesicles with Lubrol 12A9 (Asano et al., 1984b). It is not clear whether this represents an inhibition of the ability of one pool of G_s to bind nucleotide, a locking in place of bound GDP by phospholipid, or the possible occlusion of the G_s binding site from the bulk solution. We therefore assay reconstituted G_s in an incubation mixture that contains Lubrol 12A9. Because the apparent affinity of soluble G_s for GTPγS is lower than that displayed by the reconstituted receptor-stimulated binding reaction, we also increase the concentration of [^{35}S]GTPγS from 0.2 μ*M* to 10 μ*M*. We used Mg^{2+} (50 m*M*) to promote nucleotide exchange. With this assay, we found that the fraction of G$_s$ that is available for binding nucleotide in the absence of detergent can be as low as 30% or as high as 100%. Interestingly, only that pool of G$_s$ that can bind GTPγS in the absence of detergent is also accessible for receptor-stimulated binding, and we refer to this as the receptor-sensitive or coupled pool of G$_s$ (Asano et al., 1984b).

The activity of adenylate cyclase itself is also environment-sensitive. Even in the presence of Mn^{2+} or forskolin, which stimulate enzymatic activity directly, varying specific activity is found when the enzyme is assayed in Lubrol 12A9, CHAPS, Tween 60, or phospholipid vesicles (Smigel, 1986; Pfeuffer et al., 1985a,b; Ross, 1981). These difficulties in assaying reconstituted proteins must be addressed when one attempts to estimate the recovery of a specific protein during reconstitution and when one wishes to evaluate the specific activity of such a protein in vesicles as a function of some experimental manipulation.

The physical characterization of reconstituted phospholipid vesicles includes understanding the basic physical properties of their lipid components as well as the size of the vesicles, the distribution of the size, their permeability, and the orientation of the protein components within them. The techniques used for elucidating the detailed structure of phospholipid bilayers and the mobility of their protein and lipid components is beyond the scope of this review. Numerous texts and symposium volumes cover this area (Szoka and Papahadjo-poulos, 1980, for example). However, slightly more macroscopic properties are of immediate relevance. Because many signal-transducing events reflect ligand-regulated protein–protein interactions, the concentration of each protein is an important parameter. For intrinsic membrane proteins, concentration is calculated with reference to bilayer surface area rather than to volume. For example, the relevant concentrations of the βγ subunits of a G protein, adenylate cyclase, or a receptor are most reasonably expressed in moles of protein per unit amount of lipid, the lipid:protein ratio. Unless an interaction with a truly soluble component is considered, the actual concentration of protein in a suspension of vesicles is largely irrelevant, except for calculating amounts needed for a particular assay.

To continue this reasoning, the size of the vesicles into which the receptors are reconstituted and the distribution of receptors or other proteins among these vesicles also determines their ability to interact. As an example, an early preparation of receptor–G_s vesicles contained such a low concentration of beta-adrenergic receptors that, on the average, there was less than one receptor per phospholipid vesicle (Pedersen and Ross, 1982). Because the concentration of G_s was much greater than that of receptor, significant amounts of G_s were reconstituted into vesicles that had no receptor and that G_s was physically "uncoupled." This can be a general problem when the amounts of protein available for a reconstitution are low. The distribution of proteins among lipid vesicles can be estimated if one knows the number of vesicles per unit volume of suspension and the molar concentration of protein. The concentration of vesicles can be calculated using the molar concentration and the mol wt of the phospholipid, the density of the phospholipid in a bilayer, the diameter of the vesicles, and the thickness of the vesicle bilayer. One simply calculates the "mol wt" of a vesicle having dimensions of a spherical shell appropriate to the lipid (Huang and Mason, 1978, for example). The diameter of phospholipid vesicles can be estimated according to autocorrelation light scattering (Selser et al., 1976; Wong and Thompson, 1982) or gel filtration (Nozaki et al., 1982; Reynolds et al., 1983; Huang, 1969; van Renswoude et al., 1980), or from electron micrographs of the vesicles. Morphometric analysis of vesicles that have been prepared by freeze-etching yields the best results, but measurements of negatively stained preparations can also be used if attention is paid to possible swelling or shrinking artifacts that occur as results of osmotic changes or subsequent drying. A significant problem is that the size of vesicles is frequently not uniform within a preparation; receptor–G_s vesicles in our hands range in size of 400 A diameter at the smallest to 7000–8000 A diameter at the largest. Because the surface area of a vesicle (and, therefore, the volume of its spherical shell) is proportional to the square of its diameter, errors in estimating the number of vesicles in suspension propagate as the square of the errors in the determination of diameter.

A second important physical property of a reconstituted vesicle system is the orientation of individual protein molecules with respect to the surface, generally referred to their being right-side-in or inside-out. For example, the beta-adrenergic receptor is a transmembrane protein that is asymmetrically oriented to bind ligand on the outside of the cell and to bind G_s either within the bilayer or on the inner face of the membrane. G_s is a quasiperipheral protein that binds to the inner face of the membrane. Adenylate cyclase is also an integral protein that is probably transmembranous (based on its glycosylation) and presumably binds ATP and G_s only on the inner face of the membrane. For any pair of these proteins to interact productively in reconstituted vesicles, they must be oriented similarly, even if not as in the cell. Presumably, inside-out beta-adrenergic receptor could interact perfectly well with G_s bound to the outer face of a vesicle. This was probably the orientation of the receptor–G_s vesicles described by Pedersen and Ross (1982), when G_s was added to preformed receptor-containing vesicles.

In general, investigators have not probed the sidedness of reconstituted receptor–G protein–effector vesicles, but have been satisfied when reasonably efficient coupling occurs. Since there are well-documented cases where a protein will reconstitute essentially 100% inside-out, 100% right-side-in, or scrambled (randomly oriented), this pragmatic approach may cause one to overlook the presence of large amounts of topologically uncoupled proteins in one's preparation. Furthermore, where ligands must be accessible to both sides of the membrane, the orientation of the proteins determines which ligands must permeate the vesicle and which are free to act on the outside. Thus, in a reconstituted adenylate cyclase system that is oriented right-side-in, catecholamines can bind to the beta-adrenergic receptor on the outer surface of the vesicle. However, GTP, ATP, and Mg^{2+} must all reach the lumen of the vesicle. It has not been demonstrated that such a permeability barrier has limited guanine nucleotide binding or G_s-stimulated adenylate cyclase activity in vesicle preparations that have been described, perhaps suggesting that the vesicles are leaky. Alternatively, it may reflect the low concentrations of nucleotides and cations that are actually required and the large ratio of vesicle surface to luminal volume found in relatively large unilamellar vesicles.

The orientation of membrane proteins in vesicles is determined by their differential accessibility to soluble and impermeant probes.

Numerous chemical probes for protein orientation are available in addition to proteases, kinases, and antibodies. The impermeance of the probe must be demonstrated for each preparation, however. Furthermore, some knowledge of the asymmetric orientation of the protein must be known before an appropriate probe can be chosen. Thus, a GTP or ATP affinity reagent could be used to locate the binding sites of a G protein or adenylate cyclase, but their inability to cross the vesicle membrane in significant quantities would have to be demonstrated. Similarly, beta-adrenergic ligands that have been covalently coupled to macromolecules should be good probes of the ligand-binding site of the beta-adrenergic receptor (Kusiak and Pitha, 1982). In such experiments, the interaction of the impermeant probe with a reconstituted protein is measured in intact vesicles and in vesicles that have been disrupted by intense sonication, by chemical permeabilization (i.e., with alamethicin; *see* Besch et al., 1977), or by solubilization with detergent. For an interaction site that is completely shielded within the lumen, no interaction should be determined until the bilayer is disrupted. For an externally oriented site, disruption should have no effect. Intermediate results can of course be imagined, and these would blur a theoretically simple picture. They can reflect partial permeability of the probe, leakiness of some but not all of the vesicles, or partial scrambling of the protein of interest. An increasingly useful approach is to use antibodies that are specific for either the extracellular or intracellular portion of a transmembrane protein. These can be either appropriately selected monoclonal antibodies or antibodies raised against synthetic peptides representative of a sequence that faces either the inner or the outer surface of the membrane. The use of such antibodies obviates the need to determine chemically the site at which a protein may have been chemically modified or cleaved by a protease.

Unfortunately, although analytical probes of orientation of membrane proteins in reconstituted vesicles are readily available, the techniques for reconstituting a protein in a specific orientation are not well-developed. Work from the laboratories of Racker (1986 review) and Helenius et al. (1981) have indicated that specific reconstitution protocols can cause a given protein to incorporate in either the right-side-out or the inside-out configuration, and these studies have also suggested possible mechanisms for inducing asymmetry. However, predictive generalizations are not available. One useful generalization is that, when soluble proteins are mixed with preformed or "almost formed" phospholipid vesicles while detergent is being removed, they generally insert vectorally (not randomly). If one cannot reconstitute well oriented proteins, it is possible to separate inside-out from right-side-in vesicles using either affinity or immunoaffinity matrices, but such matrices may not be easy to prepare, and recoveries are not expected to be good. The provocative finding that oxidized and reduced cytochrome *c* can influence the orientation of reconstituted cytochrome oxidase also holds out a suggestion for further developments based on reconstitution in the presence or absence of ligands (Carroll and Racker, 1977).

9. Prospects

It should be clear from the tone of this review that the strategies and techniques of reconstituting receptor–effector systems are based on the mechanistic questions to be approached at least as much as they are based on a given body of experimental detail. Reconstituted assay systems are particularly well-suited to quantitative evaluation of receptor function, and questions related to the regulation of G protein coupled systems are increasingly able to be formulated in quantitative terms. The sorts of assay protocols that have helped answer questions of what receptors do should now be applicable to questions of how these activities are regulated.

Acknowledgment

Studies from the author's laboratory have been supported by NIH grant GM30355 and Welch Foundation grant I-982.

References

The Atlas HLB System. ICI, Wilmington, Delaware.

McCutcheon's Detergents and Emulsifiers, Annual North American Ed. (MC Publishing Co., Ridgewood, New Jersey.

Abou-Issa, H., and Reichert, L. E., Jr. (1977) Solubilization and some characteristics of the follitropin receptor from calf testis. *J. Biol. Chem.* **252,** 4166-4174.

Agnew, W. S. and Raftery, M. A. (1979) Solubilized tetrodotoxin binding component from the electroplax of *Electrophorus electricus*. Stability as a function of mixed lipid-detergent micelle composition. *Biochemistry* **18,**1912–1919.

Ansell, G. B., Hawthorne, J. N., and Dawson, R. M. C. (eds.) (1973) *Form and Function of Phospholipids,* Vol. 3. (Elsevier Scientific, Amsterdam.)

Asano, T., Katada, T., Gilman, A. G., and Ross, E. M. (1984a) Activation of the inhibitory GTP-binding protein of adenylate cyclase, G_i, by the β-adrenergic receptors in reconstituted phospholipid vesicles. *J. Biol. Chem.* **259,** 9351–9354.

Asano, T., Pedersen, S. E., Scott, C. W., and Ross, E. M. (1984b) Reconstitution of catecholamine-stimulated binding of guanosine 5'-O-(3-thiotriphosphate) to the stimulatory GTP-binding protein of adenylate cyclase. *Biochemistry* **23,** 5460–5467.

Asano T., and Ross, E. M. (1984) Catecholamine-stimulated guanosine 5'-O- (3-thiotriphosphate) binding to the stimulatory GTP-binding protein of adenylate cyclase: Kinetic analysis in reconstituted phospholipid vesicles. *Biochemistry* **23,** 5467–5471.

Bar-Sinai, A., Aldouby, Y., Chorev, M., and Levitzki, A. (1986) Association of turkey erythrocyte β-adrenoceptors with a specific lipid component. *EMBO J.* **5,** 1175–1180.

Baron, C., and Thompson, T. E. (1975) Solubilization of bacterial membrane proteins using alkyl glucosides and dioctanoyl phosphatidylcholine. *Biochim. Biophys. Acta* **382,** 276–285.

Becher, P. (1967) Micelle formation in aqueous and nonaqueous solutions, in Nonionic Surfactants, (Schick, M. J. , ed.), Marcel Dekker., New York. pp. 478–603.

Besch, H. R., Jr., Jones, L. R., Fleming, J. W., and Watanabe, A. M. (1977) Parallel unmasking of latent Na+, K+-ATPase and adenylate cyclase activities in cardiac sarcolemmal vesicles: A new use of the channel-forming ionophore alamethicin. *J. Biol. Chem.* **252,** 7905–7908.

Boege, F., Jürss, R., Cooney, D., Hekman, M., Keenan, A. K., and Helmreich, E. J. M. (1987) Functional and structural characterization of the two β_1-adrenoceptor forms in turkey erythrocytes with molecular masses of 50 and 40 kilodaltons. *Biochemistry* **26,** 2418–2425.

Brandt, D. R., and Ross, E. M. (1986) Catecholamine-stimulated GTPase cycle: Multiple sites of regulation by β-adrenergic receptor and Mg^{2+} studied in reconstituted receptor-G_s vesicles. *J. Biol. Chem.* **261,** 1656–1664.

Brandt, D. R., Asano, T., Pedersen, S. E., and Ross, E. M. (1983) Reconstitution of catecholamine-stimulated guanosinetriphosphatase activity. *Biochemistry* **22,** 4357–4362.

Carlson, K. E., Woolkalis, M. J., Newhouse, M. G., and Manning, D. R. (1986) Fractionation of the β subunit common to guanine nucleotide-binding regulatory proteins with the cytoskeleton. *Mol. Pharmacol.* **30,** 463–468.

Caron, M. G. and Lefkowitz, R. J. (1976) Solubilization and characterization of the β-adrenergic receptor binding sites of frog erythrocytes. *J. Biol. Chem.* **251,** 2374–2384.

Carroll, R. C. and Racker, E. (1977) Preparation and characterization of cytochrome *c* oxidase vesicles with high respiratory control. *J. Biol. Chem.* **252,** 6981–6990.

Cassel, D. and Selinger, Z. (1976) Catecholamine-stimulated GTPase activity in turkey erythrocytes. *Biochem. Biophvs. Acta* **452,** 538–551.

Cerione, R. A., Strulovici, B., Benovic, J. L., Lefkowitz, R. J., and Caron, M. G. (1983) Pure β-adrenergic receptor: The single polypeptide confers catecholamine responsiveness to adenylate cyclase. *Nature* **306,** 562–566.

Cerione, R. A., Codina, J., Benovic, J. L., Lefkowitz, R. J., Birnbaumer, L., and Caron, M. G. (1984) The mammalian β_2-adrenergic receptor: Reconstitution of functional interactions between pure receptor and pure stimulatory nucleotide binding protein of the adenylate cyclase system. *Biochemistry* **23,** 4519-4525.

Cerione, R. A., Codina, J., Kilpatrick, B. F., Staniszewski, C., Gierschik, P., Somers, R. L., Spiegel, A. M., Birnbaumer, L., Caron, M. G., and Lefkowitz, R. J. (1985a) Transducin and the inhibitory nucleotide regulatory protein inhibit the stimulatory nucleotide regulatory protein mediated stimulation of adenylate cyclase in phospholipid vesicle systems. *Biochemistry* **24,** 4499-4503.

Cerione, R. A., Gierschik, P., Staniszewski, C., Benovic, J. L., Codina, J., Somers, R., Birnbaumer, L., Spiegel, A. M., Lefkowitz, R. J., and Caron, M. G. (1987) Functional differences in the βγ complexes of transducin and the inhibitory guanine nucleotide regulatory protein. *Biochemistry* **26,** 1485–1491.

Cerione, R. A., Staniszewski, C., Benovic, J. L., Lefkowitz, R. J., Caron, M. G., Gierschik, P., Somers, R., Spiegel, A. M., Codina, J., and Birnbaumer, L. (1985b) Specificity of the functional interactions of the β-adrenergic receptor and rhodopsin with guanine nucleotide regulatory proteins reconstituted in phospholipid vesicles. *J. Biol. Chem.* **260,** 1493–1500.

Cerione, R. A., Regan, J. W., Nakata, H., Codina, J., Benovic, J. L., Gierschik, P., Somers, R. L., Speigel, A. M., Birnbaumer, L., Lefkowitz, R. J., and Caron, M. G. (1986a) Functional reconstitution of the α_2-adrenergic receptor with guanine nucleotide regulatory proteins in phospholipid vesicles. *J. Biol. Chem.* **261,** 3901–3909.

Cerione, R. A., Staniszewski, C., Gierschik, P., Codina, J., Somers, R. L., Birnbaumer, L., Spiegel, A. M., Caron, M. G., and Lefkowitz, R. J. (1986b) Mechanism of guanine nucleotide regulatory protein-mediated inhibition of adenylate cyclase. Studies with isolated subunits of transducin in a reconstituted system. *J. Biol. Chem.* **261,** 9514–9520.

Churchill, P., McIntyre, J. O., Eibl, H., and Fleischer, S. (1983) Activation of D-β-hydroxybutyrate apodehydrogenase using molecular species of mixed fatty acyl phospholipids. *J. Biol. Chem.* **258,** 208–214.

Citri, Y. and Schramm, M. (1980) Resolution, reconstitution, and kinetics of the primary action of a hormone receptor. *Nature* **287,** 297–300.

Cortese, J. D., Vidai, J. C., Churchill, P., McIntyre, J. O., and Fleischer, S. (1982) Reactivation of D-β-hydroxybutyrate dehydrogenase with short-chain lecithins: Stoichiometry and kinetic mechanism. *Biochemistry* **21,** 3899–3908.

Coussen, F., Haiech, J., D'Alayer, J., and Monneron, A. (1985) Identification of the catalytic subunit of brain adenylate cyclase: A calmodulin binding protein of 135 kDa. *Proc. Natl. Acad. Sci. USA* **82,** 6736–6740.

Cullis, P. B., Hope, M. J., and Tilcock, C. P. S. (1986) Lipid polymorphism and the roles of lipids in membranes. *Chem. Phys. Lipids* **40,** 127–144.

Darfler, F. J., Mahan, L. C., Koachman, A. M., and Insel, P. A. (1982) Stimulation by forskolin of intact S49 lymphoma cells involves the nucleotide regulatory protein of adenylate cyclase. *J. Biol. Chem.* **257,** 11901–11907.

Duzgunes, N. (1985) Membrane fusion. *Subcell. Biochem.* **11,** 195–286.

Egan, R. W., Jones, M. A., and Lehninger, A. L. (1976) Hydrophile-lipophile balance and critical micelle concentration as key factors influencing surfactant disruption of mitochondrial membranes. *J. Biol. Chem.* **251,** 4442–4447.

Eimerl, S., Neufeld, G., Korner, M., and Schramm, M. (1980) Functional implantation of a solubilized β-adrenergic receptor in the membrane of a cell. *Proc. Natl. Acad. Sci. USA* **77,** 760–764.

Engelhard, V. H., Glaser, M., and Storm, D. R. (1978) Effect of membrane phospholipid compositional changes on adenylate cyclase in LM cells. *Biochemistry* **17,** 3191–3200.

Enoch, H. G., Fleming P. J., and Strittmatter, P. (1979) The binding of cytochrome b_5 to phospholipid vesicles and biological membranes. Effect of orientation on intermembrane transfer and digestion by carboxypeptidase Y. *J. Biol. Chem.* **254,** 6483–6488.

Eytan, G. D., Matheson, M. J., and Racker, E. (1976) Incorporation of mitochondrial membrane proteins into liposomes containing acidic phospholipids. *J. Biol. Chem.* **251,** 6831–6837.

Feder, D., Arad, H., Gal, A., Hekman, M., Helmreich, E. J. M., and Levitzki, A. (1984) Resolution, reconstitution, and mode of action of the β-adrenergic receptor-dependent adenylate cyclase, in *Advances in Cyclic Nucleotide and Protein Phosphorylation Research,* Vol. 17, (Greengard, P. et al , eds.) Raven, New York.

Feder, D., Im, M-J., Klein, H. W., Hekman, M., Holzhofer, A., Dees, C., Levitzki, A., Helmreich, E. J. M., and Pfeuffer, T. (1986) Reconstitution of adrenoceptor-dependent adenylate cyclase from purified components. *EMBO J .* **5,** 1509–1514.

Feller, D. J., Talvenheimo, J. A., and Catterall, W. A. (1985) The sodium channel from rat brain. Reconstitution of voltage-dependent scorpion toxin binding in vesicles of defined lipid composition. *J. Biol. Chem.* **260,** 11542–11547.

Fleming, J. W. and Ross, E. M. (1980) Reconstitution of β-adrenergic receptors into phospholipid vesicles: restoration of [^{125}I]iodohydroxybenzyl-pindolol binding to digitonin-solubilized receptors. *J. Cyclic Nucleotide Res.* **6,** 407–419.

Florio, V. A. and Sternweis, P. C. (1985) Reconstitution of resolved muscarinic cholinergic receptors with purified GTP-binding proteins. *J. Biol. Chem.* **260,** 3477–3483.

Frye, L. D. and Edidin, M. (1970) The rapid intermixing of cell surface antigens after formation of mouse-human heterokaryons. *J. Cell Sci.* **7,** 319–335.

Fung, B. K-K. (1983) Characterization of transducin from bovine retinal rod outer segments. I. Separation and reconstitution of subunits. *J. Biol. Chem.* **258,** 10495–10502 .

Gal, A., Braun, S., Feder, D., and Levitzki, A. (1983) Reconstitution of a functional β-adrenergic receptor using cholate and a novel method for its functional assay. *Eur. J. Biochem.* **134,** 391–396.

Gaylor, J. L. and Delwiche, C. V. (1969) Removal of nonionic detergents from proteins by chromatography on Sephadex LH-20. *Anal. Biochem.* **28,** 361–368.

Gennis, R. B. and Jonas, A. (1977) Protein-lipid interactions. *Annu. Rev. Biophys. Bioeng.* **6,** 195–238.

Glavind, J. and Hartmann, S. (1955) Studies on methods for the determination of lipoperoxides. *Acta Chem. Scand.* **9,** 497–508.

Haga, K., Haga, T., Ichiyama, A., Katada, T., Kurose, H., and Ui, M. (1985) Functional reconstitution of purified muscarinic receptors and inhibitory guanine nucleotide regulatory protein. *Nature* **316,** 731–733.

Haywood, A. M. (1974a) Characteristics of Sendai virus receptors in a model membrane. *J. Mol. Biol.* **83,** 427–436.

Haywood, A. M. (1974b) Fusion of Sendai viruses with model membranes. *J. Mol. Biol.* **87,** 625–628.

Hekman, M., Feder, D., Keenan, A. K., Gal, A., Klein, H. W., Pfeuffer, T., Levitzki, A., and Helmreich, E. J. M. (1984) Reconstitution of β-adrenergic receptor with components of adenylate cyclase. *EMBO J.* **3,** 3339–3345.

Helenius, A., McCaslin, D. R., Fries, E., and Tanford, C. (1979) Properties of detergent. *Methods. Enzymol.* **56,** 734–749.

Helenius, A. and Simons, K. (1975) Solubilization of membranes by detergents. *Biochim. Biophys. Acta* **415,** 29–79.

Hjelmeland, L. M. and Chrambach, A. (1984) Solubilization of functional membrane proteins. *Methods. Enzymol.* **104,** 305–318.

Holloway, P. W. (1973) A simple procedure for removal of Triton X-100 from protein samples. *Anal. Biochem.* **53,** 304–308.

Hong, K. and Hubbell, W. L. (1973) Lipid requirements for rhodopsin regenerability. *Biochemistry* **12**, 4517–4523.

Hong, K., Knudsen, P. J., and Hubbell, W. L. (1982) Purification of rhodopsin on hydroxyapatite columns, detergent exchange, and recombination with phospholipids. *Meth. Enzymol.* **81**, 144–150.

Howlett, A. C., Sternweis, P. C., Macik, B. A., Van Arsdale, P. M., and Gilman, A. G. (1979) Reconstitution of catecholamine-sensitive adenylate cyclase: Association of a regulatory component of the enzyme with membranes containing the catalytic protein and b-adrenergic receptors. *J. Biol. Chem.* **254**, 2287–2295.

Huang, C-H. (1969) Studies on phosphatidylcholine vesicles. Formation and physical characteristics. *Biochemistry* **8**, 344–352.

Huang, C. and Mason, J. T. (1978) Geometric packing constraints in egg phosphatidylcholine vesicles. *Proc. Natl. Acad. Sci. USA* **75**, 308–310.

Hubbard, R. (1953) The molecular weight of rhodopsin and the nature of the rhodopsin-digitonin complex. *J. Gen. Physiol.* **37**, 381–399.

Im, M-J., Holzhöfer, A., Böttinger, H., Pfeuffer, T., and Helmreich, E.J.M. (1988) Interactions of pure βγ-subunits of G-proteins with purified $β_1$- adrenoceptor. *FEBS Lett.* **227**, 225–229.

Im, M-J., Holzhöfer, A., Keenan, A. K., Gierschik, P., Hekman, M., Helmreich, E. J. M., and Pfeuffer, T. (1987) The role of β,γ subunits of guanine nucleotide binding proteins in control of a reconstituted signal transmission chain containing purified components of the adenylate cyclase system. *J. Receptor Res.* **7**, 17–42.

Insel, P. A. and Koachman, A. M. (1982) Cytochalasin B enhances hormone and cholera toxin-stimulated cyclic AMP accumulation in S49 lymphoma cells. *J. Biol. Chem.* **257**, 9717–9723.

Isaacson, Y. A., Deroo, P. W., Rosenthal, A. F., Bittman, R., McIntyre, J. O., Bock, H-G., Gazzotti, P., and Fleischer, S. (1979) The structural specificity of lecithin for activation of purified D-β- hydroxybutyrate apodehydrogenase. *J. Biol. Chem.* **254**, 117–126.

Johnson, R. M. and Siddiqi, I. W. (1970) *The Determination of Organic Peroxides* (Pergamon Press, Oxford).

Kagawa, Y. (1972) Reconstitution of oxidative phosphorylation. *Biochim. Biophys. Acta* **265**, 297–338.

Kaibuchi, K., Takai, Y., and Nishizuka, Y. (1981) Cooperative roles of various membrane phospholipids in the activation of calcium-activated, phospholipid-dependent protein kinase. *J. Biol. Chem.* **256**, 7146–7149.

Kasahara, M. and Hinkle, P. C. (1977) Reconstitution and purification of the D-glucose transporter from human erythrocytes. *J. Biol. Chem.* **252**, 7384-7390.

Kassis, S. and Fishman, P. H. (1982) Different mechanisms of desensitization of adenylate cyclase by isoproterenol and prostaglandin E1 in human fibroblasts. Role of regulatory components in desensitization. *J. Biol. Chem.* **257**, 5312–5318.

Kassis, S. and Fishman, P. H. (1984) Functional alteration of the β-adrenergic receptor during desensitization of mammalian adenylate cyclase by β-agonists. *Proc. Natl. Acad. Sci. USA* **81**, 6686–6690.

Kassis, S., Henneberry, R. C. and Fishman, P. H. (1984) Induction of catecholamine-responsive adenylate cyclase in HeLa cells by sodium butyrate. *J. Biol. Chem.* **259**, 4910–4916.

Keenan, A. K., Gal, A., and Levitzki, A. (1982) Reconstitution of the turkey erythrocyte adenylate cyclase sensitivity to l-epinephrine upon reinsertion of the lubrol solubilized components into phospholipid vesicles. *Biochem. Biophys. Res. Commun.* **105**, 615–623.

Kelleher, D. J., Rashidbaigi, A., Ruoho, A. E., and Johnson, G. L. (1983) Rapid vesicle reconstitution of alprenolol-Sepharose-purified β1-adrenergic receptors. Interaction of the purified receptor with N. *J. Biol. Chem.* **258**, 12881-12885.

Kirilovsky, J., Steiner-Mordoch, S., Selinger, Z., and Schramm, M. (1985) Lipid requirements for reconstitution of the delipidated β-adrenergic receptor and the regulatory protein. *FEBS Lett.* **183**, 75–80.

Klappe, K., Wilschut, J., Nir, S., and Hoekstra, D. (1986) Parameters affecting fusion between Sendai virus and liposomes. Role of viral proteins, liposome composition, and pH. *Biochemistry* **25**, 8252–8260.

Klausner, R. D., Van Renswoude, J., and Rivnay, B. (1984) Reconstitution of membrane proteins. *Methods. Enzymol.* **104**, 340–347.

Kornberg, A. (1982) *DNA Synthesis* (W. A. Freeman, San Francisco).

Korner, M., Gilon, C., and Schramm, M. (1982) Locking of hormone in the adrenergic receptor by attack on a sulfhydryl in an associated component. *J. Biol. Chem.* **257**, 3389–3396.

Koski, G., Simonds, W. F., and Klee, W. A. (1981) Guanine nucleotides inhibit binding of agonists and antagonists to soluble opiate receptors. *J. Biol. Chem.* **256**, 1536–1538.

Kusiak, J. W. and Pitha, J. (1982) Mapping of mammalian β-adrenoreceptors by use of macromolecular alprenolol derivatives. A comparison with amphibian erythrocyte receptors. *Biochem. Pharmacol.* **31**, 2071–2076.

Kyte, J. and Doolittle, R. F. (1982) A simple method for displaying the hydropathic character of a protein. *J. Mol. Biol.* **157**, 105–132.

Larner, A. C. and Ross, E. M. (1981) Alteration in the protein components of catecholamine-sensitive adenylate cyclase during maturation of rat reticulocytes. *J. Biol. Chem.* **256**, 9551–9557.

Lefkowitz, R. J., Cerione, R. A., Codina, J., Birnbaumer, L., and Caron, M. G. (1985) Reconstitution of the β-adrenergic receptor. *J. Membrane Biol.* **87**, 1–12.

Lever, M. (1977) Peroxides in detergents as interfering factors in biochemical analysis. *Anal. Biochem.* **83**, 274–284.

Levitzki, A. (1985) Reconstitution of membrane receptor systems. *Biochim. Biophys. Acta* **822**, 127–153.

Limbird, L. E. and Lefkowitz, R. J. (1978) Agonist-induced increase in apparent-adrenergic receptor size. *Proc. Natl. Acad. Sci. USA* **75,** 228–232.

Limbird, L. E., Gill, D. M., and Lefkowitz, R. J. (1980) Agonist-promoted coupling of the β-adrenergic receptor with the guanine nucleotide regulatory protein of the adenylate cyclase system. *Proc. Natl. Acad. Sci. USA* **77,** 775–779.

May, D. C., Ross, E. M., Gilman, A. G., and Smigel, M. D. (1985) Reconstitution of catecholamine-stimulated adenylate cyclase using three purified proteins. *J. Biol. Chem.* **260,** 15829–15833.

May, D. C. and Ross, E. M. (1988) Rapid binding of guanosine-5'-0-(3-thiotriphosphate) (GTPγS) to an apparent complex of β-adrenergic receptor and the GTP-binding regulatory protein, G_s. *Biochemistry* **27,** 4888–4893.

Miller, C. (1984) Ion channels in liposomes. *Annu. Rev. Physiol.* **46,** 549–558.

Miller, C., Arvan, P., Telford, J. N., and Racker, E. (1976) Ca^{++}-induced fusion of proteoliposomes: Dependence on transmembrane osmotic gradient. *J. Membrane Biol.* **30,** 271–282.

Morris, S. J., Gibson, C. C., Smith, P. D., Greif, P. C., Stirk, C. W., Bradley, D., Haynes, D. H., and Blumenthal, R. (1985) Rapid kinetics of Ca^{2+}-induced fusion of phosphatidylserine/phosphatidylethanolamine vesicles. The effect of bilayer curvature on leakage. *J. Biol. Chem.* **260,** 4122–4127.

Moxham, C., Ross, E. M., George, S. T., and Malbon, C. C. (1988) β-Adrenergic receptors display intramolecular disulfide bridge *in situ:* Analysis by immunoblotting and functional reconstitution. *Mol. Pharmacol.* **33,** 486–492.

Mukerjee, P. and Mysels, K. J. (1971) *Critical Micelle Concentrations of Aqueous Surfactant Systems* (National Bureau of Standards, Washington, DC). Neufeld, G., Schramm, M., and Weinberg, N. (1980) Hybridization of adenylate cyclase components by membrane fusion and the effect of selective digestion by trypsin. *J. Biol. Chem.* **255,** 9268–9274.

Neufeld, G., Steiner, S., Korner, M., and Schramm, M. (1983) Trapping of the β-adrenergic receptor in the hormone-induced state. *Proc. Natl. Acad. Sci. USA* **80,** 6441–6445.

Nir, S., Klappe, K., and Hoekstra, D. (1986) Mass action analysis of kinetics and extent of fusion between Sendai virus and phospholipid vesicles. *Biochemistry* **25,** 8261–8266.

Nozaki, Y., Lasic, D. D., Tanford, C., and Reynolds, J. A. (1982) Size analysis of phospholipid vesicle preparations. *Science* **217,** 366, 367.

Orly, J. and Schramm, M. (1976) Coupling of catecholamine receptor from one cell with adenylate cyclase from another cell by cell fusion. *Proc. Natl. Acad. Sci. USA* **73,** 4410–4414.

Parente, R. A. and Lentz, B. R. (1986) Rate and extent of poly(ethylene glycol)-induced large vesicle fusion monitored by bilayer and internal contents mixing. *Biochemistry* **25,** 6678–6688.

Pedersen, S. E. and Ross, E. M. (1982) Functional reconstitution of beta-adrenergic receptors and the stimulatory GTP-binding protein of adenylate cyclase. *Proc. Natl. Acad. Sci. USA* **79,** 7228–7232.

Pedersen, S. E. and Ross, E. M. (1985) Functional activation of beta-adrenergic receptors by thiols in the presence or absence of agonists. *J. Biol. Chem.* **260,** 14150–14157.

Peretz, H., Toister, Z., Laster, Y., and Loyter, A. (1974) Fusion of intact human erythrocytes and erythrocyte ghosts. *J. Cell Biol.* **63,** 1–11.

Pfeuffer, E., Mollner, S., and Pfeuffer, T. (1985b) Adenylate cyclase from bovine brain cortex: Purification and characterization of the catalytic unit. *EMBO J.* **4,** 3675–3679.

Pfeuffer, E., Dreher, R-M., Metzger, H., and Pfeuffer, T. (1985a) Catalytic unit of adenylate cyclase: Purification and identification by affinity crosslinking. *Proc. Natl. Acad. Sci. USA* **82,** 3086–3090.

Pike, L. J., Limbird, L. E., and Lefkowitz, R. J. (1979) β-Adrenoreceptors determine affinity but not intrinsic activity of adenylate cyclase stimulants. *Nature* **280,** 502–504.

Poste, G. and Nicolson, G. L., (eds.) (1978) *Membrane Fusion* (Elsevier, Amsterdam.)

Poste, G. and Pasternak, C. A. (1978) Virus-induced cell fusion in *Membrane Fusion* (Poste, G. and Nicolson, G. L., eds.), Elsevier, Amsterdam, pp. 305-367.

Racker, E. (1973) A new procedure for the reconstitution of biologically active phospholipid vesicles. *Biochem. Biophys. Res. Commun.* **55,** 224–230.

Racker, E. (1976) *A New Look at Mechanisms in Bioenergetics* (Academic, New York).

Racker, E. (1986) *Reconstitutions of Transporters, Receptors and Pathological States* (Academic, New York).

Racker, E., Chien, T-F., and Kandrach, A. (1975) A cholate-dilution procedure for the reconstitution of the Ca^{++} pump, $^{32}P_i$-ATP exchange, and oxidative phosphorylation. *FEBS Lett.* **57,** 14–18.

Racker, E. and Hinkle, P. C. (1974) Effect of temperature on the function of a proton pump. *J. Membrane Biol.* **17,** 181–188.

Rasenick, M. M., Stein, P. J., and Bitensky, M. W. (1981) The regulatory subunit of adenylate cyclase interacts with cytoskeletal components. *Nature* **294,** 560–562.

Reynolds, J. A., Nozaki, Y., and Tanford, C. (1983) Gel-exclusion chromatography on S1000 Sephacryl: Application to phospholipid vesicles. *Anal. Biochem.* **130,** 471–474.

Rhoden, V. and Goldin, S. M. (1979) Formation of unilamellar lipid vesicles of controllable dimensions by detergent dialysis. *Biochemistry* **18,** 4173–4176.

Rivnay, B. and Metzger, H. (1982) Reconstitution of the receptor for immunoglobulin E into liposomes. Conditions for incorporation of the receptor into vesicles. *J. Biol. Chem.* **257,** 12800–12808.

Ross, E. M. (1981) Physical separation of the catalytic and regulatory proteins of hepatic adenylate cyclase. *J. Biol. Chem.* **256,** 1949–1953.

Ross, E. M. (1982) Phosphatidylcholine-promoted interaction of the catalytic and regulatory proteins of adenylate cyclase. *J. Biol. Chem.* **257**, 10751–10758.

Ross, E. M. and Gilman, A. G. (1977) Resolution of some components of adenylate cyclase necessary for catalytic activity. *J. Biol. Chem.* **252**, 6966–6969.

Ross, E. M. and Gilman, A. G. (1980) Biochemical properties of hormone-sensitive adenylate cyclase. *Annu. Rev. Biochem.* **49**, 533–564.

Ross, E. M., Howlett, A. C., Ferguson, K. M., and Gilman, A. G. (1978) Reconstitution of hormone-sensitive adenylate cyclase activity with resolved components of the enzyme. *J. Biol. Chem.* **253**, 6401–6412.

Ross E. M. and Schatz G. (1978) Purification and subunit structure of yeast cytochrome c_1. *Methods. Enzymol.* **53**, 222–229.

Rubenstein, R. C., Wong, S. K-F., and Ross, E. M. (1987) The hydrophobic tryptic core of the β-adrenergic receptor retains G_s-regulatory activity in response to agonists and thiols. *J. Biol. Chem.* **262**, 16655–16662.

Rudolph, S. A., Greengard, P., and Malawista, S. E. (1977) Effects of colchicine on cyclic AMP levels in human leukocytes. *Proc. Natl. Acad. Sci. USA* **74**, 3404–3408.

Rydstrom, J., Hoek, J. B., and Hundall, T. (1976) Selective solubilization of the components of the mitochondrial inner membrane by lysolecithin. *Biochim. Biophys. Acta* **455**, 24–35.

Sandermann, H., Jr. (1978) Regulation of membrane enzymes by lipids. *Biochim. Biophys. Acta* **515**, 209–237.

Schramm, M. (1979) Transfer of glucagon receptor from liver membranes to a foreign adenylate cyclase by a membrane fusion procedure. *Proc. Natl. Acad. Sci. USA* **76**,1174–1178.

Schramm,M., Orly, J., Eimerl, S., and Korner, M. (1977) Coupling of hormone receptors to adenylate cyclase of different cells by cell fusion. *Nature* **268**, 310–313.

Schrock, H. L. and Gennis, R. B. (1977) High affinity lipid binding sites on the peripheral membrane enzyme pyruvate oxidase. *J. Biol. Chem.* **252**,5990–5995.

Schulster, D., Orly, J., Seidel, G., and Schramm, M. (1978) Intracellular cyclic AMP production enhanced by a hormone receptor transferred from a different cell. *J. Biol. Chem.* **253**, 1201–1206.

Schwarzmeier, J. and Gilman, A. G. (1977) Reconstitution of catecholamine-sensitive adenylate cyclase activity: Interaction of components following cell-cell and membrane-cell fusion. *J. Cyclic Nucleotide Res.* **3**, 227–238.

Selser, J. C., Yeh, Y., and Baskin, R. J. (1976) A light-scattering characterization of membrane vesicles. *Biophys. J.* **16**, 337–356.

Senogles, S. E., Benovic, J. L., Amlaiky, N., Unson, C., Milligan, G., Vinitsky, R., Spiegel, A. M., and Caron, M. G. (1987) The D_2-dopamine receptor of anterior pituitary is functionally associated with a pertussis toxin-sensitive guanine nucleotide binding protein. *J. Biol. Chem.* **262**, 4860–4867.

Smigel, M. D. (1986) Purification of the catalyst of adenylate cyclase. *J. Biol. Chem.* **261**, 1976–1982.

Smith, S. K. and Limbird, L. E. (1981) Solubilization of human platelet α- adrenergic receptors: Evidence that agonist occupancy of the receptor stabilizes receptor-effector interactions. *Proc. Natl. Acad. Sci. USA* **78**, 4026–4030.

Steele, J. C. H., Jr., Tanford, C., and Reynolds, J. A. (1978) Determination of partial specific volumes for lipid-associated proteins. *Methods Enzymol.* **48**, 11–23.

Sternweis, P. C. (1986) The purified α subunits of G_o and G_i from bovine brain require βγ for association with phospholipid vesicles. *J. Biol. Chem.* **261**, 631–637.

Sternweis, P. C. and Gilman, A. G. (1979) Reconstitution of catecholamine-sensitive adenylate cyclase: Reconstitution of the uncoupled (UNC) variant of the S49 lymphoma cell. *J. Biol. Chem.* **254**, 3333–3340.

Sternweis, P. C., Northup, J. K., Smigel, M. D., and Gilman, A. G. (1981) The regulatory component of adenylate cyclase: Purification and properties. *J. Biol. Chem.* **256**, 11517–11526.

Strittmatter, S. and Neer, E. J. (1980) Properties of the separated catalytic and regulatory units of brain adenylate cyclase. *Proc. Natl. Acad. Sci. USA* **77**, 6344-6348.

Struck, D. K., Hoekstra, D., and Pagano, R. E. (1981) Use of resonance energy transfer to monitor membrane fusion. *Biochemistry* **20**, 4093–4099.

Strulovici, B., Stadel, J. M., and Lefkowitz, R. J. (1983) Functional integrity of desensitized β-adrenergic receptors. Internalized receptors reconstitute catecholamine-stimulated adenylate cyclase activity. *J. Biol. Chem.* **258**, 6410-6414.

Sweet, L. J., Wilden, P. A., Spector, A. A., and Pessin, J. E. (1985) Incorporation of the purified human placental insulin receptor into phospholipid vesicles. *Biochemistry* **24**, 6571–6580.

Szoka, F., Jr. and Papahadjopoulos, D. (1980) Comparative properties and methods of preparation of lipid vesicles (liposomes) in *Annual Review of Biophysics and Bioengineering* (Mullins, L. J. , Hagins, W. A., Newton, C., and Weber, G., eds.), Annual Reviews, Palo Alto, California, **9**, 467–508.

Takagaki, Y., Radhakrishnan, R., Gupta, C. M., and Khorana, H. G. (1983b) The membrane-embedded segment of cytochrome b_5 as studied by crosslinking with photoactivatable phospholipids. I. The transferable form. *J. Biol. Chem.* **258**, 9128–9135.

Takagaki, Y., Radhakrishnan, R., Wirtz, K. W. A., and Khorana, H. G. (1983) The membrane-embedded segment of cytochrome b_5 as studied by crosslinking with photoactivatable phospholipids. II. The nontransferable form. *J. Biol. Chem.* **258**, 9136–9142.

Tamkun, M. M., Talvenheimo, J. A., and Catterall, W. A. (1984) The sodium channel from rat brain: Reconstitution of neurotoxin-activated ion flux and scorpion toxin binding from purified components. *J. Biol. Chem.* **259**, 1676–1688.

Tanford, C. (1980) *The Hydrophobic Effect: Formation of Micelles and Biological Membranes* (Wiley, New York).

Tanford, C. and Reynolds, J. A. (1976) Characterization of membrane proteins in detergent solutions. *Biochim. Biophys. Acta* **457**, 133–170.

Tank, D. W., Miller, C., and Webb, W. W. (1982) Isolated-patch recording from liposomes containing functionally reconstituted chloride channels from *Torpedo* electroplax. *Proc. Natl. Acad. Sci. USA* **79,** 7749–7753.

Tausk, R. J. M., Karmiggelt, J., Oudshoorn, C., and Overbeek, J. T. G. (1974) Physical chemical studies of short-chain lecithin homologues. I. Influence of the chain length of the fatty acid ester and of electrolytes on the critical micelle concentration. *Biophys. Chem.* **1,** 175–183.

Toister, Z. and Loyter, A. (1973) The Mechanism of cell fusion. II. Formation of chicken erythrocyte polykaryons. *J. Biol. Chem.* **248,** 422–432.

Tsai, S-C., Adamik, R., Kanaho, Y., Hewlett, E. L., and Moss, J. (1984) Effects of guanyl nucleotides and rhodopsin on ADP-ribosylation of the inhibitory GTP-binding component of adenylate cyclase by pertussis toxin. *J. Biol. Chem.* **259,** 15320–15323.

Van Renswoude, A. J. B. M., Blumenthal, R., and Weinstein, J. N. (1980) Thin-layer chromatography with agarose gels. A quick, simple method for evaluating liposome size. *Biochim. Biophys. Acta* **595,** 151–156.

Vauquelin, G., Bottari, S., and Strosberg, A. D. (1980) Inactivation of β- adrenergic receptors by N-ethylmaleimide: Permissive role of β-adrenergic agents in relation to adenylate cyclase activation. *Mol. Pharmacol.* **17,** 163–171.

Walter, P. and Lingappa, V. R. (1986) Mechanism of protein translocation across the endoplasmic reticulum membrane. *Annu. Rev. Cell Biol.* **2,** 499–516.

Wang, C-T., Saito, A., and Fleischer, S. (1979) Correlation of ultrastructure of reconstituted sarcoplasmic reticulum membrane vesicles with variation in phospholipid to protein ratio. *J. Biol. Chem.* **254,** 9209–9219.

Welton, A. F., Lad, P. M., Newby, A. C., Yamamura, H., Nicosia, S., and Rodbell, M. (1977) Solubilization and separation of the glucagon receptor and adenylate cyclase in guanine nucleotide-sensitive states. *J. Biol. Chem.* **252,** 5947–5950.

Wickner, W. T. and Lodish, H. F. (1985) Multiple mechanisms of protein insertion into and across membranes. *Science* **230,** 400–407.

Wilschut, J. and Hoekstra, D. (1986) Membrane fusion: Lipid vesicles as a model system. *Chem. Phys. Lipids* **40,** 145–166.

Witkin, K. M. and Harden, T. K. (1981) A sensitive equilibrium binding assay for soluble β-adrenergic receptors. *J. Cyclic Nucleotide Res.* **7,** 235–246.

Wojcieszyn, J. W., Schlegel, R. A., Lumley-Sapanski, K., and Jacobson, K. A. (1983) Studies on the mechanism of polyethylene glycol-mediated cell fusion using fluorescent membrane and cytoplasmic probes. *J. Cell Biol.* **96,** 151–159.

Wong, M. and Thompson, T. E. (1982) Aggregation of dipalmitoylphosphatidyl-choline vesicles. *Biochemistry* **21,** 4133–4139.

Zimmermann, U. and Vienken, J. (1982) Electric field-induced cell-to-cell fusion. *J. Memb. Biol.* **67,** 165–182.

Zumbuehl, O. and Weder, H. G. (1981) Liposomes of controllable size in the range of 40 to 180 nm by defined dialysis of lipid/detergent mixed micelles. *Biochim. Biophys. Acta* **640,** 252–262.

CHAPTER 5

Antibodies
to Beta-Adrenergic Receptors

Craig C. Malbon, Cary P. Moxham, and Harvey J. Brandwein

1. Introduction

Contemporary biochemists and molecular biologists strive to understand the relationship between the function and the detailed chemical structure of macromolecules. Whereas chemical and direct physical analyses are employed to probe molecular structure, specific antibodies to proteins have been invaluable reagents in the determination of the fine-structure of the antigen as well as the immunologic relationship of the antigen to other proteins. Often a crowning achievement to many years of arduous work purifying and characterizing a cellular protein is the production of specific antibodies to the protein. The availability of specific antibodies then propels the direction of research into investigations of entirely new areas of protein structure, function, and regulation that could not be approached by any other route.

The β-Adrenergic Receptors Ed.: J. P. Perkins © 1991 The Humana Press Inc.

Historically, the use of antibodies to explore important questions of protein structure and regulation was first applied to the study of enzymes involved in intermediary metabolism. Methods of purification developed for many of these enzymes, which are cytosolic, globular proteins, provided a solid foundation for the successful production of specific antibodies. In the last two decades antibodies have been prepared to many structural proteins, enzymes, and cell-surface antigens. Recent advances in several converging technologies now make possible the production of highly specific antibodies to membrane-bound receptors for hormones and drugs. First, the widely applied methodologies of affinity chromatography have facilitated our ability to purify membrane-bound receptors (Cuatrecasas, 1972). These receptor proteins are often of very low abundance but can be rapidly purified by taking advantage of a specific interaction between the receptor and a high-affinity ligand immobilized on an insoluble matrix. Utilized in tandem with ion-exchange, hydrophobic, and high-performance liquid chromatography systems, affinity chromatography has been employed as a means to purify membrane proteins that often constitute less than $10^{-4}\%$ of cellular protein (*see* Chapter 2 this vol.; Bahouth et al., 1988). Second, advances in microsequencing of proteins (Hunkapiller and Hood, 1983) as well as solid-phase peptide synthesis (Marglin and Merrifield, 1970) now make it possible to provide primary sequence information from picomole quantities of purified proteins as well as the automated synthesis of peptides (ranging up to 30 amino acid residues in length) for use as defined antigens. Molecular cloning and the application of the polymerase chain reaction (PCR) to isolate and amplify regions of DNA displaying sequences of homology are but two prime examples of how molecular biology has provided powerful new tools for the identification of low-abundance membrane proteins based upon scant protein or DNA sequence information. Finally, hybridoma-monoclonal antibody techniques (Kohler and Milstein, 1975) permit the production of many different, yet individually-homogeneous antibodies to single antigenic sites (epitopes) of proteins and synthetic peptides. In concert, these techniques have revolutionized the immunochemical approaches employed to generate antibodies capable of probing the fine-structure and function of membrane proteins.

Although the widespread use of immunochemical approaches to study membrane proteins is still in its infancy, there already exist outstanding examples in which the careful and thorough application of these methodologies has yielded major advances in our knowledge of the structure and biology of membrane receptors. The acetylcholine receptor (Gullick et al., 1981; Tzartos and Lindstrom, 1980), the insulin receptor (Herrera et al., 1985; Jacobs et al., 1978; Kull et al., 1982; Morgan et al., 1986; Roth et al., 1982), the low density lipoprotein receptor (Beisiegel et al., 1981a,b), and the epidermal growth factor receptor (Schreiber et al.,1981,1983) are but a few examples of a growing list of cell surface receptors whose detailed structural features have been illuminated using such advanced techniques. The receptors for beta-adrenergic catecholamines have not been as thoroughly investigated as the aforementioned receptors. It is the goal of this chapter to review past efforts focused on the production of antibodies specific for beta-adrenergic receptors and to highlight the current and future directions of efforts to apply immunochemical strategies to the analysis of these important receptor molecules.

2. Approaches
to Production of Antibodies to Receptors

Currently there are two basic, fundamentally different approaches to producing antibodies to receptor molecules. The first of these is what might be considered the "classical" approach of producing polyclonal antibodies in rabbits, goats, rats, or mice by immunization of these animals with essentially pure receptor protein isolated from tissues and cells on a large scale (Fig. 1). In recent years, this strategy has been employed successfully to generate specific antibodies to several membrane proteins found in relatively high abundance in cells. The success of the approach critically depends on the availability of relatively large quantities (0.1–1.0 mg protein) of highly pure membrane protein to be used as the "antigen." Immunization of animals with impure receptor preparations should be avoided when using this approach, since antibodies will be generated against "contaminants" as well as the receptor proteins used as immunogens. This potential

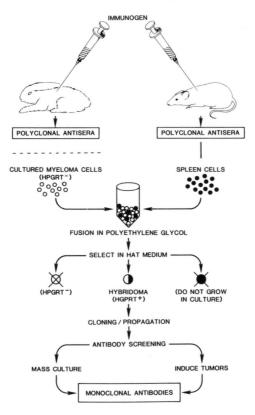

Fig. 1. Classical and hybridoma methods for the production of antibodies. Polyclonal antibodies are produced by the "classical" approach of immunizing rabbits or mice with antigen and analyzing the serum for antibodies to the antigen. For the production of monoclonal antibodies, spleen cells producing antibodies are isolated from the immunized mouse and fused in the presence of polyethylene glycol to myeloma cells that are deficient in the enyzme HGPRT. When placed in medium containing hypoxanthine-aminopterine-thymidine, the myeloma cells are unable to utilize the hypoxanthine and thymidine for the *de novo* synthesis of DNA that is required in the presence of the drug aminopterine. Spleen cells from the mice fail to thrive under culture conditions. The hybridomas, by virtue of the contributions from both cell types, grow in HAT medium. Upon cloning and propagation, the hybridoma supernatants are screened for the presence of antibody to the antigen. The positive clones secreting a monospecific antibody to a single epitope of the antigen are cultured in mass or used to induce tumors and accumulation of ascites fluid. For details, *see* the text.

problem becomes even more formidable if the receptor protein proves to be a relatively poor antigen in comparison to other proteins in the mixture. Failure to appreciate the possible ramifications of this problem at an early stage usually compromises the value of any antibodies produced to impure preparations of receptor.

The isolation of milligram quantities of essentially pure receptor from tissues in which the receptor is found at relatively high abundance is an arduous and demanding task. Many receptors, however, are present on sensitive cells in very low abundance, less than 5000 receptors/cell. The task of isolating membrane receptors of very low abundance to apparent homogeneity on a large scale is an even more formidable task that often yields only microgram quantities of protein when recoveries are high (20–50%). Mammalian beta-adrenergic receptors exemplify low-abundance membrane receptors that are difficult to purify (Benovic et al., 1984; Cubero and Malbon, 1984; George and Malbon, 1985; Graziano et al., 1985; Homcy et al., 1983). Consequently, progress in the elucidation of the fine-structure and biology of beta-adrenergic receptors using immunologic tools has not been as rapid as that achieved in the study of other membrane-bound receptors. Many "tricks" specifically designed to avoid the need to isolate nanomole quantities of pure receptor have been attempted in the last five years. Immunizations have been performed with partially purified preparations of receptor and with receptor-bearing whole cells. Although antisera have been developed using these approaches, none of these attempts has been truly successful in terms of generating specific antisera or antibodies that have been useful as reagents for further immunochemical analysis of the structure and biology of beta-adrenergic receptors. Moreover, the ability to characterize as well as produce antibodies to beta-receptors by the "classical" approach requires the availability of microgram quantities of pure receptor protein or synthetic peptide. For this reason, "tricks" employed to avoid the isolation of essentially pure receptor for immunization are often doomed to failure in the screening process because of a lack of purified receptor. When the receptor of interest is of the same relative abundance as the beta-adrenergic receptor, is at a much lower abundance than the beta-adrenergic receptor, or cannot be isolated to apparent homogeneity in at least microgram quantities, the use of the "classical" approach is often precluded.

An alternative to the classical approach to the production of antireceptor antibodies is the antiidiotype approach. This strategy obviates, at least in theory, the need to purify receptor for use as the immunogen (Fig. 2). Consequently, the antiidiotype approach has sparked the interest of biochemists and molecular biologists who seek to investigate the nature of low-abundance membrane receptors. The antiidiotype strategy is based on Jerne's theory of immune networks (Jerne, 1974), which predicts the appearance of antibodies (antiidiotypes) directed against other antigen-induced antibodies (idiotypes) as a natural mechanism for controlling the temporal regulation of the immune response to an insult by an antigen. In this way, it is envisioned that some subpopulation of antiidiotype antibodies will be directed against the specific antigen-combining site of the primary, idiotype antibodies. Only those antiidiotype antibodies that are generated against the antigen-combining site of idiotypes raised against a ligand would be expected to bind to the cellular receptor for that ligand (Fig. 2). Antiidiotypes against domains other than the ligand-binding domain would not be expected to recognize and to bind to the cellular receptor for the ligand. Thus it is possible, at least in theory, to obtain antibodies to a membrane receptor without first purifying the receptor for use as an immunogen. Antibodies are prepared to a ligand that binds with high specificity to the receptor of interest (Fig. 2, step 1). The antiligand antibody (idiotype) is then isolated and used as an immunogen to prepare antibodies (antiidiotypes) to the idiotype (Fig. 2, step 2). Analysis of the ability of the antiidiotypes to bind to the cellular receptor will differentiate those that have structural homology with the ligand and recognize the membrane receptor from those that have been generated against other domains of the idiotypes and fail to recognize the receptor (Fig. 2, analysis step). Theoretically, antiidiotypes to receptors should arise in the immune response of a single animal immunized with a ligand to the receptor, eliminating the need for two independent animals. For the sake of clarity, the scheme shown in Fig. 2 uses two separate rats. Antiidiotype antibodies have been generated in animals immunized with a ligand (or ligand conjugate) as

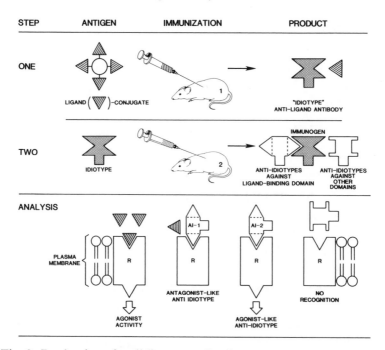

Fig. 2. Production of antiidiotype antibodies to receptors. Ligand coupled to a conjugate is used to immunize a mouse (mouse 1). An idiotype antibody directed against the ligand is generated in mouse 1, and this idiotype antibody is used to immunize a second animal (mouse 2). Antiidiotype antibodies against the idiotype antibodies will be generated in mouse 2. Included in these antiidotype populations of antibodies will be some antibodies that were made to the antigen (ligand)-combining site that will also recognize the receptor for the ligand. Antiidiotypes against other domains of the idiotype will be generated also, but these fail to recognize the receptor for the ligand. The analysis of the antiidiotype antibodies produced by this protocol reveals three possibile outcomes: Antiidiotype antibody AI-1 recognizes the receptor, and its binding to the receptor produces no biological response. In the presence of this antibody, the ligand would not have access to the receptor. This antibody, AI-1, is thus "antagonist-like." Antiidiotype antibody AI-2 also recognizes and binds to the receptor for the ligand. In this case, AI-2 is capable of stimulating a biological response and is thus "agonist-like."

Antiidiotypes to other domains of the idiotype do not recognize the receptor.

For details of the application of this protocol to production of antibodies to beta-adrenergic receptors, consult the text.

well as in animals immunized with antiligand (idiotype) antibodies raised in separate animals (as depicted in Fig. 2). Description of the antiidiotype and classical approaches used to prepare antibodies to membrane-bound receptors in general and beta-adrenergic receptors specifically is presented below.

2.1. Antiidiotype Antibody Production

2.1.1. Antiidiotype Antibodies to Membrane-Bound Receptors

Aside from the important immunoregulatory significance of antiidiotypes in the immune networks as proposed by Jerne (1974), the ability to produce antiidiotype antibodies that, like the ligand used to prepare the idiotype antibody, also bind specifically to a cellular receptor of very low abundance provides a "short cut" to the preparation of antireceptor antibodies. Sege and Peterson (1978) provided experimental evidence in support of Jerne's theory and the general approach of antiidiotype antibody production. These researchers succeeded in generating a polyclonal antiidiotype antibody against antiinsulin antibodies that blocked the ability of radioiodinated insulin to bind to receptors on rat epididymal fat cells. The antiidiotype antibodies were capable of immunoprecipitating radiolabeled Fab fragments of the antiinsulin, idiotype antibodies (Sege and Peterson, 1978). Shechter et al. (1982) showed the occurrence of biologically active antiidiotype antibodies to the insulin receptor in mice that had been immunized with bovine or porcine insulin alone. The initial generation of anti-insulin antibodies was presumably followed by the "network" generation of antiidiotypes that bind to the insulin receptor (Sege and Peterson, 1978).

Another excellent example of this approach is the work of Farid and colleagues, who first generated antibodies against the pituitary hormone TSH in rats and then used the anti-TSH antibodies to immunize New Zealand White rabbits (Farid et al., 1983; Farid and Lo, 1985; Islam et al., 1983a,b). The antisera from rabbits immunized with anti-TSH antibodies contained antiidiotype antibodies to the thyroid TSH receptor. The antiidiotype antisera inhibited the binding of radioiodinated TSH to plasma membranes prepared from thyroid tissue, stimulated GTP-dependent adenylate cyclase activity in thyroid plasma membranes like TSH, and stimulated the uptake of iodide by isolated thyroid cells like TSH (Islam et al., 1983a). The antiidiotype antibodies were also shown to bind the same 197kDa protein of thyroid membranes as does TSH (Islam et al., 1983b). These studies provide compelling evidence to support the notion that antiidiotype antibodies have the same "internal image" as the TSH ligand itself (Strosberg, 1984). Graves' disease, typified by the appearance of autoantibodies that interact with TSH receptors in the thyroid, may be a pathological correlate to the antiidiotype antibodies to the TSH receptor produced in the studies of Farid and coworkers (Farid et al., 1983).

Wasserman et al. (1982) provided an additional example of experimentally induced production of antiidiotype antibodies that provoked many features of another pathological autoimmune condition, myasthenia gravis. These investigators immunized rabbits with a potent analog of acetylcholine, trans-3,3'-Bis[alpha-(trimethylamino) methyl]azobenzenebiomide (Bis-Q), and found circulating antiidiotypes that recognized the acetylcholine receptor of several species, including rat, *Torpedinidae californum*, and *Electrophorus electricus* (Wasserman et al., 1982). Two of the animals immunized with Bis-Q displayed the transient muscle weakness that is observed in experimentally induced myasthenia gravis in animals immunized with purified acetylcholine receptor (Cleveland et al., 1983; Patrick and Lindstrom, 1973).

Antiidiotype antibodies to other cell surface receptors have also been reported (for review, *see* Gaulton et al., 1985). Greene and coworkers, by immunization of mice with the hemagglutinin protein of reovirus serotype 3, have succeeded in obtaining antibodies that recognize the mammalian reovirus binding protein (Nepom et al., 1982; Noseworthy et al., 1983). The results of these interesting studies will be described later in this chapter. Antiidiotype antibodies have also been employed to study the presence of idiotypic determinants on T cells (Dietz et al., 1981; Owen et al., 1977). Marasco and Becker (1982) provided several independent lines of evidence to support their claim that antiidiotype antibodies specific for the receptor involved in chemotaxis could be prepared by immunization of animals with antibodies to f-Met-Leu-Phe peptide. Gramsch et al. (1988), too, have

190   Malbon, Moxham, and Brandwein

successfully approached the problem of generating antibodies specific to opioid receptor via the antiidiotype route. Thus, the antiidiotypic approach can be profitable for producing antireceptor antibodies without first isolating the receptor.

2.1.2. Antiidiotype Antibodies to Beta-Adrenergic Receptors

Three research groups have reported the successful production of antiidiotype antibodies that bind to beta-adrenergic receptors (Homcy et al., 1983; Schreiber et al., 1980; Itami et al., 1987). Schreiber et al. (1980) first prepared antibodies against a conjugate of the beta-adrenergic antagonist (–)alprenolol to bovine serum albumin (BSA) (Hoebeke et al., 1977), and then used the IgG fraction of the primary antiserum as the immunogen for a second set of rabbits. Following immunization with antialprenolol/BSA antibodies once weekly for four consecutive weeks, the appearance of antibodies with the capability of inhibiting the binding of a tritiated, high-affinity beta-adrenergic antagonist, dihydroalprenolol, to membranes of turkey erythrocytes was detected. Immune IgG at 5 mg/mL inhibited binding of the radioligand to the membranes by nearly 70%. Furthermore, these researchers reported that the antiidiotype antibodies produced a small stimulation of adenylate cyclase as well as an enhancement of epinephrine-stimulated adenylate cyclase activity. Turkey erythrocytes replete in beta-adrenergic receptors, but not erythrocytes of human or sheep origin that display few beta-adrenergic receptors, were agglutinated by the antiidiotype antibodies. Preincubation with antialprenolol/BSA antibodies blocked binding of the antiidiotypes to the turkey erythrocyte (Schreiber et al., 1980). Homcy et al. (1982) later reported production of antiidiotypic antibodies using as an immunogen antialprenolol antibodies (Rockson et al., 1980) that were isolated from bulk rabbit antisera by affinity chromatography on an acebutolol-immobilized affinity resin. Six months after the primary immunization, with a booster immunization at every 3–4 wk, one rabbit antiserum displayed antiidiotypic activity capable of inhibiting the binding of tritiated dihydroalprenolol to membranes prepared from turkey erythrocytes, dog lung, and rat reticulocytes (Homcy et al., 1982). Unlike the antiidiotype antibodies prepared by Schreiber et al. (1980), those prepared by Homcy et al. (1982) inhibited rather than enhanced the ability of isoproterenol to activate adenylate cyclase activity. Nonspecific inhibition of "specific" radioligand binding to membranes, as well as inhibition of adenylate cyclase activity of membranes by either crude antisera or purified IgG fractions generated against unrelated antigens is commonly observed when these immunological reagents are used at very high concentrations. The lack of purified beta-adrenergic receptors with which to establish the nature of these antiidiotype antisera hampered these early efforts.

Guillet et al. (1984) reported the use of hybridoma cells that bear monoclonal antibodies against (–) alprenolol as an immunogen to raise monoclonal antiidiotype antibodies to beta-adrenergic receptors. Of the 23 hybridoma supernatants that were found to recognize the idiotype, six inhibited hapten binding, and three of these were found to bind to membranes of murine mastocytoma and turkey erythrocytes adsorbed onto poly-L-lysine-treated polystyrene plates as well as to human A431 epidermoid cells grown in microtiter culture wells and then fixed with glutaraldehyde. Antiidiotype monoclonal antibodies reacted with membranes from these cells that are replete with beta-adrenergic receptors, but not with membranes from a rabbit B cell line, reported by these workers to be devoid of beta-adrenergic receptors (Guillet et al., 1984).

One of the monoclonal antiidiotype antibodies, mAb2B4, identified in these earlier studies (Guillet et al., 1984) was later shown to recognize beta-adrenergic receptors using several additional criteria (Guillet et al., 1985). Tritiated dihydroalprenolol binding of digitonin extracts of A431 cells could be immunoprecipitated by mAb2B4. Cyclic AMP accumulation by A431 cells was stimulated by mAb2B4 purified from ascites fluid (50 µg/mL) and this response, unlike that provoked by the rabbit antiidiotype antibody (Schreiber et al., 1980), could be blocked by the beta-adrenergic antagonist propranolol (Guillet et al., 1985). Immunoblots of membranes from A431 cells subjected to sodium dodecyl sulfate-polyacrylamide gel electrophoresis (SDS-PAGE) under reducing conditions, transferred to nitrocellulose and probed with mAb2B4, revealed a band of strong immunoreactivity with 55kDa (Guillet et al., 1985). Under conditions that chemically reduce disulfide bonds and in the presence of sodium dodecyl sulfate, essentially pure mammalian beta-1 as well as beta-2 adrenergic receptors from several sources have been shown to display an electrophoretic mobility of 65–67kDa (Benovic et al., 1984; Cubero and Malbon, 1984; George and Malbon, 1985; Graziano et al., 1985). More recently, the beta-2 adrenergic receptors of A431 cells have been shown to display two different types of oligosaccharides, providing an explanation for the 55kDa species (Kaveri et al., 1987; Cervantes-Olivier et al., 1988). In addition, we have demonstrated the existence of disulfide bridges in purified beta-1 and beta-2 adrenergic receptors that are essential for the ligand-binding property of these receptors (George and Malbon, 1985; Moxham and Malbon, 1985; Moxham et al., 1988). When subjected to electrophoresis in the absence of chemical reduction, these receptors containing intramolecular disulfide bridges display greater electrophoretic mobility with 55kDa (George and Malbon, 1985; Moxham and Malbon, 1985), even *in situ* (Moxham et al., 1988). Thus alterations in glycosylation or incomplete reduction of the A431 beta-adrenergic receptors may provide an explanation for the prominent 55kDa band on the immunoblots shown by Guillet et al. (1985).

Efforts to produce antiidiotype antibodies to beta-adrenergic receptors are laudable, since they represent alternatives to the classical approach of producing antibodies to these low-abundance membrane proteins. The theoretical merit of the approach is considerable, especially for peptide hormones that have been sequenced and synthesized and whose receptors appear to be of very low abundance. The recent extension of monoclonal antibody techniques to the production of stable hybridomas, each secreting a single type of antiidiotype antibody, is a real improvement over the transient and low-titer antiidiotype responses observed in polyclonal sera from whole animals. However, success of this approach is critically dependent upon the availability of a panel of sensitive and reliable assays for screening the antibodies. The criteria employed to identify antiidiotype antibodies to receptors must be rigorous enough to eliminate false positives as well as experimental artifacts, such as steric obstruction of ligand-binding domains of receptors by the binding of putative "antiidiotype" antibodies to adjacent membrane sites. The work of Chuang (1985) provides an excellent example in which antibodies to a membrane component other than the receptor can "interact" with the beta-adrenergic receptor. Antiidiotype antibodies to beta-adrenergic receptors have provided and will continue to provide additions to our repertoire of immunologic reagents for the study of these receptors.

2.2. The "Classical" Approach to Antireceptor Antibody Production

2.2.1. Polyclonal Antibodies to Membrane-Bound Receptors

The classical approach to preparing antibodies against a membrane-bound receptor or any other protein antigen is to immunize animals repeatedly with sufficient quantities of an essentially pure preparation of protein that has been carefully and thoroughly characterized (Fig. 1). Historically, the animals of choice for immunization were rabbits or goats. More recently mice have been used, particularly to produce monoclonal antibodies. As alluded to earlier, the straightforward, classical approach has been most successful in cases where a highly abundant source of the target protein exists and an efficient high-yield purification strategy has been developed. Perhaps the best example of a membrane-bound receptor for which both of these criteria have been met is the nicotinic acetylcholine receptor that can be isolated from the receptor-replete membranes of the electric-organ of *Torpedinidae californum* or *Electrophorus electricus* by affinity chromatography and standard biochemical techniques (Conti-Tronconi and Raftery, 1982). In the studies of the acetylcholine receptor (Conti-Tronconi et al., 1982), the availability of milligram quantities of the purified receptor as well as of its individual subunits was critical to the screening and characterization of both polyclonal (for review, *see* Conti-Tronconi and Raftery, 1982) and monoclonal (Tzartos and Lindstrom, 1980) antibodies. Other examples of the use of the classical approach are the work on the low-density lipoprotein receptor (Beisiegel et al., 1981a) and the insulin receptor (for review, *see* Czech, 1985). In all of these instances, the most significant fact is that, when finally generated after considerable investment of time and effort, antireceptor antibodies became invaluable reagents for the study of the structure and biology of the receptors. Biochemical characterization, quantification of peptide, elucidation of the biosynthesis of the receptor, and direct visualization of the receptor by immunocytochemical methods are several of the new directions that research can assume when specific antibodies become available.

Successful production of polyclonal antibodies using the classical approach demands the isolation of microgram–milligram quantities of purified receptor, both for immunizations as well as for screening procedures. However, this "clean in/clean out" approach is not the only strategy that can be employed to produce antireceptor antibodies. Hybridoma antibody production offers some real advan-tages over the conventional approach of harvesting antisera from animals that have been immunized with essentially pure antigen (for review, *see* Fraser and Lindstrom, 1984). In addition to providing a continuous, unlimited supply of monoclonal antibodies directed at single antigenic determinants, the hybridoma technology does not require the availability of pure antigen for the immunization. This "dirty in/clean out" approach, however, does require pure receptor at some point to rigorously establish the true antigen for the monoclonal antibodies that are produced. Advances in the production of antibodies to beta-adrenergic receptors by the "classical" approach are described first, followed by a discussion of hybridoma antibody production and its application to the generation of monoclonal antibodies to beta-adrenergic receptors.

2.2.2. Polyclonal Antibodies to Beta-Adrenergic Receptors

Several groups have utilized the classical approach to produce polyclonal antibodies to beta-adrenergic receptors. Immunization with only partially pure preparations of beta-adrenergic receptors was attempted early. Wrenn and Haber (1979) immunized rabbits with 0.2 nmol of a high-affinity binding site for propranolol obtained from dog heart and isolated by affinity chromatography on columns of Sepharose to which norepinephrine had been immobilized. The specific activity of the partially purified receptor preparation was reported to be approx 1.1 nmol of ligand bound/mg of protein (Wrenn and Haber, 1979). It is well established now by biochemical analysis and molecular cloning that mammalian beta-adrenergic receptors are single-chain polypeptides with 65–67kDa when subjected to SDS-PAGE in the presence of an agent capable of reducing disulfide bridges (Benovic et al., 1984; Cubero and Malbon, 1984; George and Malbon, 1985; Graziano et al., 1985). Assuming a single binding site per polypeptide, the theoretical specific activity of the pure receptor should be approx 12 nmol of ligand bound/mg of protein. Thus, receptor preparations estimated to be less than 10% of theoretical purity were used in this early work (Wrenn and Haber, 1979). Antiserum raised against the partially purified propranolol binding site of cardiac membranes did effectively inhibit both the binding of propranolol to, as well as the stimulation of adenylate cyclase by, catecholamines in cardiac membranes. The immune serum was found to inhibit stimulation of cardiac adenylate cyclase activity in response to isoproterenol, but not in response to either the GTP analog Gpp(NH)p or sodium fluoride. Interestingly, this immune serum did not inhibit isoproterenol-stimulated adenylate cyclase activity in liver membranes. Subsequent studies of beta-adrenergic receptors using this rabbit polyclonal antiserum have not appeared.

Couraud et al. (1981) reported the production of three mouse polyclonal antisera raised against beta-adrenergic receptors purified by affinity chromatography some 20,000-fold from digitonin extracts of turkey erythrocyte membranes. The receptor used in these studies was composed of two major polypeptides with 60 and 70kDa, and three smaller peptides with 42, 33, and 30kDa (Couraud et al., 1983). Mice were immunized with 1–2 pmol (50–100 ng protein) in complete Freund's adjuvant and boosted monthly with the same quantity of receptor in incomplete Freund's adjuvant. At a fivefold dilution, the antisera immunoprecipitated 30% of radioiodinated receptor; at 60-fold dilution, none of the receptor was immunoprecipitated. Adenylate cyclase activity of turkey erythrocyte membranes was stimulated by IgG fractions purified from the crude antisera, at immunoglobulin concentrations from 0.2 to 1.4 mg/mL (Couraud et al., 1981). Interestingly, IgG did not alter the ability of epinephrine to stimulate the adenylate cyclase activity in the membranes, suggesting that the antibodies recognize domains of the receptor other than the ligand-binding domain. Later these researchers reported on the use of indirect immunofluorescence and one of the antisera to visualize putative beta-adrenergic receptors of P815 mastocytoma cells (Couraud et al., 1983). Autoradiography of immunoprecipitates using this antiserum and radioiodinated beta-adrenergic receptor purified from turkey erythrocytes revealed an intense band of radioactivity with 60kDa (Couraud et al., 1981). The purified beta-adrenergic receptor isolated from turkey erythrocytes has been reported to be composed of polypeptides with 40 and 45kDa (Shorr et al., 1982), and more recently as a polypeptide with 53kDa (Yarden et al., 1986). The molecular cloning of the beta-adrenergic receptor from turkey erythrocytes (Yarden et al., 1986) predicts that this beta-adrenergic receptor is composed of 483 amino acids and has a mol wt of 54000, although a major proteolysis product (40kDa) and the peptides with 53kDa are observed in sodium dodecyl sulfate-polyacrylamide gels of receptor purified on a large scale from these erythrocytes (Yarden et al., 1986). The differences in the M_r between the immunoprecipitated receptor in the studies of Couraud et al. (1983) and the purified receptor (Shechter et al., 1982; Yarden et al., 1986) or the predicted mass of the cloned protein (Yarden et al., 1986) may be more apparent than real. However, these differences must be reconciled if these antibodies are to be used with confidence in other pursuits.

Strader et al. (1983) reported the successful production of a polyclonal antiserum to beta-adrenergic receptors using the "classical" approach and beta-2 adrenergic receptors that were 50–80% pure, isolated from frog erythrocytes. A New Zealand White rabbit was immunized by subcutaneous injection of 0.1 mg (approx 1.7 nmol) of receptor antigen over a period of several months. The antiserum obtained from this rabbit immunoprecipitated radiolabel from incubations of radioiodinated, partially purified receptor with the immune antiserum diluted up to 100-fold. This antiserum also immunoprecipitated radiolabel from incubations of purified receptor that had first been tagged with radioiodinated cyanopindolol, a high-affinity beta-adrenergic antagonist ligand. Interestingly, beta-2 adrenergic receptors from several sources showed various degrees of immunologic crossreactivity, whereas several beta-1 adrenergic receptors failed to show any crossreactivity with this antiserum raised against the frog beta-2 adrenergic receptor. The antiserum did not alter the ability of the receptor to bind antagonist ligands, but did attentuate stimulation of adenylate cyclase activity by isoproterenol in frog erythrocyte membranes. Immunocytochemical analysis of rat and frog brain using this antiserum suggested a postsynaptic localization of beta-adrenergic receptors (Strader et al., 1983). Aoki et al. (1987) employed this immune serum to conduct ultrastructural localization of receptor-like immunoreactivity in the cortex and neostriatum of rat brain.

The studies described above provide a framework for a discussion of the optimal methods for the production of polyclonal antireceptor antibodies. Emphasis is placed on facets of the classical approach that most often provide serious obstacles to the preparation of useful antireceptor antibodies. Methods that we have recently developed in our laboratory for the production of high-quality, useful antibodies to beta-adrenergic receptors are highlighted (Moxham and Malbon, 1985; Moxham et al., 1985a,1986a,b).

2.2.3. Techniques

2.2.3.1. SELECTION OF ANIMAL TO BE IMMUNIZED. For producing polyclonal antisera, one may generally consider goats or rabbits as the most practical choices. Although goats have the advantage of producing large volumes of antiserum (50–200 mL/bleed), they are considerably larger animals and will often require substantial amounts of receptor immunogen (1–10 mg) for effective immunizations. In situations where the amount of receptor antigen to be used is in scarce supply and quite difficult to obtain, rabbits or mice may be the animals of choice for immunization. We (Moxham and Malbon, 1985; Moxham et al., 1985a,b;1986a,b,c) have reported, as have others (Couraud et al., 1981; Strader et al., 1983), the generation of polyclonal antisera against beta-adrenergic receptors using rabbits as well as mice. Nonetheless, for some protein antigens there may well be interspecies differences in the immune response, and goats are often tried if sufficient antigen is available.

2.2.3.2. PREPARATION OF IMMUNOGEN AND IMMUNIZATION SCHEDULE. Details of large-scale purification of beta-adrenergic receptors are provided elsewhere (Malbon, 1990) and in this volume, and the reader is advised to consult this work for detailed discussion of the topic (see Chapter 2). In the context of antibody production, the source from which the receptors are to be purified and the purification technique to be employed are the most important considerations. Careful attention should be paid to each of these items. In real terms, the selection of tissues for starting material from which nmol quantities of beta-adrenergic receptors can be obtained is rather limited. Membranes prepared from guinea pig (Benovic et al., 1984) or hamster (Benovic et al., 1984; Dixon et al., 1986) lung as well as basal membranes prepared from human placenta (Bahouth et al., 1986; Kelley et al., 1983) possess an abundance of beta-adrenergic receptors. Basal membranes of placenta display specific activities of up to 5 pmol receptor/mg of membrane protein, and a single human placenta from a full-term birth contains approx 2 nmol of these receptors (Kelley et al., 1983). We have successfully utilized rat fat cells (Cubero and Malbon, 1984), rat liver (Graziano et al., 1985), human placenta (Bahouth et al., 1986), guinea pig lung (Moxham et al., 1986a), as well as S49 mouse lymphoma cells grown in mass cultures (George and Malbon, 1985) as sources for the isolation of mammalian beta-1 and beta-2 adrenergic receptors. Each of these sources offers some distinct advantage over others. Most notably we have used the guinea pig lung and S49 cells as mammalian sources of strictly beta-2 adrenergic receptors and the rat fat cells as a mammalian source for strictly beta-1 adrenergic receptors for the isolation of receptor to be used as an immunogen (Moxham and Malbon, 1985; Moxham et al., 1985a; 1986a,b).

Several approaches to the purification of beta-adrenergic receptors have appeared in the literature. All of these approaches rely heavily on affinity chromatography as an early, high-efficiency step in the purification. The purity and recovery of the receptor must be sufficient to allow isolation of nmol quantities (0.05–0.1 mg protein) of essentially pure receptor. Antibodies have been prepared with receptor preparations that varied in purity from 5% (Caron et al., 1979) or 10% (Wrenn and Haber, 1979) to 50–90% (Couraud et al., 1981,1983; Moxham and Malbon, 1985; Moxham et al., 1985a,1986; Strader et al., 1983).

Figure 3 displays several methods of immunogen preparation that have been employed to generate antibodies to beta-adrenergic receptors. In one case at least, receptor was apparently administered in the absence of an adjuvant to obtain antibodies to the turkey erythrocyte beta-adrenergic receptor (Strader et al., 1983). In most cases, however, receptor is emulsified with Freund's complete adjuvant for the primary immunization. Subsequent booster injections of receptor are administered in incomplete Freund's adjuvant. Subcutaneous and intradermal injections are often the route of choice for administration of receptor antigen.

We have made two modifications that improved our rate of success in generating high-titer antisera to beta-adrenergic receptors in rabbits and mice. Our experience suggests that the pure beta-adrenergic receptor is a poor immunogen. The considerable hydrophobicity of these receptors as well as their conservation between avian (Yarden et al., 1986) and mammalian (Dixon et al., 1986) species likely contributes to the lack of antigenicity that we have observed. However, we found that the antigenicity of the receptor can be markedly enhanced when the purified receptor is applied as a coating to glutaraldehyde-fixed cells (Moxham et al., 1985a). Beta-2 adrenergic receptors purified from guinea pig lung as described by Benovic et al. (1984) and determined to be 50–90% pure (Moxham et al., 1986a,b) were coupled to glutaraldehyde-fixed SP 2/0 mouse myeloma cells and used as an alternative to Freund's complete adjuvant (Fig. 3). This receptor conjugate containing approx 0.1 nmol of receptor was used to immunize New Zealand White rabbits. Recently we have also been successful in raising antibodies to antigens emulsified with trehalose dimycolate-monophosphoryl lipid A, a synthetic adjuvant.

A second modification that improves the immunization protocol involves the route of administration. Without exception, we found superior immune responses if antigens are administered into the popliteal lymph node. Primary and booster immunizations into the popliteal lymph node on the hind legs of rabbits and mice can be performed easily and painlessly to the animal by experienced animal technicians. This immunization protocol has been gaining widespread use when the amounts of purified antigens are limited. Discussion of this technique for applying antigens more directly to the immune system is provided elsewhere (Sigel et al., 1983).

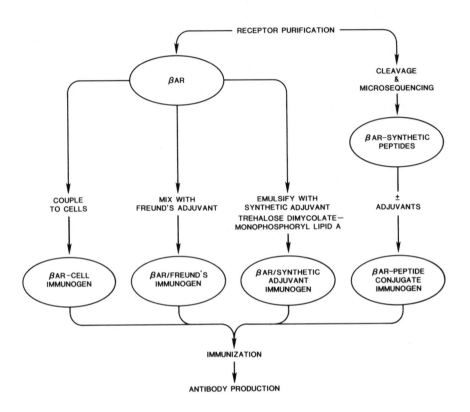

Fig. 3. Beta-adrenergic receptor immunogen preparations. Purified beta-adrenergic receptor is a poor immunogen alone. To enhance the immune response in the production of antibodies to beta-adrenergic receptor (βAR), these receptors can be coupled to cells, used in combination with Freunds' adjuvants, or used in combination with synthetic adjuvants. Alternatively, receptor is subjected to microsequencing, and synthetic peptides are designed based on the primary sequence of the receptor. These peptides are coupled to a larger molecule (such as bovine serum albumin or keyhole limpet hemacyanin), and this conjugate is used as the immunogen. Advantages of each of these approaches are discussed in the text.

With this combination of adjuvant-immunization methods, we generated an antiserum that specifically recognized beta-adrenergic receptors. In preparing such antisera, approx 20 μg of the pure receptor protein was used to immunize each animal. Isolating this quantity of the beta-adrenergic receptor is a considerable amount of work (equivalent to the isolation of beta-adrenergic receptor from the fat pads of 400 rats [Cubero and Malbon, 1984] or from 50–60 L of S49 mouse lymphoma cells at a density of 5–6 million cells/mL [George and Malbon, 1985]). Isolation of milligram and submilligram quantities of a protein antigen typically used for the successful generation of antibodies in rabbits is not feasible. Indeed, the amounts of receptor required for successful immunization of rabbits using the improvements described above are closer to the amounts of protein antigen typically used to produce an antibody response in mice.

Based on these observations, it would seem most profitable, when trying to produce a useful rabbit (or goat) antiserum against small quantities of a pure receptor protein, to investigate the following: the use of adjuvants, such as Freunds' and synthetic adjuvants, as well as the creation of a more antigenic vehicle by coupling the antigen to a cell surface (Fig. 3); the use of more than one route of immunization, in particular immunization into the lymph system (Sigel et al., 1983); the use of different schedules of immunization; and the use of more than one type of animal for the production of antibodies. Less-pure immunogens may be more antigenic and thereby may require less "boosting" of the immune response with adjuvant mixtures. Immunization with impure antigens should be adopted as a strategy only when the options outlined above are not available. The antisera generated in response to an impure mixture of antigens, however, must be screened for the presence of *bona fide* antibodies to the antigen of interest.

2.2.3.3. INITIAL CHARACTERIZATION OF POLYCLONAL ANTIRECEPTOR ANTISERA. Depending on the purity and quantity of receptor available, there are a number of possible methods of screening polyclonal antisera for antireceptor antibodies. Unlike working with hybridoma antibodies, when one must sometimes simultaneously screen large numbers (hundreds) of test supernatants, working with polyclonal sera generally requires small numbers (6–10) of test bleeds in which one can and is obliged to use multiple methods of screening. A useful approach is immunoprecipitation of radiolabeled pure receptor or of receptor that has been tagged with a specific radioligand. However, identification of true antireceptor antibodies requires the application of multiple screening techniques in order to eliminate troublesome false positives. Examples of such multiple, mutually confirmatory screens might be a solid-phase ELISA against purified receptor or cell membranes, immunoprecipitation of purified receptor protein as well as immunoprecipitation of receptor binding activity, immunoblotting of membrane proteins and pure receptor, and ability of test antibody to interfere with the binding of radioligand to receptor in membranes or to purified receptor. Each screen contributes information that, when compiled with data from other methods of analysis, gives a more complete picture of the specificity of the antibody.

2.3. Monoclonal Antibody Production

2.3.1. Monoclonal Antibodies to Membrane-Bound Receptors

The pioneering work of Kohler and Milstein (1975) in developing techniques to immortalize individual, antibody-producing B lympho-cytes through fusion to tumorigenic, rapidly-dividing myeloma cells has added much to the study of membrane receptors in recent years. Figure 1 depicts the hybridoma methodology that several laboratories have been using for the production of mouse monoclonal antibodies to purified beta-adrenergic receptors. The first step in the process is identical to that used in the "classical" approach. Mice, rather than rabbits, are repeatedly immunized with purified antigen or an antigen-containing mixture until a high-titer, polyclonal antiserum specific for the antigen is detected in test bleeds from the animals. Spleen cells are removed from mice scored "positive" for high-titer antiserum and fused with cultured myeloma cells in the presence of polyethylene glycol (PEG). The critical feature of the protocol is the selection for true hybridomas, the fused hybrids of splenocytes with myeloma cells. Initial selection of the fusion products is performed after the cell mixture has been plated in 96-well microtiter plates (600–1000 wells/fusion is average). The selection is performed in a culture medium containing hypoxanthine, aminopterine, and thymidine (HAT medium). Unfused spleen cells from the immunized mouse fail to survive under these culture conditions. The unfused myeloma cells, lacking hypoxanthine-guanine ribosyltransferase (HGPRT⁻) activity, cannot survive in HAT medium. Only true hybridomas, which are HGPRT⁺ from the spleen cell contribution and are now immortalized by virtue of their fusion with the myeloma cells, will grow in the HAT selection medium. Each antibody-producing hybridoma secretes an antibody to a single epitope (antigenic site) (Kohler and Milstein, 1975). The ability to produce homogeneous populations of hybridoma cells, each secreting large (milligram) quantities of a unique antibody directed against a single epitope on a receptor molecule, is a powerful new approach to probing receptor structure and function. The production of monoclonal antibodies for the acetylcholine receptor (Tzartos and Lindstrom, 1980), low-density lipoprotein receptor (Beisiegel et al., 1981b), and insulin receptor (Kull et al., 1982; Morgan et al., 1986) has provided important new reagents for the study of these receptors.

2.3.2. Monoclonal Antibodies to Beta-Adrenergic Receptors

In 1980, Fraser and Venter reported the production of monoclonal antibodies to beta-adrenergic receptors of turkey erythocytes. These investigators used isoelectric focusing of digitonin extracts from turkey erythrocyte ghosts to isolate beta-adrenergic receptors. About 50% of the membrane protein of the turkey erythrocyte ghosts was extracted by digitonin. Beta-adrenergic receptors in these crude extracts were found to migrate in a broad band ranging from pH 4.5 to pH 7.5 when subjected to isoelectric focusing on a gradient of pH 3–10 (Venter and Fraser, 1981). Beta-adrenergic receptors of calf lung were extracted in Triton X-100 and partially purified by gel exclusion chromatography for use in these studies also (Fraser and Venter, 1980). The specific activities of the preparations as well as the amount of receptor used in the immunizations were not reported (Fraser and Venter, 1980; 1981). Four of eight hybridoma obtained from a single fusion plated out into 196 wells were identified as "positives" (Fraser and Venter, 1980). Clones scored "positive" were those whose conditioned media inhibited the amount of tritiated dihydroalprenolol binding in crude extracts after immunoprecipitation. A single "positive" hybridoma was the product of a fusion that generated hybridoma in 72 of 196 wells (Venter and Fraser, 1981). One monoclonal antibody (FV-104) was reported to inhibit specific binding of radioiodinated hydroxy-benzylpindolol to detergent-dispersed and membrane-bound beta-adrenergic receptors. Inhibition was observed only at ascites fluid dilutions of less than 1 to 20.

In the same studies, Fraser and Venter (1980) also reported the novel purification of the turkey beta-adrenergic receptor using immunoaffinity adsorption by the FV-104 monoclonal antibodies immobilized to a resin. A single peptide (70kDa) of turkey erythrocyte extracts was specifically eluted from the immunoaffinity matrix. The M_r for the turkey beta-adrenergic receptor purified in this manner (Fraser and Venter, 1980), however, differs significantly from that reported by others based on large-scale purification (Shorr et al., 1982; Yarden et al., 1986), and more recently derived from molecular cloning data (Yarden et al., 1986). Beta-2 adrenergic receptors of membranes from cultured human lung (VA4) cells were purified using immunoaffinity chromatography on FV-104-Sepharose 4B and reported by Venter and Fraser (1981) to be a single peptide monomer with 47kDa. Immunoaffinity purification of canine lung beta-2 adrenergic receptors, too, was reported by Fraser and Venter (1982). The subunit mol wt of 59 was determined for this purified receptor by SDS-PAGE of the purified protein (Fraser and Venter, 1982). Benovic et al. (1984) reported that the M_r of the lung beta-2 adrenergic receptor, like that of the mammalian liver (Graziano et al., 1985) as well as S49 mouse lymphoma cells (George and Malbon, 1985), was 65–67. Differences in electrophoretic separation protocols may play a role in these discrepancies.

In addition to the production of polyclonal antiidiotype antibodies that recognize beta-adrenergic receptors (Schreiber et al., 1980), Strosberg and colleagues succeeded in preparing both antireceptor (Couraud et al., 1983; Kaveri et al., 1987) and antiidiotype monoclonal antibodies (Guillet et al., 1984,1985). These antibodies have been characterized using a combination of techniques, including immunoprecipitation, ELISA, and immunoblotting (Guillet et al., 1985; Kaveri et al., 1987). The availability of a panel of hybridomas that secrete antibodies of a constant nature in theoretically unlimited quantities, and recognize well-defined epitopes of beta-adrenergic receptors, remains an important but elusive goal.

Recently, we have used highly purified beta-adrenergic receptor from rat fat cells to immunize Balb/C mice and to prepare antisera specific for these receptors (Moxham et al., 1985a,1986a). Our efforts to produce a panel of stable hybridomas from mice displaying high-titer antireceptor antisera have achieved only limited success. Site-directed antipeptide antibody production to defined regions of the mammalian beta-adrenergic receptor has been far more fruitful (George et al., 1988; Wang et al., 1988a,b,c; Bahouth et al., 1988; Hadcock et al., 1989).

2.3.3. Techniques

2.3.3.1. IMMUNOGENS AND IMMUNIZATIONS. As discussed above (*see* Section 2.2.3.2.), several different immunogens can be employed, ranging from crude, receptor-bearing cell membranes, through detergent-solubilized partially purified receptor preparations, to essentially pure receptor proteins. The use of receptor-bearing whole cells to obtain monoclonal antibodies to a single, specific membrane receptor was elegantly demonstrated by Schreiber et al. (1981) for the receptor for epidermal growth factor, but has not yet been successfully applied to beta-adrenergic receptors. Recently we have succeeded in transfecting Chinese hamster ovary (CHO) cells with a cDNA encoding the beta-2 adrenergic receptor under the control of the SV40 early promoter and have isolated several clones expressing $0.5–2.0 \times 10^6$ receptors/cell (George et al., 1988). These high-expressing CHO cells may prove to be sufficiently antigenic to obtain antireceptors monoclonal antibodies. More recently, we have succeeded in the high-efficiency expression of mammalian beta-adrenergic receptors in baculovirus-infected insect cells (*Spodoptera frugiperda*), providing cells that express 15–60 million receptors per cell (George et al., 1989).

Specific, unambiguous screens useful for the rapid testing of supernatants from several hundred or thousand hybridomas are also required. As for the use of partially purified receptors as immunogens, we have already noted the earlier reports from two laboratories, where turkey erythrocyte beta-adrenergic receptor was being used as immunogen. Fraser and Venter (1980) first immunized 6-wk-old Balb/C mice intraperitoneally with an unreported amount of partially purified, turkey erythrocyte beta-adrenergic receptor emulsified in Freund's incomplete adjuvant. A booster immunization was also given intraperitoneally four weeks later. Only a small number of hybridomas grew from fusions of these immunized spleens (8 wells of 196 wells seeded [Fraser and Venter, 1980]), which may be a reflection of the poor antigenicity of this crude immunogen.

Couraud et al. (1983) used 1–2 pmol (50–100 ng) of a more highly purified (20,000-fold) preparation of turkey beta-adrenergic receptor to immunize Balb/C mice subcutaneously with Freund's complete adjuvant. This was followed with a second booster immunization to produce a mouse antiserum with relatively low titer in immunoprecipitation and adenylate cyclase assays. In this study, the use of a very small quantity of purified receptor may have limited the potential success of producing high-titer, high-affinity antibody. Moreover, these observations agree well with our own that suggest beta-adrenergic receptors are weak immunogens. Kaveri et al. (1987) obtained monoclonal antibodies from mice immunized with beta-2 adrenergic receptor isolated from A431 cells by affinity chromatography on a BSA–alprenolol conjugate. These antibodies have already demonstrated their utility in a variety of applications (Kaveri et al., 1987; Cervantes-Olivier et al., 1988; Ventimiglia et al., 1987).

We have dealt with the relatively poor antigenicity of beta-adrenergic receptors by immunizing animals with microgram quantities of highly (50–90%) purified receptor covalently attached to glutaraldehyde-fixed mouse myeloma cells. The use of a highly purified receptor protein as a starting material maximizes the chances for generating specific antibodies. Moreover, we have found that the use of the protein–cell conjugate immunogens, as originally described by Relyveld and Ben-Efraim (1981), greatly increases the immune response to 2–5 µg of purified receptor protein when administered intraperitoneally (or in the popliteal lymph node) as primary as well as booster immunizations at 3–4 wk intervals for 3–4 mo. Other synthetic, well-defined adjuvants, such as trehalose dimycolate-monophosphoryl Lipid A (*see* Fig. 3), have been tested in our laboratory and appear to be effective in increasing the immune response to small quantities of purified protein antigens.

2.3.3.2. HYBRIDOMA PRODUCTION. Since the original publication of Kohler and Milstein (1975), several related methodologies for producing hybridomas by PEG-mediated fusion of mouse spleen and myeloma cells have been well described. In our laboratory, where monoclonal antibodies to several protein antigens are being developed, a modification (Brandwein et al., 1981) of the original procedure of Gefter et al. (1977) is utilized. In this protocol, 37% PEG-1000 is added slowly to the mixed splenocyte–myeloma cell pellet over a 1–2 min period, and then slowly diluted with dropwise addition of calcium-free and serum-free media over a 4–5 min period, followed by a low-speed centrifugation of the mixture to collect the cells. Spleen cells of Balb/C mice displaying high-titer antisera are fused with the non-Ig secreting SP2/0 myeloma cell line. After the fusion and subsequent washes, the cells are suspended in a selective culture medium containing hypoxanthine, aminopterin, and thymidine (HAT medium) and distributed into the wells of ten microtiter culture plates of 96 wells each. Under these conditions, normal splenocytes and myelomas die, and only fused hybridomas grow. After 2–3 wk of culture, wells with hybridomas can be screened for receptor-specific antibody by one or more of the screening assays described below. Positive wells can then be subcloned in 0.5% agarose with complete medium. After 10 d, discrete clones are identified, picked from soft agar, and transferred to liquid medium for further growth in microwells. The resultant stable monoclonal populations can be injected into mice primed with 0.5 mL Pristane one wk prior to the intraperitoneal injection of approx 5–6 million hybridoma cells. One to two weeks later, ascites fluid is obtained and tested for antibody titer.

2.3.3.3. HYBRIDOMA SCREENING ASSAYS. Typically, a successful fusion will produce hybridomas in 40–100% of the wells seeded. Thus some 400–900 hybridoma supernatants must be screened for antibody activity when 960 wells (ten plates 96 wells each) are seeded from a single spleen. Testing this number of wells dictates the use of an initial screening method that can be performed quickly on a large number of samples. If not subcloned expeditiously, precious antibody-secreting clones can be overgrown by nonsecreting clones. The solid-phase ELISA methods in which antibodies are screened for their ability to bind purified receptor immobilized on microtiter wells has become the method of choice (*see* Section 3.3.). These ELISA methods are rapid and convenient, and when used with essentially pure antigen, are unambiguous. Screening hybridoma supernatants requires the isolation of considerable quantities of purified antigen, which may not be possible in all instances. In such cases there are several other methods discussed below that can be used to initially screen monoclonal antibodies.

3. Screening Antibodies
Against Beta-Adrenergic Receptors

The successful production of antibodies to membrane-bound receptors depends on not only the intrinsic antigenicity of the receptor molecule, but also the development of useful screening techniques capable of detecting and characterizing antibodies specifically directed against the receptor. A variety of methods are available for the screening of antibodies directed against specific antigens, including the use of membrane-bound hormone receptors. The section that follows describes methods useful for screening antisera and hybridoma supernatants and their application to the identification of antibodies to beta-adrenergic receptors and other membrane-bound receptors. The advantages, disadvantages, and limitations of each method will be highlighted.

3.1. Biology-Based Assays

Initial characterization of antisera or hybridoma supernatants for antireceptor antibody activity often includes an analysis of the effects of these immunologic preparations on a biological response mediated by the receptor of interest. These biology-based assays for antibodies have been employed to demonstrate the presence of antibodies directed against several cell surface, membrane-bound receptors, including beta-adrenergic receptors (Table 1, *see* pp. 210 and 211). Schreiber et al. (1981,1983) reported the use of biology-based assays to identify monoclonal antibodies against the receptor for epidermal growth factor in A431 cells. Antibodies to the receptor displayed many of the biological effects of epidermal growth factor itself in these cells (Schreiber et al., 1981,1983). Polyclonal antibodies to insulin receptors also have been identified that, like insulin, display the ability to stimulate glucose oxidation in intact cells (Jacobs et al., 1978). Mouse monoclonal antibodies against the insulin receptor have been shown to inhibit 2-deoxyglucose uptake in rat fat cells, receptor kinase activity, and the ability of insulin to stimulate maturation of *Xenopus* oocytes (Morgan et al., 1986). Monoclonal antibodies to the low-density lipoprotein receptor have been shown to block the binding, internalization, and degradation of LDL (Beisiegel et al., 1981b). Antibodies to the acetylcholine receptor of *Torpedinidae californum* have been shown to inhibit receptor activation by acetylcholine (Tzartos and Lindstrom, 1980).

Several groups have adopted a biology-based strategy to detect antibodies against beta-adrenergic receptors (Table 1). Cyclic AMP accumulation or activation of adenylate cyclase activity of cell membranes in response to stimulation by beta-adrenergic agonists can be measured with high sensitivity and reliability. Measurement of the ability of antisera or hybridoma supernatants to alter either of these parameters provides a biology-based assay for the detection of antibodies against beta-adrenergic receptors. Antibodies capable of stimulating adenylate cyclase activity in cells and membranes in an agonist-like fashion have been reported (Couraud et al., 1981; Fraser and Venter, 1980; Guillet et al., 1985). Fraser and Venter (1984) reported the identification, in the serum of asthmatic patients, of autoimmune antibodies to beta-adrenergic receptors. These antibodies inhibited agonist stimulation of cyclic AMP accumulation in cells. Strader et al. (1983) observed an inhibition of isoproterenol-stimulated adenylate cyclase activity of frog erythrocyte membranes by a polyclonal antibody raised against purified beta-adrenergic receptors. We too have observed that antisera from mice immunized with pure beta-adrenergic receptor isolated from rat fat cells inhibit isoproterenol-stimulated cyclic AMP accumulation by rat fat cells (Moxham et al., 1985a).

Table 1
Biological-Based Analysis of Antireceptor Antibodies

Antibody	Immunogen	Biological assay	Effect	Reference
Mouse mAb	A-431 Cells	EGF receptor clustering and internalization short delayed growth	+ + +	Schreiber et al., 1981
Mouse mAb	A-431 Cells	EGF receptor kinase [³H]thymidine incorporation	+	Schreiber et al., 1983
Rabbit pAb	Purified human placental insulin receptor	Glucose oxidation	+	Jacobs et al., 1978
Mouse mAb	IM-9 lymphocytes	2-Deoxyglucose uptake	−	Roth et al., 1982
Rabbit pAb	Insulin receptor synthetic peptide	Insulin receptor auto-phosphorylation and exogeneous substrate phosphorylation	− −	Herrera et al., 1985
Mouse mAb	Purified human placental insulin receptors	Insulin-stimulated receptor kinase and maturation of *Xenopus* oocytes	−	Morgan et al., 1986

Mouse mAb	Partially purified LDL receptor	Binding, degradation, and internalization of LDL	−	Beisiegel et al., 1981b
Mouse mAb	Purified torpedo AcHR	Experimental animal myasthenia gravis	+	Tzartos and Lindstrom, 1985
Mouse pAb	Purified turkey B_1AR	Adenylate cyclase	+	Couraud et al., 1981
Rabbit pAb	Alprenolol-BSA	Adenylate cyclase	+	Schreiber et al., 1980
Human aAb	?	Cyclic AMP accumulation	+	Fraser and Venter, 1984
Mouse mAb	Hybridoma bearing anti-ALP antibodies	Cyclic AMP accumulation	+	Guillet et al., 1985
Rabbit pAb	Purified frog B_2AR	Adenylate cyclase	−	Strader et al., 1983
Mouse pAb	Purified fat cell B_1AR	Cyclic AMP accumulation	−	Moxham et al., 1985

The abbreviations used are as follows: EGF, epidermal growth factor; LDL, low-density lipoprotein; AcHR, acetylcholine receptor; BAR, beta-adrenergic receptor; mAb, monoclonal antibody; pAb, polyclonal antibody; aAb, autoantibody.

Although biology-based assays may provide supporting evidence that antireceptor antibodies are present in an antiserum or hybridoma supernatant, their utility is compromised to a large extent by their sensitivity to nonspecific effects of other hormones, fatty acids, and unidentified serum factors. The potential for unpredictable as well as artifactual responses of cells or cell membranes treated with antisera or hybridoma supernatants cannot be overstated. Over the past several years, we have observed nonspecific stimulatory as well as inhibitory effects of antisera in biology-based assays. Investigations of the effects of putative antireceptor antibodies on cyclic AMP accumulation by intact cells (or on adenylate cyclase activity of cell membranes) should include an analysis of their effects on the responses to stimulation by isoproterenol, by other hormones, and by the diterpene forskolin. Sera of animals immunized with antigens not related to beta-adrenergic receptors often display an ability to inhibit the cyclic AMP response to isoproterenol or forskolin. Couraud et al. (1981) reported that a polyclonal serum to beta-adrenergic receptors stimulated adenylate cyclase activity even in the presence of maximal concentrations of isoproterenol. Thus, interpretation of data from biology-based assays can be complicated by the nature of the antisera, the nature of antireceptor antibodies present in the sera, and an incomplete understanding of the biology in question. When used in combination with other methods of screening, biology-based assays have a great potential to provide useful information. When used as the sole means of establishing the presence of putative antireceptor antibodies, however, they display an equally great potential to provide "false" positives or misinformation.

3.2. Radioligand Binding-Based Assays

Antibodies to the ligand-binding domain of a receptor often are detected by the use of classical radioligand binding assays with receptor preparations. Monoclonal antibodies to mammalian LDL receptors were identified using the binding of radiolabeled LDL to receptors of cell membranes as an assay for antibody activity (Beisiegel et al., 1981a). Antibodies to insulin receptors have been identified using this same strategy (Roth et al., 1982). As expected, antibodies to the ligand-binding domain of a receptor score as "positives" in biology-based assays of receptor-mediated events (Beisiegel et al., 1981b; Roth et al., 1982).

Research in the field of beta-adrenergic receptors benefits more than most fields of receptor research from the availability of a wide spectrum of well-characterized high-affinity antagonist and agonist ligands. Radiolabeled antagonists of high specific activity are available commercially and have become the mainstay for much of the research in this area. Tritiated dihydroalprenolol and radioiodinated cyano-pindolol, in particular, are extremely useful probes of beta-adrenergic receptors both in membranes and in detergent-dispersed preparations. Cyanopindolol with its high specific activity offers an advantage when the absolute amount of receptor being measured is low, i.e., less than 10 fmol/assay. For the screening of antibodies, the amounts of receptor as well as antisera or hybridoma supernatant employed in the assays can be minimized by the use of radiolabeled cyanopindolol. Dihydro-alprenolol is useful in assay systems where the more hydrophobic cyanopindolol cannot be employed because of high levels of nonspecific (nonreceptor) binding. As discussed with the biology-based assays, the complex nature of sera or hybridoma supernatants dictates that due caution be employed when using these radioligand binding assays. Antibodies against the ligand-binding domain or any epitope capable of influencing access to or the integrity of the ligand-binding domain will score "positive" in these analyses.

Several laboratories have reported the effects of antibodies to beta-adrenergic receptors on radioligand binding to these receptors. In most cases, the antibodies decreased the amount of specific binding of radioligand beta-adrenergic receptors in membranes, without altering the apparent affinity of the receptors for the ligand (Fraser and Venter, 1980; Guillet et al., 1985; Moxham et al., 1985a). Homcy et al. (1982) reported, in contrast to these observations, that antiidiotype antibodies to beta-receptors altered only the affinity of the receptors for radioligand. Other reports (Couraud et al., 1981), like that of Strader et al. (1983), identified antibodies that failed to alter the ability of the receptor to bind radioligand. The lack of an effect of an antireceptor antibody to alter ligand binding indicates only that the antibodies recognize domains of the molecule not related to the ligand-binding domain.

Polyclonal antisera that we have prepared against purifed beta-1 and beta-2 adrenergic receptors effectively inhibit the binding of radioligand by receptor. Antireceptor antibodies generated against pure beta-1 receptor of rat fat cells inhibit the specific binding of cyanopindolol to rat fat cell membranes (Fig. 4). The equilibrium bind-ing data have been transformed to a Scatchard-Rosenthal plot. Preimmune serum or serum from animals immunized with unrelated proteins does not alter the interaction of the radioligand with the beta-adrenergic receptors in these membranes. Similar data are obtained using the radioligand dihydroalprenolol and cell membranes from S49 mouse lymphoma cells. The reduction in radioligand binding by the high-titer antisera or purified IgG fractions is nearly complete, even when used at dilutions of 1:400. As shown in Fig. 4, the loss in binding can be accounted for by a reduction in the maximal binding activity with no significant change in the affinity of the remaining sites for the radioligand. It is reasonable to suspect that polyclonal antisera to beta-adrenergic receptors also contain populations of antibodies against domains of the receptor other than those that constitute or somehow affect the ligand-binding domain.

Antibodies to beta-adrenergic receptors also can be screened using detergent-dispersed, rather than membrane-bound, receptor preparations. The analysis can be performed in detergent extracts after immunoprecipitation with the antisera, using the loss of residual bind-ing of the extract as an index of specific immunoprecipitation of the receptor (Fraser and Venter, 1982). Alternatively, the ability of the antibody to block binding to detergent-dispersed receptor preparations can be assayed directly. Our experience suggests that, unless detergent concentration and ionic strength of the mixtures are properly controlled, this approach is particularly prone to experimental artifacts. Cyano-pindolol is the radioligand of choice for these purposes. However, the concentration of digitonin must be reduced to 0.05% or less in extracts of membranes if cyanopindolol is to be employed. This restriction generally precludes the use of crude extracts with specific activities less than 0.5 pmol receptor/mg protein in 1% digitonin, because the signal-to-noise ratio is not favorable in these extracts diluted more than 1:20.

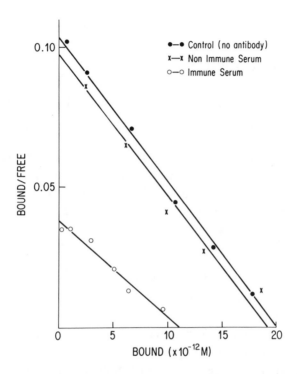

Fig. 4. Radioligand binding-based assay of antibodies to beta-adrenergic receptors. Antibodies from mice immunized with pure beta-adrenergic receptors were incubated with rat fat cell membranes replete with beta-adrenergic receptors, and the membranes then were probed with the high-affinity beta-adrenergic antagonist radioligand, [^{125}I] cyanopindolol. Binding was assayed at concentrations of radioligand from 1–1500 pmol, and the data are transformed to a Scatchard-Rosenthal plot. The slopes of the lines are the negative reciprocals of the affinity constants (K_a) for the interaction between the radioligand and the receptor. The intercept on the X-axis is equal to the maximal binding capacity of the receptors in the preparation being analyzed. Note that the immune serum specifically inhibits binding of the radioligand to the receptors in these membranes. The data transformation shows that this decline in binding is a result of the loss of receptor sites, with no apparent change in the affinity of the remaining receptors for the radioligand.

The availability of several high-affinity radiolabeled photoaffinity ligands for beta-adrenergic receptors provides an interesting alternate approach to analysis of putative antireceptor antibodies. (The utlility or the photoaffinity labeling strategy is discussed in detail elsewhere in this volume.) For routine detection of antireceptor antibodies, this approach offers little advantage over those highlighted above. However, when multiple molecular forms of receptor are identified by photoaffinity labeling (Bahouth et al., 1986) this approach may be profitable for probing novel receptor species with antireceptor antibodies well-characterized in other systems. Study of the immunologic crossreactivity of antireceptor antibodies for various forms of the receptor or of receptor fragments retaining the ligand-binding domain may provide a means of mapping the ligand-binding domain of beta-adrenergic receptors using photoaffinity labeling.

3.3. Enzyme-Linked Immunosorbent Assay (ELISA)

The use of enzyme-linked immunosorbent assay (ELISA) for the detection of antibodies offers several distinct advantages over the other methods discussed in this section. The conventional protocol (*see* Fig. 5A) immobilizes the antigen to a solid phase, such as polystyrene or polypropylene surfaces of wells of microtiter plates. After immobilization of the antigen, the surfaces are "blocked" with a solution containing a protein, such as bovine serum albumin, to coat any remaining "sticky" surfaces that might nonselectively bind the antibody used to probe the wells in the next step. The immobilized antigen is first probed with the primary antiserum or hybridoma supernatant under conditions that permit antibody–antigen complex to form (Fig. 5, step 2). The presence of an antibody to the antigen is detected by the use of a second antibody directed to the constant region of the primary antibodies. The second antibody is coupled to an enzyme, such as horseradish peroxidase or bovine alkaline phosphatase (Johnstone and Thorpe, 1982). After incubation with the second anti-body, the wells are washed extensively and a solution containing a colorless substrate of the enzyme is added. The enzyme catalyzes the generation of a colored or fluoroscent product localized in "positive" wells that contain antigen-antibody complexes (Fig. 5, step 3).

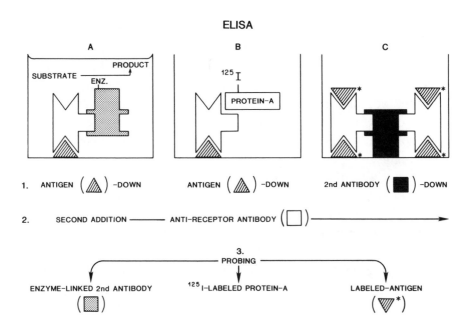

Fig. 5. Analysis of antireceptor antibodies using ELISA. (A) In the conventional ELISA, the antigen (receptor in this case) is immobilized to the well of a microtiter plate (step 1) and probed first with a primary antisera or hybridoma supernatant (step 2). The analysis of an immune complex is performed using a second antibody that is directed against the constant region of the primary antibodies and is coupled to an enzyme. The presence of the enzyme is detected by enzyme-catalyzed conversion of a colorless substrate to a colored or fluorogenic product. (B) An alternative to the use of second antibody coupled to enzyme is the use of radiolabeled Protein-A. Protein-A also recognizes the constant region of the primary antibody, and its presence can be detected by simple measure of radioactivity in the microtiter well following the wash procedures (step 3). (C) The "piggyback" assay is performed by immobilizing the second antibody to the well (step 1). Antibodies in the primary antisera or hybridoma supernatants are immobilized through their constant regions to the plate. The plates are then probed with radiolabeled antigen (receptor in this case) to identify "positives."

Foremost among the advantages of the ELISA is its high sensitivity, which reflects the amplification of signal produced by the
enzyme linked to the second antibody. Another advantage of the
ELISA is its format, which can be readily applied to the screening of
several hundred hybridoma supernatants in an afternoon. Finally, with
the use of pure antigen, the ELISA provides high specificity and a
reliable method of identification of antibodies to a particular antigen.
This lack of ambiguity greatly simplifies the further characterization
of sera or hybridoma supernatants, particularly when a large number
of samples must be screened. For these reasons, conventional ELISA
is the mainstay of current hybridoma technology.

The utility of the ELISA is dependent upon efficient immobilization of antigen to a solid support (Fig. 5). Lack of uniform, efficient coating of the solid support with antigen almost precludes successful
use of this technique. For beta-adrenergic receptors, immobilization
by conventional means has been a problem. We have used polypropylene and polystyrene plates as the solid phase for ELISA. Following
the procedures recommended by the commercial vendors, we were
unsuccessful in achieving uniform, high-efficiency coating of the wells.
The presence of digitonin, required for solubility of these membrane
proteins, appears to interfere with the protein–surface interaction.
Success was achieved using polypropylene plates (Immulon, Dynatech)
and a protocol described by Beisiegel et al. (1981a) in which the receptor
is fixed with 50% methanol (Moxham et al., 1985a, 1986a,b).
Alternatively, the wells of the microtiter plates can first be treated with
glutaraldehyde or coated with poly-L-lysine. This enhances the coating
of the wells with antigen. Optimal results with beta-adrenergic receptors
in digitonin have been obtained, however, using methanol fixation as
described above. Nitrocellulose has also been used as a matrix for the
immobilization of membrane proteins for ELISA (Roof et al., 1985).
Commercially available manifolds make the immobilization of soluble
antigens to nitrocellulose by vacuum filtration feasible for up to 96
samples at a time.

We have employed ELISA techniques to screen hybridoma
supernantants as well as antisera for antibodies to beta-adrenergic
receptors (Fig. 6). Antisera were raised in mice immunized with pure

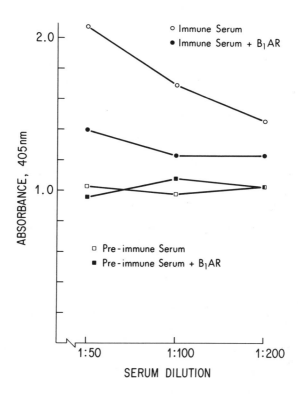

Fig. 6. ELISA of antibodies to beta-adrenergic receptors. Purified beta-adrenergic receptor was immobilized to wells of microtiter plates with 50% methanol. The wells were incubated with serum from preimmune mice and mice immunized with pure beta-1 adrenergic receptor. Competition for the antibody interaction with immobilized receptor was performed by prior incubation of the sera with purified beta-1 adrenergic receptor prior to addition of this antisera to the wells. Goat antimouse second antibody coupled to horseradish peroxidase was added, and the product formed from added substrate was determined at 405 nm. Note that immune serum specifically recognizes the immobilized receptor. Prior incubation of the immune serum with purified antibody reduces the positive signal from the ELISA. For further details consult the text.

Malbon, Moxham, and Brandwein

beta-1 adrenergic receptors administered as an antigen–cell conjugate with SP2/0 cells. The antiserum from an early bleed was screened using purified beta-adrenergic receptor isolated from rat fat cells and immobilized to microtiter plates. Goat antimouse antibody coupled to horseradish peroxidase was used to identify antireceptor antibodies. Product formation was measured spectrophotometrically. Serum from mice immunized with receptor, but not the serum from preimmune mice or mice immunized with an unrelated antigen (not shown in Fig. 6) gave a strong positive signal in the ELISA. The signal obtained from this antiserum of an early bleed was observed up to a serum dilution of 1:200. A competitive ELISA, in which the ability of pure antigen to compete with the immobilized antigen for antireceptor antibody, was also employed to test further the specificity of a mouse antiserum (Fig. 6). The signal generated by the antireceptor antiserum could be inhibited by treating the antiserum with pure receptor prior to the ELISA; the signal resulting from preimmune serum was not altered by prior treatment with pure antigen (Fig. 6).

A major advantage in the hybridoma approach to the production of monoclonal antireceptor antibodies discussed earlier is the elimination of need for essentially pure antigen for the immunizations. The screening of hybridoma by ELISA using purified, immobilized antigen distinguishes those clones producing antibodies to the receptor from those producing unwanted antibodies to contaminants present in the immunogen mixture. This "dirty in/clean out" strategy relies critically on the ability to identify true "positives" rapidly from a large number of samples. A typical fusion using a single spleen from an immunized animal would normally produce about 1000 clones. The conventional design of the plates used in an ELISA is a three-well array. Antigen is included in the first two wells, providing duplicate determinations. A third well, to which no antigen is added, serves as a control for nonspecific adsorption of antibody to the wells. This procedure would be routinely performed at least twice as the first screen of the products of the entire fusion. Clones identified as "positive" would be subcloned and expanded (Fig. 1) and assayed by ELISA and competitive ELISA at each step. Consequently, this protocol requires enough antigen to coat 2000–4000 wells. Although we have had success using as little as 5 ng protein of pure receptor to coat the wells, 50-ng quantities are preferred when supernatants of hybridoma that may have low titer are to be screened. Thus, under ideal conditions, 100–200 μg of pure receptor are required to screen a single fusion. Minimally 10–20 μg of pure receptor would be required in total when operating near the limit of sensitivity and reliability. Since success at producing useful hybridoma is a "numbers game," that is, to obtain a panel of monoclonal antibodies to a receptor may require the screening of products of 10 or more separate fusions (Roth et al., 1982), 50–100 μg protein (1–2 nmol) is not an overly conservative estimate of the amount of pure receptor needed.

Utilization of less-pure preparations of receptor for ELISA screening compromises the "dirty in/clean out" strategy. Unless sensitive, rapid, and reliable assays of antireceptor antibody activities can be performed using some other approach, such as radioligand- or biology-based analyses, the ELISA strategy often only protracts rather than avoids the need for pure antigen. Guillet et al. (1984,1985) reported the application of ELISA-like techniques to the identification of antibodies to beta-adrenergic receptors using glutaraldehyde-fixed human A431 epidermoid cells as well as membranes of murine mastocytoma cells and turkey erythrocytes as the immobilized antigens. The approach of these investigators would be useful if the antigen employed to immunize the mice were essentially pure. The approach would be of little value if impure antigen were used as an immunogen, since the cell membranes used in the screening protocol are likely to contain many of the contaminants present in the immunogen. Selecting receptor-specific antibodies under these conditions is virtually impossible.

Methods other than the conventional ELISA also can be used to screen antisera and hybridoma supernatants for antibody activity (Fig. 4B). The use of radiolabeled Protein-A as an alternative to an enzyme-coupled second antibody offers some advantages. Proteins can be radioiodinated to a high specific activity. Protein A recognizes the constant region of most, but not all, immunoglobulins. If onsite radioiodination poses some special problem, high-quality radioiodinated Protein-A prepared especially for this purpose can be obtained from commercial suppliers. Antibodies specific for the receptor can be probed with labeled Protein-A (Fig. 4B, step 2) and detected either by autoradiography of the intact plate or by gamma-counting of the excised wells (Fig. 4B, step 3).

Another technique we have employed to screen hybridoma supernatants uses a "piggyback" approach (Fig. 4C). According to this protocol, a second antibody directed against the constant region of the primary antibodies to be screened is first immobilized to the microtiter plates (Fig. 4C, step 1). (A second antibody can be efficiently immobilized to these surfaces following the procedures recommended by the commercial vendors.) Hybridoma supernatants or antisera are then incubated with the immobilized second antibody (Fig. 4C, step 2). Detection of the antireceptor antibody activity is performed using radiolabeled antigen, receptor in this case (Fig. 4C, step 3). This technique provides amplification of the signal, based upon the polyvalent nature of the primary as well as secondary antibodies. The amounts of pure antigen required for screening can be significantly reduced by use of this approach. However, successful use of this technique depends on defining conditions that optimize the signal from the true antireceptor antibodies over the background noise resulting from nonspecific adsorption of such hydrophobic, radiolabeled molecules to other proteins and surfaces.

3.4. Analysis by Immunoprecipitation

The techniques described above for detecting antireceptor antibodies are invaluable and, when used in concert, provide a starting point for the analysis of antisera as well as hybridoma supernatants. However, each technique is limited in the sense that the nature of the antigen recognized by the antibodies cannot be described in molecular terms. Immunoprecipitation and immunoblotting provide the means for direct analysis of the molecular nature of the antigen. Immunoprecipitation provides the means of selectively isolating the antigen from a mixture of proteins that has first been radiolabeled; SDS-PAGE of the immunoprecipitate in combination with autoradiography provides the means of determining the M_r of the antigen(s). The specificity of the antibodies can be ascertained once the nature of the labeled antigen(s) that are specifically immunoprecipitated has been established. When performed under conditions of equivalence, immunoprecipitation also can provide quantitative analysis of the amount of antigen contained in a sample mixture. Subjecting immunoprecipitates to isoelectric focusing and SDS-PAGE (two-dimensional analysis) provides the isoelectric point (pI) as well as the M_r of a labeled antigen, two physicochemical properties specific for a given protein. These analyses cannot be applied conveniently to the task of screening large numbers of antisera or hybridoma supernatants. As the means to further characterize a "positive" antiserum or hybridoma supernatant in greater detail, however, immunoprecipitation analyses are indispensible.

Immunoprecipitation of antigen by monoclonal antibodies is not obtained routinely. Although capable of crosslinking antigens to form dimers, monoclonal antibodies fail to form the antigen–antibody lattice critical for efficient immunoprecipitation reactions. Polyclonal antisera, in the absence of a second antibody, also may not provide efficient immunoprecipitation of antigen. Addition of a second antibody directed against the constant region of the primary antibodies readily overcomes this problem and generates immunoprecipitates that can be collected by centrifugation.

As discussed earlier, antibodies to beta-adrenergic receptors have been detected by measuring the loss of receptors from a detergent-dispersed preparation following immunoprecipitation (Fraser and Venter, 1980; Guillet et al., 1985; Venter et al., 1984). Loss of receptor from a detergent extract by immunoprecipitation with antibodies not directed to the receptor, but rather to antigens that may specifically or nonselectively bind the beta-adrenergic receptor, however, cannot be distinguished from loss of receptor resulting from immunoprecipitation by true antireceptor antibodies. The alternative strategy, of measuring the amount of receptor immunoprecipitated by first labeling the receptor with a radioligand and then measuring the amount of radioactivity in the immunoprecipitate (Chuang, 1985; Strader et al., 1983), likewise suffers from this same potential problem.

Strader et al. (1983) and Couraud et al. (1981) reported the use of radiolabeled, purified beta-adrenergic receptors for immunoprecipitation reactions with antireceptor antibodies. Whereas Couraud et al. (1981) used excess second antibody to provide optimum immunoprecipitation, Strader et al. (1983) used ammonium sulfate precipitation to isolate immune complexes. Evaluation of antibody titer was performed by measuring the amount of radioactivity immunoprecipitated (Couraud et al., 1981; Strader et al., 1983). The success of this approach is critically dependent on both the use of pure receptor and the removal of residual radioiodinated digitonin from the receptor after the labeling reaction and prior to use of receptor in the immunoprecipitation reactions. In the absence of an additional step, such as gel electrophoresis to isolate the receptor from the immunoprecipitate, the problem of labeled digitonin being trapped in the immunoprecipitate cannot be solved and therefore severely compromises the utility of this approach. SDS-PAGE of immunoaffinity purified receptor provides another solution to this problem (Fraser and Venter, 1982).

Couraud et al. (1981) first reported the use of immunoprecipitation and gel electrophoresis in tandem to identify the nature of the antigen recognized by antibodies generated against beta-adrenergic receptors. We have used this approach to establish the specificity of antibodies raised in mice and rabbits that have been immunized with purified beta-adrenergic receptor of a single subtype. Antisera raised in mice as well as rabbits specifically immunoprecipitate the purified beta-adrenergic receptor subtypes with which these animals were immunized (Fig. 7). Pure mammalian beta-1 and beta-2 adrenergic receptors are single-chain polypeptides displaying 65–67kDa following chemical reduction of disulfide bridges and separation by SDS-PAGE (Benovic et al., 1984; Cubero and Malbon, 1984; George and Malbon, 1985; Graziano et al., 1985). The immunoprecipitates of each antireceptor antiserum display a prominent peptide with M_r identical to that of the pure receptor (Fig. 7). Analysis of immunoprecipitates of labeled antigens by simple measurements of radioactivity in the pellets must be approached with caution. In our experience, differences in the amount of label in the crude immunoprecipitates, as measured by

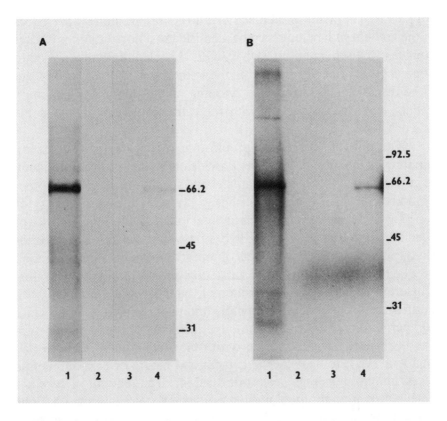

Fig. 7. Antireceptor antisera immunoprecipitate purified beta-adrenergic receptors. (A) Radioiodinated, beta-1 adrenergic receptor isolated from rat fat cells (lane 1) was incubated in buffer in the absence (lane 2) or presence of sera (1:500-fold dilution) from control mice (lane 3) or that from mice immunized with beta-1 adrenergic receptor (lane 4). Goat antimouse second antibody and control mouse ascites fluid were then added to each sample, and the resultant immunoprecipitates collected and prepared for gel electrophoresis. (B) Radioiodinated, beta-2 adrenergic receptor isolated from S49 mouse lymphoma cells (lane 1) was incubated in buffer in the absence (lane 2) or presence of sera (1:100-fold dilution) from preimmune rabbits (lane 3) or that from rabbits immunized with beta-2 adrenergic receptor (lane 4). Goat antirabbit second antibody was then added to each and the resultant immunoprecipitates collected and prepared for gel electrophoresis. The numbers on the right margins of each panel indicate the apparent mol mass (kDa) of protein standards. For further details consult the text.

simple measurements of radioactivity, do not often parallel differences in the amount of labeled receptor identified in autoradiograms of precipitates first separated by SDS-PAGE. Labeled digitonin and other nonreceptor molecules trapped in the immunoprecipitates account for these discrepancies. Repeated washing of the immunoprecipitation pellets, even in conjunction with sonication, does not always solve this problem.

The ability to immunoprecipitate receptors from complex mixtures of membrane proteins provides new opportunities to study in greater detail not only the character and specificity of antibody preparations, but also, more importantly, the structure and biology of receptor molecules. Using immunoprecipitation analysis, the titer of specific antibodies in antisera, hybridoma supernatants, or in ascites fluid can be assayed. The effects of repeated immunization of a rabbit producing anti-beta-2 receptor antisera is shown in Fig. 8. The final antiserum obtained from this animal, which was immunized with purified beta-2 adrenergic receptor, immunoprecipitates pure beta-1 adrenergic receptors in a quantitative fashion. This same antiserum has been used in conjunction with other techniques (Cubero and Malbon, 1984) to analyze the content and nature of beta-adrenergic receptors through large-scale purification. The amount of receptor detected in immunoprecipitations of receptor-containing fractions eluted from the Sepharose–alprenolol affinity matrix with (–)alprenolol agrees well with the amount of receptor determined by radioligand binding analysis of these fractions (not shown).

In combination with metabolic labeling, immunoprecipitation provides the means to investigate the biosynthesis, processing, expression, and metabolism of receptors. This approach is best exemplified by its application to the study of the metabolism of insulin receptors (for review *see* Czech, 1985). Partial purification by affinity chromatography of receptors that have been metabolically labeled is often required prior to immunoprecipitation in order to reduce the amount of background (Kasuga et al., 1983; Van Obberghen et al., 1981). In addition to insulin receptors, the receptors for LDL (Beisiegel et al., 1981a,b) and epidermal growth factor (EGF) (Schreiber et al., 1981, 1983), to name just a few, have been studied in great detail using this

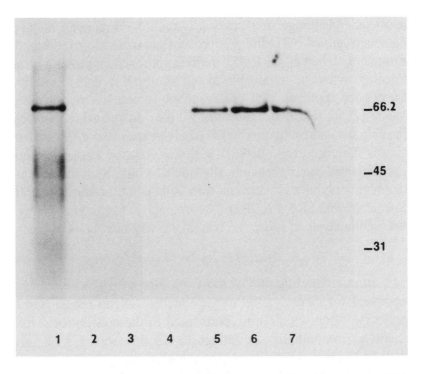

Fig. 8. Immunoprecipitation of purified beta-1 adrenergic receptor with anti-beta-2 antisera: Effects of repeated immunization on antibody titer. Anti-beta-2 antisera were obtained from a rabbit immunized repeatedly with beta-2 adrenergic receptors. Pure beta-1 adrenergic receptor isolated from rat fat cells was incubated with antisera obtained from the rabbit 3 d after the fifth booster immunization (lane 4) and at day 3, 4, and 5 after the sixth booster immunization (lanes 5, 6, and 7, respectively). Goat antirabbit second antibody was added and the resultant immunoprecipitate collected and prepared for gel electrophoresis. Lanes were arranged as follows: Lane 1— pure receptor; lane 2—receptor incubated with second antibody; lane 3— receptor incubated with control serum and then second antibody. The amount of beta-1 adrenergic receptor immunoprecipitated in each lane was determined by excising the bands and quantifying the amount of radioactivity by gamma counting. The values obtained were as follows: Lane 1—set as 100%, lane 2—7%, lane 3—7%, lane 4—20%, lane 5—68%, lane 6—100%, lane 7—76%. The numbers on the right margin of each panel indicate the apparent mol mass (kDa) of protein standards. Consult the text for further details.

approach. Although much new information on receptor metabolism has been gained by indirect approaches emphasizing the kinetics of turnover (Mahan et al., 1987), a direct approach employing metabolic labeling would be invaluable to our study of receptor metabolism. Successful application of this approach to beta-adrenergic receptors has not been reported. Recently we have succeeded in metabolically labeling beta-adrenergic receptors of high-expressing CHO transfectant clones (George et al., 1988). Cells expressing about one million receptors/cell can be metabolically labeled with [^{35}S]methionine and the detergent-solubilized receptors then isolated by immunoprecipitation in tandem with SDS-PAGE and fluorography. Employing this system, detailed analysis of receptor metabolism will now be possible.

3.5. Analysis by Immunoblotting

Immunoblotting or "Western" blotting provides a second direct approach for identification of the specific antigen recognized by an antibody. Immunoblotting, performed in the conventional manner, involves immobilizing an antigen to a solid support, such as nitro-cellulose paper, and then probing the nitrocellulose with an antibody preparation (Bers and Garfin, 1985). Like the ELISA techniques (Figs. 4A,B), the antibody–antigen complex is detected on immunoblots by the use of a second antibody that has been coupled to an enzyme. Immune complexes are made visible by the enzyme-catalyzed formation of an insoluble, colored product that precipitates on the blot. Autoradiography, in combination with the use of radiolabeled, second antibody directed against the constant region of the primary antibody or in some cases with the use of radiolabeled Protein-A, provides an alternative means for detection of immune complexes of antibody with immobilized antigen.

For most applications, the proteins applied to nitrocellulose are first separated by SDS-PAGE. The resolved proteins are transferred to the nitrocellulose passively, by diffusion, or more commonly under the influence of an electric field. This latter technique of "electroblotting" is rapid, and the apparatus required to perform the transfer is available commercially (BioRad Laboratories, Hoeffer). Several solid supports, in addition to nitrocellulose, have been developed for immunoblotting work.

Zeta-Probe® nylon membranes (BioRad Laboratories) and diazoben-zyloxymethyl (DBM)-derived cellulose membranes offer some advantages. Proteins transferred to DBM-cellulose paper, for example, can be covalently attached to this support. Treating nitrocellulose blots with glutaraldehyde likewise can improve the retention of protein by the blot during the blocking, probing, and extensive washing steps. For work with beta-adrenergic receptors, electrophoretic transfer by the method of Towbin et al. (1979), as modified by Erickson et al. (1982), onto nitrocellulose (Schleicher and Schuell, Inc., Keene, NH) offers high-efficiency transfer of the protein and acceptable levels of background for immunoblotting. To accurately monitor the efficiency of the transfer of an antigen from a gel to a blot, trace amounts of radiolabeled antigen are added to samples prior to electrophoresis on polyacrylamide gels. Autoradiography of both the blot and the residual gel after electroblotting provides the means to determine the efficiency of the transfer. Once the protein has been immobilized to the blot, as in the ELISA protocol, the blot is "blocked" using a solution containing BSA to coat any "sticky" surfaces that might indiscriminantly bind the antibodies used to probe the blot in the next step.

Nonionic detergents are used commonly to increase the stringency of the interaction between antibodies and proteins immobilized on nitrocellulose. The use of nonionic detergents poses a particular problem when the antigen being probed is a hydrophobic membrane protein like the beta-adrenergic receptor. Treatment with nonionic detergents for extended periods of time (overnight incubations), even at low concentrations, often can result in significant loss of the receptor from the blot. Air drying the blot prior to blocking and probing reduces the loss of antigen from the blot and improves the signal-to-noise ratio. Careful attention should be given to defining optimal conditions for both the blocking and probing steps, which require the presence of a nonionic detergent.

For establishing the molecular nature of the antigens recognized by an antibody, immunoblotting provides a more rapid means of analysis than immunoprecipitation used in tandem with gel electrophoresis and autoradiography. Immunoprecipitation requires a soluble or detergent-dispersed antigen for reaction. Immunoblotting, in contrast, is capable of probing antigens in crude membranes or lysates of cells separated by SDS-PAGE. Immunoblotting also affords the opportunity to manipulate the conditions of the electrophoretic separation without necessarily modifying those steps distal to the transfer to nitrocellulsose. Thus, proteins can be subjected to gel electrophoresis under denaturing as well as nondenaturing conditions. SDS-PAGE can be performed on proteins in the absence or presence of prior reduction of disulfide bridges. For beta-adrenergic receptors, this capability has been used profitably. Having established the existence of disulfide bridges in the fat cell beta-1 adrenergic receptor that is essential for the integrity of the ligand-binding domain (Moxham and Malbon, 1985), we have used immunoblotting to define the generality of this observation with regard to the structure of mammalian receptors (Moxham et al., 1986a,b). When subjected to SDS-PAGE without reduction of disulfide bridges, both beta-1 as well as beta-2 adrenergic receptors display 55kDa (Fig. 9). When either subtype of beta-adrenergic receptor was first treated with disulfide bridge reducing agents and then SDS-PAGE run under disulfide bridge reducing conditions, an 65–67kDa was observed in autoradiograms (Fig. 9A) or immunoblots (Fig. 9B). In addition, beta-adrenergic receptors *in situ* behave as species with 55kDa (Moxham et al., 1988). The role of disulfide bridge/free sulfhydryl exchange in the process of receptor "activation" by agonist is presently under intense investigation in several laboratories (Moxham and Malbon, 1985; Pedersen and Ross, 1985; Malbon et al., 1987; Moxham et al., 1988; Dixon et al., 1987; Fraser et al., 1989; Liggett et al., 1989; Karnik et al., 1988).

Immunoblotting has been used to analyze structural features of membrane-bound receptors. Beisiegel et al. (1981a,b) employed immunoblotting techniques to establish the isoelectric point of the receptor for low-density lipoproteins in cell lines with alterations in cholesterol metabolism. We have examined the immunologic cross-reactivity of antisera raised against a beta-adrenergic receptor of one subtype for receptors of the other, pharmacologically distinct subtype (Moxham et al., 1986a). Antisera raised specifically against either beta-1 or beta-2 adrenergic receptors immunoprecipitate receptors of

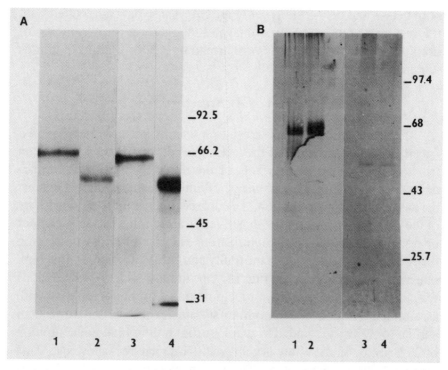

Fig. 9. Immunoblotting identifies beta-adrenergic receptors in the presence and absence of chemical reduction of disulfide bridges. (A) Radioiodinated, pure beta-1 adrenergic receptor isolated from rat fat cells and radioiodinated, pure beta-2 adrenergic receptor isolated from S49 mouse lymphoma cells were subjected to gel electrophoresis after treatment with (lanes 1 and 3) or without (lanes 2 and 4) 20 mM dithiothreitol. (B) Beta-1adrenergic receptors purified from rat fat cells (100 ng/lane) and beta-2 adrenergic receptors purified from S49 mouse lymphoma cells (100 ng/lane) were subjected to gel electrophoresis after treatment with (lanes 1 and 2) or without (lanes 3 and 4) 20 mM dithiothreitol. The resolved proteins were then electrophoretically transferred from the gels to nitrocellulose. The nitrocellulose blots were incubated with rabbit anti-beta-2 antiserum, washed and then incubated with goat antirabbit antiserum coupled to alkaline phosphatase. Bands were visualized by incubation with enzyme substrate. The numbers on the right margin of each panel indicate the apparent mol mass (kDa) of protein standards. For further details consult the text.

both beta-1 and beta-2 subtypes (Fig. 10). These data agree well with the results of other structural analyses, suggesting a high degree of homology between these two pharmacologically distinct peptides (Graziano et al., 1985).

3.6. Purification of Antireceptor Antibodies

Efforts to further characterize antisera that have been scored "positive" by one or more of the techniques discussed earlier are often frustrated by nonspecific effects of the normally occurring components of whole sera or the presence of antibodies not directed against the antigen under investigation. The degree of reliability and sensitivity of the assays used to identify "positive" antisera often dictates whether or not further detailed characterization is a useful pursuit. Having made the decision to improve the utility and purity of a whole antiserum, one can proceed as shown in Fig. 11. One of most widely used methods for such enrichment employs precipitation of immunoglobulins with a saturated solution of ammonium sulfate. A solution of ammonium sulfate (45% w/v) is added to whole serum until a final concentration of 27% ammonium sulfate is achieved. This mixture is allowed to stand overnight at room temperature. The immunoglobulin-enriched pellet thus formed is collected by centrifugation and then suspended in phosphate-buffered saline. Exhaustive dialysis of this solution against phosphate-buffered saline is required to eliminate residual ammonium sulfate. Some antibodies will be inactivated by this treatment, but some improvement in the purity of the immunoglobulin fraction is achieved.

Adsorption of whole antisera or immunoglobulin fractions to Protein-A immobilized to Sepharose provides further enrichment of IgG fractions. Chromatography of sera on DEAE-AffiGel Blue matrix provides IgG-enriched fractions with excellent recoveries and little manipulation. Either of these approaches can provide antibodies of increased purity that can then be further analyzed with regard to their specificity and titer by techniques described above.

Affinity chromatography of antibodies on antigens immobilized to insoluble matrices provides rapid, highly selective enrichment of the antibodies directed against the antigen. In practice, construction

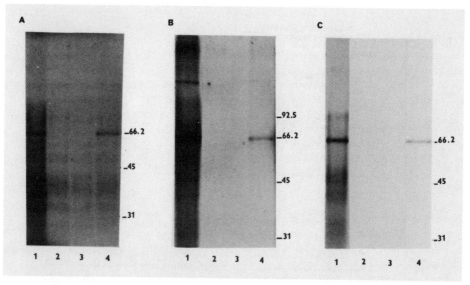

Fig. 10. Antisera raised specifically against either beta-1 or beta-2 adrenergic receptors immunoprecipitate both receptor subtypes. Pure beta-2 adrenergic receptors isolated from rat liver (A) as well as from S49 mouse lymphoma cells (B) were incubated with the mouse anti-beta-1 antiserum. Pure beta-1 adrenergic receptor isolated from rat fat cells was incubated with rabbit anti-beta-2 antisera (C). Goat antimouse or goat antirabbit second antibody were added and the resultant immunoprecipitates collected and prepared for gel electrophoresis. Lane arrangement is as follows: Lane 1; pure receptor, lane 2; receptor incubated with second antibody, lane 3; receptor incubated with control serum and then second antibody, lane 4; receptor incubated with immune serum and then second antibody. The numbers on the right margin of each panel indicate the apparent mol mass (kDa) of protein standards. Consult the text for further details.

of even a small amount of affinity matrix requires the availability of at least 0.1-mg quantities of antigen. The antigens can be coupled to cyanogen bromide-activated derivatives of Sepharose. Some epitopes may be lost in this process through participation in chemical coupling to the matrix, through chemical inactivation, or through steric hindrance by the matrix. Antibodies to these epitopes will be lost, i.e., these antibodies will not adsorb to the matrix. In the case of beta-adrenergic receptors, most researchers are reluctant to commit nanomole

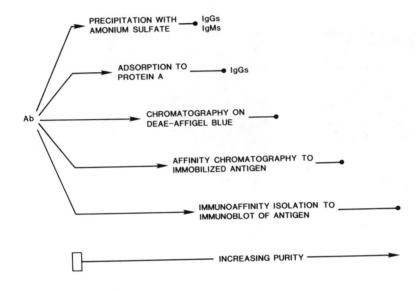

Fig. 11. Approaches to the purification of antibodies to receptors. For details about these approaches, consult the text.

quantities of pure receptor to this pursuit. Synthetic peptides of sequences used to generate antipeptide antibodies, in contrast, can be prepared, immobilized to a suitable matrix, and employed for affinity-purification of these antibodies *(see below)*. A useful alternate approach, and one we have used with varying success, is the immunoaffinity isolation of antigens from immunoblots. We used a modification (Smith and Fisher, 1984) of the original method of Olmsted (1981) to adsorb and then release antibodies to beta-adrenergic receptors from whole serum. Pure beta-adrenergic receptor (20–50 µg) subjected to SDS-PAGE is electrophoretically transferred to nitrocellulose. The band of resolved receptor is identified and excised, and used as a template for adsorption of antibodies from antiserum. Following extensive washing steps, the antibodies are released from the nitrocellulose by transient exposure to low pH (Smith and Fisher, 1984). The antibody fraction is rapidly returned to neutral pH. Antibodies specific for antigenic sites lost by denaturation of the receptor protein as well as those blocked by the interaction of the receptor with the nitrocellulose will not be purified by this technique. In our experience, the blots of receptor can be utilized several times without significant loss or inactivation of receptor.

4. Immunologic Analyses of Receptor Structure and Biology

Opportunities for analysis of the fine-structure and the biology of the beta-adrenergic receptors has expanded with the availability of well-defined, highly specific antireceptor antibodies. Monoclonal and polyclonal antibodies raised against pure beta-adrenergic receptors have provided opportunities to analyze biochemical and molecular details of receptors when the large-scale purification of receptor for this purpose is not feasible. The nature (M_r) and steady-state levels of beta-adrenergic receptors can be adequately defined by immunoblotting in cell membrane containing as little as 5 ng equivalent (<100 fmol of binding) of receptor protein, depending upon the quality of antibodies employed and the degree of homology between the receptor immunogen and the receptor transferred to the blot (Moxham et al., 1986a). Using this approach, Kaveri et al. (1987) and later Cervantes-Olivier et al. (1988) were able to define the oligosaccharides differentially expressed on beta-adrenergic receptors in human epidermoid A431 cells. Moxham et al. (1988) succeeded in demonstrating that beta-adrenergic receptors display intramolecular disulfide bridges *in situ*. Analysis of the influence of permissive hormones on the status of these receptors in fat cells was accomplished using this same approach (Ros et al., 1988). These are but a few examples of how new immunological reagents have had an impact upon our knowledge of receptor structure and metabolism.

In tandem with epifluorescence and electron microscopy, anti-receptor antibodies permit the first detailed analysis of receptors in intact cells and avoid the technical liability of cell fractionation. Indirect immunofluorescence (Couraud et al., 1981; Guillet et al., 1985; Ventimiglia et al., 1987; Strader et al., 1987; Kaveri et al., 1987; George et al., 1988; Wang et al., 1989a,b,c) and immunocytochemical (Strader et al., 1983; Aoki et al., 1987; Alho et al., 1988; Itami et al., 1987; Strader et al., 1987; Wanaka et al., 1989a,b; Ishimoto et al., 1989; Takano et al., 1989; Sato et al., 1989) analyses have provided fundamental information on receptor localization and receptor biology. Much of this new information could not have been obtained by any other means. In the case of desensitization of receptors in response to chronic stimulation of agonists, recent work has revealed receptor translocation in some systems (Alho et al., 1988; Kaveri et al., 1987), but not in all (Wang et al., 1989a,b). We have demonstrated that, although not capable of binding radiolabeled antagonist, receptors of desensitized A431 cells are available for binding by antireceptor antibodies in intact cell preparations (Wang et al., 1988a,b). Thus receptor translocation is not obligate for densensitization.

Most interesting was the observation that beta-adrenergic receptors, stained and observed by indirect immunofluorescence, appear as punctiform patterns that are not homogeneously distributed in cells (Strader et al., 1987; George et al., 1988; Wang et al., 1989a,b,c). These punctate structures do not appear to represent single receptors, but rather clusters of receptors (Fig. 12; George et al., 1988; Wang et al., 1989a,b,c). Immunogold staining in tandem with electron microscopy (EM), though in its infancy, will likely provide new insights into the nature, distribution, and perhaps biological significance of these receptor clusters.

Antibodies reactive with a native molecule can be raised against chemically synthesized peptides corresponding to defined regions of proteins (Lerner, 1982; Sutcliffe et al., 1983). Our understanding of the antigenic profile of proteins that emerged in the 1970s suggested that most globular proteins contain fewer than five antigenic sites (approx one site/5000 dalton of mass) and that these sites were generated by tertiary structure of both continuous and discontinuous regions of the molecule. Recent studies employing site-directed antipeptide antibodies has expanded this conceptual framework. Eighteen of 20 chemically synthesized peptides covering 75% of influenza virus hemagglutinin HAI chain proved to be immunogenic, generating antibodies that recognized the native molecule (Lerner, 1982; Sutcliffe

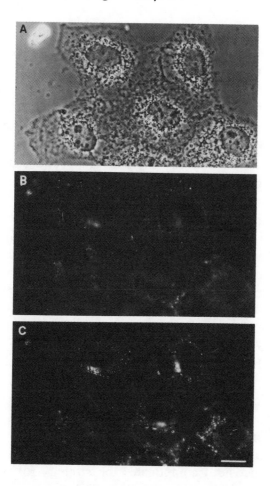

Fig. 12. Indirect immunofluorescence of beta-adrenergic receptors on
A431 cells stained with antireceptor antiserum: Analysis at different depths
of focus. Phase contrast (A) and epifluorescence (B and C) are shown.
A431 cells were grown on glass slides, fixed with 3% paraformaldehyde
and probed with anti-beta-adrenergic receptor antiserum diluted 1:50.
Rhodamine conjugated goat antirabbit IgG was used as a second antibody
diluted 1:1000. The epifluorescence image of beta-adrenergic receptors on
fixed, intact A431 cells were obtained at different depths of focus with a
63X Zeiss planapochromat objective lens (numerical aperture = 1.40). (B)
The focal plane is on the basal level of the cell surface. (C) The focal plane
is on the opical regions of the cell surface. Bar = 25 µm.

et al., 1983). The immunogenicity of synthetic peptides of regions of a protein may often be greater than that of the intact molecule. Antisera raised against the intact molecule may not recognize the synthetic peptides; thus these peptides can be used to generate antibodies that cannot be obtained by immunization with the native protein.

Peptides can be chemically synthesized according to the primary sequence obtained by direct chemical sequencing of the native protein or fragments derived from it. Alternatively, peptides can be synthesized according to the deduced amino acid sequence predicted from nucleic acid sequence information. Several protocols have been published that permit the isolation of nanomoles of beta-adrenergic receptor from tissues (Benovic et al., 1984) or cells in culture (George and Malbon, 1985) that are rich in receptor content. Gas-phase microsequencing techniques now permit the analysis of peptides with a sensitivity in the 1–10 pmol range (Hunkapiller and Hood, 1983). Based on these two facts, one would have expected that a great deal of the primary sequence of beta-adrenergic receptors would have been elucidated prior to 1986. Unfortunately, real progress in the direct chemical sequencing of these receptors has been frustratingly slow.

Sequencing low-abundance membrane proteins is a formidable challenge for several reasons. Not only is the isolation of receptor a difficult and demanding task, but the fact that membrane proteins must be purified in the presence of a detergent or surfactant (such as digitonin or octylglucoside) complicates the strategy for preparing the molecule for chemical sequencing. These detergents must be removed prior to chemical bonding of the protein to the matrix. In addition, observations from our laboratory and others suggest that the N-terminus of the beta-receptor is chemically blocked (Dixon et al., 1986; Malbon et al., 1986; Yarden et al., 1986). Thus the protein must be fragmented by enzymatic (proteinases) or chemical (cyanogen bromide, diphenylamine, and so on) means in order to uncover a free amino group required for the Edman degradation procedure. Unfortunately, the presence of detergents may inactivate proteinases, and the solubility of these membrane proteins is very low under the conditions required for chemical cleavage.

In spite of the difficulties associated with obtaining direct primary sequence information on low-abundance membrane proteins, two groups succeeded in obtaining a primary sequence of fragments of beta-adrenergic receptors. Using CNBr digestion, Dixon et al. (1986) obtained nine specific absorbance peaks on HPLC separation of the digests. Five of the peaks yielded amino-acid sequence upon N-terminal sequence analysis. The peptides, ranging in length from 9 to 34 amino acids (Dixon et al., 1986), represent about 15% of the native receptor protein. Yarden et al. (1986) employed chemical cleavage and proteolysis to fragment the turkey erthyrocyte beta-adrenergic receptor. Five peptides, comprising about 25% of the protein and ranging in length from 8 to 29 amino acids, were successfully sequenced (Yarden et al., 1986).

Once the sequence of peptide fragments from the receptor has been obtained, these peptides can be synthesized by conventional methods employing the solid-phase approach developed by Merrifield and coworkers (Marglin and Merrifield, 1970). It is important to keep in mind that the synthetic peptides alone will not likely generate acceptable titers of antibodies. Chemical coupling of the synthetic peptides to a larger molecule, such as bovine serum albumin, keyhole limpet hemocyanin, or soybean trypsin inhibitor, will generate a more suitable antigen. With this feature in mind, the addition of a cysteine residue to the N-terminus of the peptide during synthesis will provide a useful position for chemical coupling with m-maleimidobenzoyl-N-hydroxysuccinimide ester, as described by Green et al. (1982). Tyrosyl residues may also be added during the synthesis to provide a site for radioiodination if no such site exists in the peptide fragment. Yarden et al. (1986) used this strategy to prepare antibodies to synthetic fragments of the avian beta-adrenergic receptors. The antisera recognized the peptides synthesized according to predicted or chemically-determined sequence and also reacted specifically with beta-adrenergic receptors (Yarden et al., 1986).

An alternative approach to chemically coupling receptor peptides to a larger "carrier" molecule has been reported in which the immunogen for antipeptide antibodies was expressed in *Escherichia coli* (Dixon et al., 1986). A 19-amino-acid fragment of the hamster-lung beta-adrenergic receptor was isolated from a CNBr digest and subjected to *N*-terminal sequence analysis. Two oligonucleotides encoding the peptide were synthesized and cloned into a plasmid containing the gene for yeast RAS^{SC1} protein SC1. *Escherichia coli* transformed with the resultant plasmid overexpress a fusion protein (23kDa) in which the receptor peptide (2kDa) has been fused to the *N*-terminal domain of the yeast RAS^{SC1} protein (Temeles et al., 1985). Rabbits injected with the isolated fusion protein displayed antibodies that reacted with the immunogen as well as pure beta-adrenergic receptor, as determined by immunoprecipitation and immunoblotting (Dixon et al., 1986).

Cloning the gene and the complementary DNA for beta-adrenergic receptors are pursuits that both benefit from and are of benefit to the immunologic approaches discussed above (Fig. 13). Antibodies to receptors and to fragments derived from the receptor (either by cleavage of native receptor or by chemical synthesis of defined as well as predicted primary sequence) are indispensible reagents in closing the "open loop" in molecular cloning. Dixon et al. (1986) by evaluating the ability of antibodies raised against synthetic peptides to recognize pure beta-adrenergic receptor, confirmed that the determined amino acid sequences were in fact those of the receptor peptide. Oligonucleotides complementary to the DNA encoding the amino acid sequence of this peptide provided the probes necessary for the molecular cloning of the gene and the cDNA for the hamster-lung beta-adrenergic receptor (Dixon et al., 1986). Yarden et al. (1986) designed and employed oligonucleotides as hybridization probes to screen a λ gt 10 cDNA library prepared from erythrocytes of fetal turkeys. To further confirm that isolated clones carried the cDNA of the beta-adrenergic receptor, antibodies were raised against synthetic peptides that were encoded by the cDNA of interest. Antisera raised against the predicted peptides specifically recognized beta-adrenergic receptor or CNBr-generated fragments of the receptor (Yarden et al., 1986).

We have used polyclonal antibodies raised against beta-adrenergic receptors as primary reagents in the molecular cloning of G-protein-linked receptors. In 1983, Young and Davis reported on the construction of a λ phage expression vector, gt 11, in which complimentary DNA is inserted into the lacZ β-galactosidase gene and

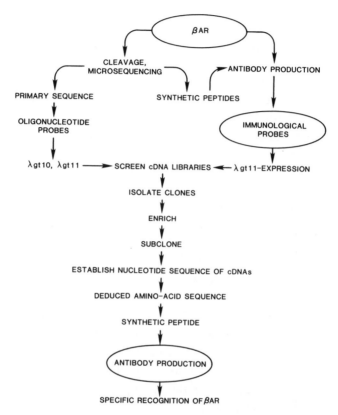

Fig. 13. Structural and molecular biology of beta-adrenergic receptors: Role of immunologic probes. This figure depicts the roles that antibodies play in the molecular cloning of beta-adrenergic receptors (βAR). Purified beta-adrenergic receptors or synthetic fragments designed from their primary structure are used as immunogens to produce antireceptor antibodies. These immunologic probes are used to screen gt 11 cDNA libraries in which the cDNAs are expressed as fusion proteins with beta-galactosidase. Positive clones are identified by immunoblotting of expressed proteins. These clones, as well as those identified by hybridization assays using oligonucleotide probes with designs based on the primary sequence information of the receptor, are enriched, subcloned, and then sequenced. Synthetic peptides based on the deduced amino acid sequence of the cloned cDNAs are prepared, coupled to a protein conjugate, and used as an immunogen. Antipeptide antibody recognition of the receptor provides compelling evidence that the isolated cDNAs encode the authentic receptor sequence. *See* the text for further details.

expressed as a fusion protein (Young and Davis, 1983). Efficient isolation of genes could be achieved by screening λ gt 11 expression libraries with antibody probes (Young and Davis, 1983). Proteins expressed in this fashion can be transferred to nitrocellulose membranes and probed using immunoblotting techniques and polyclonal or monoclonal antibodies (Fig. 13). Basal plasma membranes prepared from human placenta display the highest specific activity of both beta-1 and beta-2 adrenergic receptors reported, 4–5 pmol receptor/mg of membrane protein (Bahouth et al., 1986). A λ gt 11 expression library (Clontech, Palo Alto, CA) from human placenta containing 1.2×10^6 independent clones with an inset size range of 0.8–3.6 kilobases was screened using two polyclonal antibodies that recognize mammalian beta-2 adrenergic receptors (Malbon et al., 1986). Fifteen enriched positive clones were identified. Five of those clones display immunologic crossreactivity with more than one of the antibodies used to screen the library. In addition, these clones were shown to hybridize under high stringency with oligonucleotide probes designed from primary sequence data. These cDNAs were then subcloned into M13. The inserts were excised and sequenced (Malbon et al., 1986). The molecular cloning identified a human protein with structural features similar to those of members of the rhodopsin/beta-adrenergic receptor family (Rapiejko et al., 1988).

The molecular cloning of the beta-adrenergic receptors (Dixon et al., 1986; Yarden et al., 1986; Chung et al., 1987; Kobilka et al., 1987; Emorine et al., 1987) provided a wealth of new information on the primary sequence and probable organization of these membrane proteins. Based on the amino acid sequence data deduced from molecular cloning, it has been possible to prepare synthetic peptides corresponding to precise regions of these molecules. A wide array of research opportunities now exists for the application of well-characterized antibodies against beta-adrenergic receptors and synthetic receptor peptides. The focus of the remainder of this section will be on the fundamental questions of receptor structure and biology that challenge researchers and for which antireceptor and site-directed antipeptide antibodies are proving to be indispensible.

Struck by the functional similarities of the photopigment of bovine-rod outer segments, rhodopsin, and beta-adrenergic receptor (Cerione et al., 1985), we investigated the extent of structural similarity between these two G-protein-linked receptors using site-directed antipeptide antibodies. The strategy employed prior to the availability of molecular cloning data for beta-adrenergic receptors was to generate antibodies to the cytoplasmic domains of the rhodopsin molecule and to probe beta-adrenergic receptors for crossreactivity. We hypothesized that, if there were structural domains common to these receptor proteins, they would likely reside in the cytoplasmic domains. The first predicted cytoplasmic loop (loop 1–2) of rhodopsin is highly conserved in mammals (Nathans and Hogness, 1983), insects (Zucker et al., 1985), and in the three human color-sensitive photopigments (Nathans et al., 1986). Based upon the demonstration that rhodopsin and beta-adrenergic receptors do not maintain strict functional fidelity in vitro and can activate G-proteins other than those through which each is believed to operate in vivo (Cerione et al., 1985), we speculated that this conserved loop in rhodopsin is a G-protein binding domain of the receptor. Polyclonal antibodies were made against synthetic peptides (chemically coupled to keyhole limpet hemacyanin) that correspond to four domains of rhodposin predicted to be cytoplasmic. These antipeptide antibodies recognized rhodopsin in immunoblots. Antibodies directed against loop 1–2 and against the serine-threonine-rich region of the COOH terminus of rhodospin also recognized purified mammalian beta-2 and beta-1 adrenergic receptors (Weiss et al., 1987). Antibodies raised against membrane-associated rhodopsin, like two of the antipeptide antibodies, recognized the beta-adrenergic receptor in immunoblots of pure receptor or of crude receptor-bearing membranes. When the primary sequence of one of the beta-adrenergic receptors was revealed later, we were excited to find that sequence homologies between rhodopsin and the receptor were not colinear for the regions, but rather were made obvious only by Edmunson helical wheel projections of α-helixes of the beta-adrenergic receptor (Weiss et al., 1987). The data showed that at least two structural domains of these distinct receptors are shared.

Weiss and coworkers (1988) have extended these studies by using
antipeptide antibodies to cytoplasmic regions of rhodopsin to define
the binding domains for the G-protein, transducin. These reports
highlight the potential of using site-directed antipeptide antibodies to
map the topology of membrane receptors. Using ligand binding asaays
and defined reconstitution systems to evaluate antipeptide antibodies
to the beta-adrenergic receptor may identify specific domains of the
molecule that participate either directly or indirectly with ligand binding
or G-protein recogniton. Antibodies that we raised against the predicted
extracellular and cytoplasmic domains do not appear to influence the
ability of the receptor to bind either antagonist or agonist ligands (Wang
et al., 1989c). Mapping of the G-protein recognition domain of the
receptor by biochemical (Rubenstein et al., 1987) and recombinant
DNA (Dixon et al., 1987) techniques reveal interesting differences.
As in the case of rhodopsin (Weiss et al., 1988), site-directed antipeptide
antibodies may prove useful to mapping the G-protein recognition
domains.

The degree of structural similarity between two apparently
unrelated proteins can be investigated using the same strategy outlined
above. A fascinating example of the application and potential pitfalls
of this approach is the work of Greene and coworkers (Co et al., 1985).
These investigators prepared monoclonal and polyclonal antiidiotype
antibodies to mammalian reovirus receptor (Nepom et al., 1982;
Noseworthy et al., 1983) and examined the crossreactivity of these
antibodies with beta-adrenergic receptors (Co et al., 1985). Based on
immunological and structural data, they proposed that the two
membrane proteins were very similar, if not identical (Co et al., 1985).
Both proteins exhibited identical masses and isoelectric points; trypsin
digests of the proteins displayed identical patterns upon SDS-PAGE;
purified reovirus receptors bind beta-adrenergic ligands, and the
binding is sensitive to competition by isoproterenol; and finally and
most importantly, antireovirus receptor antibodies immunoprecipitated
putative beta-adrenergic receptor isolated from calf lung (Co et al., 1985).
As compelling as these results are for a high degree of structural similarity, if not identity, subsequent reports from the same group (Sawutz
et al., 1987) and an independent group (Choi and Lee, 1988) demon-

strate that the reovirus receptor is not the beta-adrenergic receptor and that the two proteins are structurally dissimilar (Graziano et al., 1985). Molecular cloning data likewise fail to provide a strong case to support the original observations (Dixon et al., 1986).

The primary sequence of the hamster beta-2 adrenergic receptor is displayed in Fig. 14. We have used this information to generate site-directed antipeptide antibodies to each of the domains predicted to be extracellular or cytoplasmic (Dixon et al., 1986). In most cases, immunization of 3–6 rabbits with peptide-KLH antigen was sufficient to obtain a high-titer antibody, specific to the peptide, that also recognized the intact receptor. In some cases, immunization of 10–12 animals was required to obtain a suitable antibody. These antibodies have been exhaustively characterized (Wang et al., 1989) and employed subsequently in indirect immunofluorescence studies (Wang et al., 1989a,b,c). Comparison of the staining of beta-adrenergic receptors of human epidermoid A431 cells with antisera against the holoprotein and a synthetic peptide corresponding to the first extracellular loop (*see* Fig. 14) is displayed in Fig. 15. Note that the punctiform patterns of epifluorescence are visible in cells stained with either antibody. Figure 16 shows the images of mouse L-cells stained with antireceptor antibodies. Mouse L-cells display very low levels of beta-adrenergic receptors (as determined by radioligand binding). In agreement with these data, the specific epifluorescence staining with antireceptor antibodies was found to be low (Fig. 16).

The availability of a cell line that expresses high levels of beta-adrenergic receptors would be a tremendous asset for many applications, including indirect immunofluorescence. CHO cells transfected with an expression vector harboring the cDNA for the hamster receptor under the control of the SV40 early promoter have been isolated in our laboratory. These cells express, in a stable fashion, levels of receptor from several hundred to several million receptors per cell (George et al., 1988). The expressed receptor displays normal coupling to adenylate cyclase, even for the very high-expressing clones. Indirect immunofluorescence of two of these CHO clones is displayed in Fig. 17. The cells were stained with antibodies to either native or synthetic receptor peptides. Comparison of the images obtained from

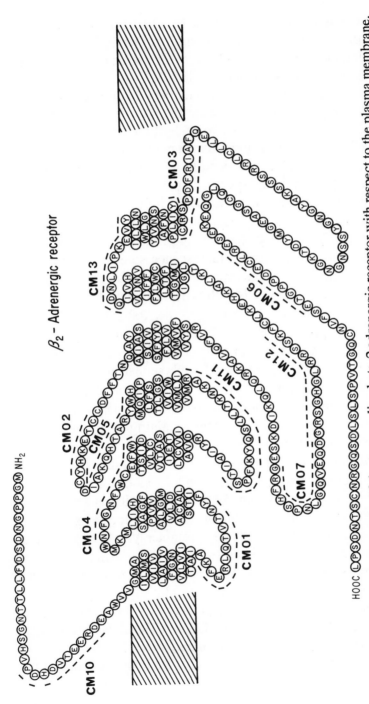

Fig. 14. Defined topography of the mammalian beta-2 adrenergic receptor with respect to the plasma membrane. Transmembrane-spanning regions are numbered sequentially from the *N*- to C-termini from one to seven (left to right). The assignment of tertiary structure is arbitrary. The primary sequence was taken from Dixon et al. (1986), based upon a model of Yarden et al. (1986) made by analogy to the known orientation of opsin. Antipeptide antibodies prepared to peptides corresponding to the regions indicated were used to establish the topography of this member of the G-protein-linked receptor family (Wang et al., 1989).

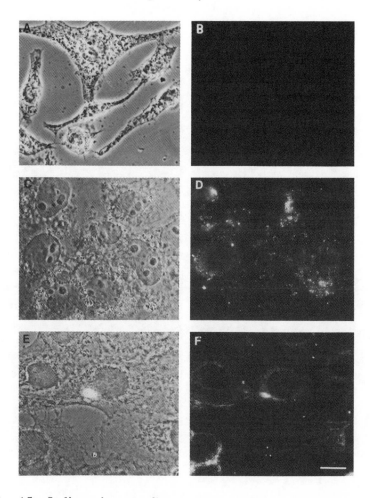

Fig. 15. Indirect immunofluorescence of A431 cells stained with antireceptor antisera. Phase-contrast (A, C, and E) and epifluorescence (B, D, and F) micrographs are shown. A431 cells were grown on glass slides, fixed with 3% paraformaldehyde and probed with preimmune serum, anti-beta-adrenergic receptor antiserum or antipeptide of beta-adrenergic receptor antiserum diluted 1:50. Rhodamine conjugated goat antirabbit IgG was used as a second antibody diluted 1:1000. (A) and (B), fixed and intact A431 cells were probed with preimmune serum. (C) and (D), fixed and intact A431 cells were probed with anti-beta-adrenergic receptor antiserum (6EN-1) (E) and (F) fixed and intact A431 cells were probed with antipeptide of beta-adrenergic receptor antiserum (CM02).

248   *Malbon, Moxham, and Brandwein*

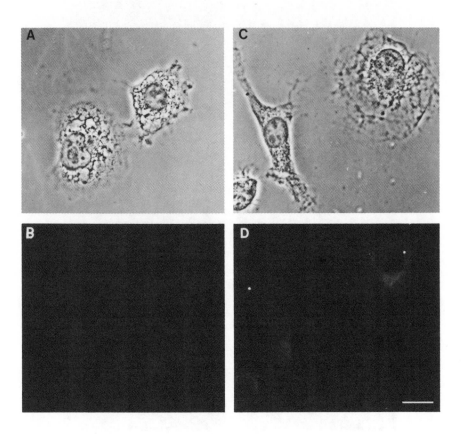

Fig. 16. Indirect immunofluorescence of mouse L cells stained with antireceptor antisera. Phase-contrast (A and C) and epifluorescence (B and D) micrographs are shown. L cells were grown on glass slides, fixed, permeabilized with detergent, and probed with preimmune serum or anti-beta-adrenergic receptor antiserum (6EN-1) diluted 1:100. Rhodamine-conjugated goat antirabbit IgG was used as a second antibody diluted 1:1000. (A and B): Fixed and permeabilized L cells were probed with preimmune serum. (C and D): Fixed and permeabilized L cells were probed with anti-beta-adrenergic receptor antiserum (6EN-1). Bar = 25 μm.

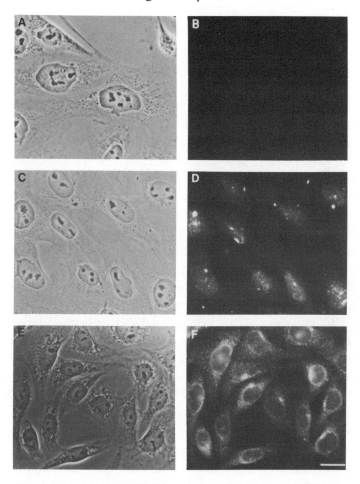

Fig. 17. Indirect immunofluorescence of beta-adrenergic receptors on CHO cells exhibiting stable expression of a cDNA that encodes the beta-2 adrenergic receptor. Phase-contrast (A, C, and E) and epifluorescence (B, D and F) micrographs. Transfected CHO clones 32d that expresses beta-adrenergic receptors were grown on glass slides, fixed, permeabilized with detergent, and probed with preimmune serum, anti-beta-adrenergic receptor antiserum or antipeptide of beta-adrenergic receptor antisera diluted 1:100. Rhodamine-conjugated goat antirabbit IgG was used as a second antibody, diluted 1:1000. (A and B): Fixed and permeabilized transfected CHO clones were probed with anti-beta-adrenergic antiserum (6EN- 1). (E and F): Fixed and permeabilized transfected CHO clones 32e were probed with antipeptide antibody CM02. Bar = 25 μm.

the CHO transfectant clones and those from A431 cells (that display one of the highest levels of beta-adrenergic receptors) highlights the advantages of the high-expressing clones for immunocytochemical analysis of the receptor.

Immunoprecipitations and immunoblotting provide an approach to comparing the structural features of beta-adrenergic receptors of a single subtype (beta-1 vs beta-2) isolated from different species and tissues, or to comparing beta-adrenergic receptors of different subtypes (beta-1 vs beta-2). Antibodies raised against beta-2 adrenergic receptors isolated from frog erythrocytes were shown to have no cross-reactivity with beta-1 adrenergic receptors of turkey erythrocytes (Strader et al., 1983). Autoantibodies for beta-2 adrenergic receptors isolated in humans also were reported to lack crossreactivity with beta-1 adrenergic receptors (Venter et al., 1980). Based on these and other data, the concept was advanced that beta-1 and beta-2 adrenergic receptors were structurally and immunologically quite distinct (Venter and Fraser, 1983). Our data suggest that, contrary to this earlier speculation about beta-1 as compared to beta-2 adrenergic receptors (Fraser and Venter, 1983), mammalian beta-1 and beta-2 adrenergic receptors are immunologically and structurally quite similar to each other (Graziano et al., 1985; Moxham et al., 1985a,1986a,b). However, using antibodies raised against peptides corresponding to sequences of the mammalian beta-2 adrenergic receptor, we were able to identify antibodies capable of differentiating the two subtypes (Bahouth et al., 1988). The basis for the subtype selectivity of several of the antipeptide antibodies was clarified when the sequences of the mammalian beta-1 adrenergic receptors (deduced by molecular cloning [Frielle et al., 1987]) were compared to those of the beta-2 adrenergic receptor used to design the synthetic peptide antigens. The corresponding sequences of the beta-1 adrenergic receptor were found to be quite dissimilar (Wang et al., 1989c). The subtype-selective antipeptide antibodies to beta-adrenergic receptors provide new reagents for identification and quantification of receptor species in systems bearing both subtypes.

Central to our understanding of beta-adrenergic receptors at the molecular level is information on the organization and topology of the receptor in the membrane. Several speculative models have been proposed, including one model in which the receptor displays seven transmembrane-spanning regions (Yarden et al., 1986) and a second with only five such domains (Chung et al., 1987). Both of these models picture the receptor as a "Chinese dragon" weaving through the membrane, head distant from tail. In contrast, based on our analysis of the status of disulfide bridges in the molecule, we proposed a "bagel" model in which the transmembrane-spanning domains 1 and 2 are in close proximity to domains 6 and 7 (Malbon et al., 1987). To test the basic assumptions of these models (which have since been applied to a large number of G-protein-linked hormone receptors), antipeptide antibodies and indirect immunofluorescence used in tandem are invaluable. Using this approach we have established, for the first time, the topography of a G-protein-linked receptor—the beta-adrenergic receptor (Wang et al., 1989c). Thus, the availability of site-directed antipeptide antibodies to the beta-adrenergic receptor provides oppotunities to analyze receptor organization and topology that cannot be easily addressed by any other current technology.

Antireceptor antibodies could provide a new set of therapeutic agents for specific uses where delivery of an antibody is feasible. Delivery via nasal spray or inhalants may provide a means of administering antibodies to the respiratory tract to treat asthma and other conditions that typically respond to treatment with beta-adrenergic ligands. Although only speculations, these are but a few areas for new drug development that may emerge from immunologic analysis of beta-adrenergic receptors.

5. Conclusions

It would have been difficult if not impossible to imagine only five years ago that our knowledge of beta-adrenergic receptors would include that obtained from the large-scale purification of both mammalian beta-1 and beta-2 adrenergic receptors as well as the isolation of the cDNAs and the gene encoding these receptors. Production of antibodies to beta-adrenergic receptors has been an arduous task, complete with many false starts. The availability of monoclonal as well as polyclonal antibodies to beta-adrenergic receptors provides us with a new set of tools and new opportunities for investigation of the fine-structure and biology of these important receptors. Our charge is now to apply these tools to the formidable problems that lie ahead with the same rigor and perseverance that has brought us to this threshold.

Acknowledgments

This work was supported by United States Public Health Service grant DK25410 from the National Institutes of Health, by Genetic Diagnostics Corporation, and by the Center for Advanced Technology, State University of New York at Stony Brook. C. C. M. is the recipient of Career Development Award K04-00786 from the National Institute of Arthritis, Metabolism, Diabetes, Digestive and Kidney Diseases, NIH. The authors would like to express their appreciation to June Moriarty for her expert assistance in preparing this manuscript and to Karen Henrickson for her expert medical illustrations that appear in this work.

References

Alho, H., Dillion-Carter, O., Moxham, C. P., Malbon, C. C., and Chuang, D.-M. (1988) Changes in immunohistochemical properties of β-adrenergic receptors in frog erythrocytes by isoproterenol-induced desensitization. *Life Sciences* **42,** 321–328.

Aoki, C., Joh, T. H., and Pickel, V. M. (1987) Ultrastructural localization of β-adrenergic receptor-like immunoreactivity in the cortex and neostriatum of rat brain. *Brain Res.* **437,** 264–282.

Bahouth, S. W., Kelley, L. K., Smith, C. H., Arbabian, M. A., Ruoho, A. E., and Malbon, C. C. (1986) Identification of a novel 76-kDa form of β-adrenergic receptors. *Biochem. Biophys. Res. Comm.* **141,** 411–417.

Bahouth, S. W. and Malbon, C. C. (1987) Human β-adrenergic receptors: Simultaneous purification of β_1- and β_2-adrenergic-receptor peptides. *Biochem. J.* **248,** 557–566.

Bahouth, S. W., Berrios, M., George, S. T., Hadcock, J. R., Wang, H. -S., and Malbon, C. C. (1988) β-adrenoceptors: New advances in purification and analysis, in *Progress in Catecholamine Research*, Liss, pp. 157–165.

Beisiegel, U., Kita, T., Anderson, R. G. W., Schneider, W. J., Brown, M. S., and Goldstein, J. L. (1981a) Immunologic crossreactivity of the low density lipoprotein receptor from bovine adrenal cortex, human fibroblasts, canine liver and adrenal gland, and rat liver. *J. Biol. Chem.* **256,** 4071–4078.

Beisiegel, U., Schneider, W. J., Goldstein, J. L., Anderson, R. G. W., and Brown, M. S. (1981b) Monoclonal antibodies to the low density lipoprotein receptor as probes for study of receptor-mediated endocytosis and the genetics of familial hypercholesterolemia. *J. Biol. Chem.* **256,** 11923–11931.

Benovic, J. L., Shorr, R. G. L., Caron, M. G., and Lefkowitz, R. J. (1984) The mammalian β_2-adrenergic receptor: Purification and characterization. *Biochemistry* **23,** 4510–4518.

Bers, G. and Garfin, D. (1985) Protein and nucleic acid blotting and immunochemical detection. *BioTechniques* **3,** 276–287.

Brandwein, H., Lewicki, J., and Murad, F. (1981) Production and characterization of monoclonal antibodies to soluble rat lung guanylate cyclase. *Proc. Natl. Acad. Sci. USA* **78,** 4241–4245.

Caron, M. G., Srinivasan, Y., Snyderman, R., and Lefkowitz, R. J. (1979) Antibodies raised against purified β-adrenergic receptors specifically bind β-adrenergic ligands. *Proc. Natl. Acad. Sci. USA* **76,** 2263–2267.

Cerione. R. A., Staniszewski, C., Benovic, J. L., Lefkowitz, R. J., Caron, M. G., Gierschick, P., Somers, R., Spiegel, A. M., Codina, J., and Birnbaumer, L. (1985) Specificity of the functional interactions of the β-adrenergic receptor and rhodopsin with guanine nucleotide regulatory proteins reconstituted in phospholipid vesicles. *J. Biol. Chem.* **260,** 1493–1500.

Cervantes-Olivier, P., Delavier-Klutchko, C., Durieu-Trautmann, O., Kaveri, S., Desmandril, M., and Strosberg, A. D. (1988) The β_2-adrenergic receptors of human epidermoid carcinoma cells bear two different types of oligosaccharides which influence expression on the cell surface. *Biochem. J.* **250,** 133–143.

Choi, A. H. C., and Lee, P. W. K. (1988) Does the β-adrenergic receptor function as a reovirus receptor? *Virology* **163,** 193–197.

Chuang, D. M. (1985) A monoclonal antibody to a membrane component that interacts with the β-adenergic receptor. *J. Cyclic Nucleotide Protein Phosphor. Res.* **10,** 281–292.

Chung, F. -Z., Lentes, K. -U., Gocayne, J., Fitzgerald, M., Robinson, D., Kerlavage, A. R., Fraser, C. M., and Venter, J. C. (1987) Cloning and sequence analysis of the human brain β-adrenergic receptor. *FEBS Lett.* **211,** 200–206.

Cleveland, W. L., Wassermann, N. H., Sarangarajan, R., Penn, A. S., and Erlanger, B. F. (1983) Monoclonal antibodies to the acetylcholine receptor by a normally functioning auto-antiidiotypic mechanism. *Nature* **305,** 56–57.

Co, M. S., Gaulton, G. N., Tominaga, A., Homcy, C. J., Fields, B. N., and Greene, M. I. (1985) Structural similarities between the mammalian β-adrenergic and reovirus type 3 receptors. *Proc. Natl. Acad. Sci. USA* **82,** 5315–5318.

Conti-Tronconi, B., Hunkapiller, M., Lindstrom, J., and Raftery, M. (1982) Subunit structure of the acetylcholine receptor from *Electrophorous electricus*. *Proc. Natl. Acad. Sci. USA* **79,** 6489–6493.

Conti-Tronconi, B. M. and Raftery, M. A. (1982) The nicotinic cholinergic receptor: Correlation of molecular structure with functional properties. *Annu. Rev. Biochem.* **51,** 491–530.

254   *Malbon, Moxham, and Brandwein*

Couraud, P. O., Delavier-Klutchko, D., Durieu-Trautmann, O., and Strosberg, A. D. (1981) Antibodies raised against β-adrenergic receptors stimulate adenylate cyclase. *Biochem. Biophys. Res. Commun.* **99,** 1295–1302.
Couraud, P. O., Lu, B. Z., Schmutz, A., Durieu-Trautmann, O., Klutchko-Delavier, C., Hoebeke, J., and Strosberg, A. D. (1983) Immunological studies of β-adrenergic receptors. *J. Cell. Biochem.* **21,** 187–193.
Cuatrecasas, P. (1972) Affinity chromatography and purification of the insulin receptor of liver cell membranes. *Proc. Natl. Acad. Sci. USA* **69,** 1277–1281.
Cubero, A. and Malbon, C. C. (1984) The fat cell β-adrenergic receptor: Purification and characterization of a mammalian β_1-adrenergic receptor. *J. Biol. Chem.* **259,** 1344–1350.
Czech, M. P. (1985) The nature and regulation of the insulin receptor: Structure and function. *Annu. Rev. Physiol.* **47,** 357–381.
Dietz, M. H., Sy, M. -S., Benacerraf, B., Nisonoff, A., Greene, M. I., and Germain, R. H. (1981) Antigen- and receptor-driven regulatory mechanisms. *J. Exp. Med.* **153,** 450–463.
Dixon, R. A. F., Kobilka, B. K., Strader, D. J., Benovic, J. L., Dohlman, H. G., Frielle, T., Bolanowski, M. A., Bennett, C. D., Rands, E., Diehl, R. E., Mumford, R. A., Slater, E. E., Sigal, I. S., Caron, M. G., Lefkowitz, R. J., and Strader, C. D. (1986) Cloning of the gene and cDNA for mammalian β-adrenergic receptor and homology with rhodopsin. *Nature* **321,** 75–79.
Dixon, R. A. F., Sigal, I. S., Rands, E., Register, R. B., Candelore, M. R., Blake, A. D., and Strader, C. D. (1987) Ligand binding to the β-adrenergic receptor involves its rhodopsin-like core. *Nature* **326,** 73–77.
Dixon, R. A. F., Sigal, I. S., Candelore, M. R., Register, R. B., Scattergood, W., Rands, E., and Strader, C. D. (1987) Structural features required for ligand binding to the β-adrenergic receptor. *EMBO J.* **6,** 3269–3275.
Emorine, L. J., Marullo, S., Delavier-Klutchko, C., Kaveri, S. V., Durieu-Trautmann, O., and Strosberg, A. D. (1987) Structure of the gene for human β-adrenergic receptor: Expression and promoter characterization. *Proc. Natl. Acad. Sci. USA* **84,** 6995–6999.
Erickson, P. F., Minier, L. N., and Lasher, R. S. (1982) Quantitative electrophoretic transfer of polypeptides from SDS polyacrylamide gels to nitrocellulose sheets: A method for their reuse in immunoautoradiographic detection of antigens. *J. Immunol. Methods* **51,** 241–249.
Farid, N. R., Briones-Urbina, R., and Bear, J. C. (1983) Graves' disease—The thyroid stimulating antibody and immunological networks. *Mol. Asp. Med.* **6,** 355–457.
Farid, N. R. and Lo, T. C. Y. (1985) Antiidiotypic antibodies as probes for receptor structure and function. *Endocr. Rev.* **6,** 1–23.
Fraser, C. M. (1989) Site-directed mutagenesis of β-adrenergic receptors. *J. Biol. Chem.* **264,** 9266–9270.
Fraser, C. M. and Lindstrom, J. (1984) The use of monoclonal antibodies in recep-

tor characterization and purification, in *Molecular and Chemical Characterization of Membrane Receptors.* Alan R. Liss, New York, pp. 1–30.

Fraser, C. M. and Venter, J. C. (1980) Monoclonal antibodies to β-adrenergic receptors: Use in purification and molecular characterization of β receptors. *Proc. Natl. Acad. Sci. USA* **77,** 7034–7038.

Fraser, C. M. and Venter, J. C. (1982) The size of the mamamlian lung β$_2$-adrenergic receptor as determined by target size analysis and immunoaffinity chromatography. *Biochem. and Biophys. Res. Commun.* **109,** 21–29.

Fraser, C. M. and Venter, J. C. (1984) Antireceptor antibodies in human disease. *J. Allergy and Clin. Immun.* **74,** 661–673.

Frielle, T., Collins, S., Daniel, K. W., Caron, M. G., Lefkowitz, R. J., and Kobilka, B. K. (1987) Cloning of the cDNA for the human β-adrenergic receptor. *Biochemistry* **84,** 7920–7924.

Gaulton, G. N., Co, M. S., Royer, H. D., and Greene, M. I. (1985) Antiidiotypic antibodies as probes of cell surface receptors. *Mol. Cell Biochem.* **65,** 5–21.

Gefter, M. L., Margulies, D. H., and Scharff, M. D. (1977) A simple method for polyethylene glycol-promoted hybridization of mouse myeloma cells. *Somatic Cell Genet.* **3,** 231–236.

George, S. T. and Malbon, C. C. (1985) Large-scale purification of β-adrenergic receptors from mammalian cells in culture. *Prep. Biochem.* **15,** 349–366.

George, S. T., Berrios, M., Hadcock, J. R., Wang, H.-Y., and Malbon, C. C. (1988) Receptor density and cyclic AMP accumulation: Analysis in CHO cells exhibiting stable expression of a cDNA that encodes the β$_2$-adrenergic receptor. *Biochem. Biophys. Res. Commun.* **150,** 665–672.

George, S. T., Arbabian, M. A., Ruoho, A. E., Kiely, J., and Malbon, C. C. (1989) High-effeciency expression of mammalian β-adrenergic receptors in baculovirus-infected insect cells. *Biochem. Biophys. Res. Comm.* **163,** 1265–1269.

Gramsch, C., Schulz, R., Kosin, S., and Herz, A. (1988) Monoclonal antiidiotypic antibodies to opioid receptors. *J. Biol. Chem.* **263,** 5853–5859.

Graziano, M. P., Moxham, C. P., and Malbon, C. C. (1985) Purified rat hepatic β$_2$-adrenergic receptor. *J. Biol. Chem.* **260,** 7665–7674.

Green, N., Alexander, H., Olson, A., Alexander, S., Shinnick, T. M., Sutcliffe, J. G., and Lerner, R. A. (1982) Immunogenic structure of the influenza virus hemagglutinin. *Cell* **28,** 477–487.

Guillet, J. G., Chamat, S., Hoebeke, J., and Strosberg, A. D. (1984) Production and detection of monoclonal antiidiotype antibodies directed against a monoclonal anti-beta-adrenergic ligand antibody. *J. Immunol. Methods* **74,** 163–171.

Guillet, J. G., Kaveri, S. V., Durieu, O., Delavier, C., Hoebeke, J., and Strosberg, A. D. (1985) β-adrenergic agonist activity of a monoclonal antiidiotypic antibody. *Proc. Natl. Acad. Sci. USA* **82,** 1781–1784.

Gullick, W., Tzartos, S., and Lindstrom, J. (1981) Monoclonal antibodies as probes of acetylcholine receptor structure. *Biochemistry* **20,** 2173–2180.

Hadcock, J. R., Wang, H.-Y., and Malbon, C. C. (1989) Agonist-induced destabilization of β-adrenergic receptor mRNA. *J. Biol. Chem.* **264,** 19928–19933.

Herrera, R., Petruzzelli, L., Thomas, N., Bramson, H. N., Kaiser, E. T., and Rosen,
O. M. (1985) An antipeptide antibody that specifically inhibits insulin receptor
autophosphorylation and protein kinase activity. *Proc. Natl. Acad. Sci. USA*
82, 7899–7903.

Hoebeke, J., Vauquelin, G., and Strosberg, A. D. (1977) The production and characterization of antibodies against β-adrenergic antagonists. *Biochem.
Pharmacol.* **27,** 1527–1532.

Homcy, C. J., Rockson, S. G., and Haber, E. (1982) An antiidiotypic antibody that
recognizes the β-adrenergic receptor. *J. Clin. Invest.* **69,** 1147–1154.

Homcy, C. J., Rockson, S. G., Countaway, J., and Egan, D. A. (1983) Purification
and characterization of the mammalian β₂-adrenergic receptor. *Biochemistry* **22,** 660–668.

Hunkapiller, M. W. and Hood, L. E. (1983) Protein sequence analysis: Automated
microsequencing. *Science* **219,** 650–659.

Ishimoto, I., Kiyama, H., Malbon, C. C., Iwahashi, Manabe, R., and Tohyama, M.
(1989) Localization of adrenergic receptors in the rat retina: An immuno-
cytochemistry study. *Neurosci. Res.* in press.

Islam, M. N., Pepper, B. M., Briones-Urbina, R., and Farid, N. R. (1983a) Biological activity of anti-thyrotropin antiidiotypic antibody. *Eur. J. Immunol.* **13,**
57–62.

Islam, M. N., Briones-Urbina, R., Bako, G., and Farid, N. R. (1983b) Both TSH
and thyroid-stimulating antibody of Graves' disease bind to a M_r 197,000
holoreceptor. *Endocrinology* **113,** 436–438.

Itami, S., Kino, J., Halprin, K. M., and Adachi, K. (1987) Immunohistochemical
study of β-adrenergic receptors in the psoriatic epidermis using an anti-
alprenolol antiidiotypic antibody. *Arch Dermatol. Res.* **279,** 439–443.

Jacobs, S., Chang, K. J., and Cuatrecasas, P. (1978) Antibodies to purified insulin
receptor have insulin-like activity. *Science* **200,** 1283–1284.

Jerne, N. K. (1974) Towards a network theory of the immune system. *Ann. Immunol.
(Inst. Pasteur)* **125C,** 373–388.

Johnstone, A. and Thorpe, R. (1982) *Immunochemistry in Practice* (Blackwell
Scientific Publishers, London).

Karnik, S. S., Sakmar, T. P., Chen, H.-B., and Khorana, H. G. (1988) Cysteine
residues 110 and 187 are essential for the formation of correct structure in
bovine rhodopsin. *Proc. Natl. Acad. Sci. USA* **85,** 8459–8463.

Kasuga, M., Hedo, J. A., Yamada, K. M., and Kahn, C. R. (1983) The structure of
insulin receptor and its subunits. *J. Biol. Chem.* **257,** 10392–10399.

Kaveri, S. V., Cervantes-Olivier, P., Delavier-Klutchko, C., and Strosberg, A. D.
(1987) Monoclonal antibodies directed against the human A431 β₂-adrenergic receptor recognize two major polypeptide chains. *Eur. J. Biochem.* **167,**
449–456.

Kelley, L. K., Smith, C. H., and King, B. F. (1983) Isolation and partial characterization of the basal cell membrane of human placental trophoblast. *Biochem. Biophys. Acta* **734**, 91–98.

Kohler, G. and Milstein, C. (1975) Continuous cultures of fused cells secreting antibody of predefined specificity. *Nature* **256**, 495–497.

Kull, F. C., Jr., Jacobs, S., Su, Y.-F., and Cuatrecasas, P. (1982) A monoclonal antibody to human insulin receptor. *Biochem. Biophys. Res. Commun.* **106**, 1019–1026.

Lerner, R. A. (1982) Tapping the immunological repertoire to produce antibodies of predetermined specificity. *Nature* **299**, 592–596.

Liggett, S. B., Bouvier, M., O'Dowd, B. F., Caron, M. G., Lefkowitz, R. J., and DeBlasi, A. (1989) Substitution of an extracellular cysteine in the β_2-adrenergic receptor enhances agonist-promoted phosphorylation and receptor desensitization. *Biochem. Biophys. Res. Commun.* **165**, 257–263.

Mahan, L. C., McKernan, R. M., and Insel, P. A. (1987) Metabolism of alpha- and beta-adrenergic receptors *in vitro* and *in vivo*. *Annu. Rev. Pharmacol. Toxicol.* **27**, 215–235.

Malbon, C. C. (1990) Purification of β-adrenergic receptors: Isolation of mammalian β_1- and β_2-subtypes, in *Receptor Purification* (Litwack, G., ed., Humana Press) in press.

Malbon, C. C., George, S. T., and Moxham, C. P. (1987) Intramolecular disulfide bridges: Avenues to receptor activation? *Trends Biochem. Sci.* **12**, 172–175.

Malbon, C. C., Moxham, C. P., Rapiejko, P. J., Bahouth, S. W., Brandwein, H., and George, S. T. (1987) The structure and biology of β-adrenergic receptors: Analysis by biochemical, immunologic, and molecular biological approaches, in *Synaptic Transmitters and Receptors* (John Wiley and Sons, New York), 239-248.

Marasco, W. A. and Becker, E. L. (1982) Antiidiotype as antibody against the formyl peptide chemotaxis receptor of the neutrophil. *J. Immunol.* **128**, 963–968.

Marglin, A. and Merrifield, R. B. (1970) Chemical synthesis of peptides and proteins. *Annu. Rev. Biochem.* **39**, 841–866.

Morgan, D. O., Ho, L., Korn, L. J., and Roth, R. A. (1986) Insulin action is blocked by a monoclonal antibody that inhibits the insulin receptor kinase. *Proc. Natl. Acad. Sci. USA* **83**, 328–332.

Moxham, C. P. and Malbon, C. C. (1985) Fat cell β_1-adrenergic receptor: Structural evidence for existence of disulfide bridges essential for ligand binding. *Biochemistry* **24**, 6072–6077.

Moxham, C. P., Cubero, A., Brandwein, H., and Malbon, C. C. (1985a) Murine polyclonal antibodies to the fat cell β_1-adrenergic receptor. *Biophys. J.* **47**, 200a.

Moxham, C. P., Graziano, M. P., Brandwein, H., and Malbon, C. C. (1985c) Mammalian β_1- and β_2-adrenergic receptors: Structural and immunological comparisons. *Fed. Proc.* **44**, 1795.

Moxham, C. P., George, S. T., Graziano, M. P., Brandwein, H., and Malbon, C. C. (1986a) Mammalian β_1- and β_2-adrenergic receptors: Immunologic and structural comparisons. *J. Biol. Chem.* **261,** 14562–14570.

Moxham, C. P., George, S. T., Brandwein, H., and Malbon, C. C. (1986b) Mammalian β-adrenergic receptors: Immunolgical analysis of native forms in membranes. *Fed. Proc.* **45,** 1569.

Moxham, C. P., Ross, E. M., George, S. T., and Malbon, C. C. (1988) β-adrenergic receptors display intramolecular disulfide bridges *in situ*: Analysis by immunoblotting and functional reconstitution. *Mol. Pharmacol.* **33,** 486– 492.

Nathans, J. and Hogness, D. S. (1983) Isolation, sequence analysis, and intron-exon arrangement of the gene encoding bovine rhodopsin. *Cell* **34,** 807–814.

Nathans, J., Thomas, D., and Hogness, D. S. (1986) Molecular genetics or human color vision: The genes encoding blue, green, and red pigments. *Science* **232,** 193–202.

Nepom, J. T., Tardieu, M., Epstein, R. L., Noseworthy, J. H., Weiner, H. L., Gentsch, J., Fields, B. N., and Greene, M. I. (1982) Virus-binding receptors: Similarities to immune receptors as determined by antiidiotypic antibodies. *Surv. Immunol. Res.* **1,** 255–261.

Noseworthy, J. H., Fields, B. N., Dichter, M. A., Sobotka, C., Pizer, E., Perry, L. L., Nepom, J. T., and Greene, M. I. (1983) Cell receptors for the mammalian reovirus. I. Syngeneic monclonal antiidiotypic antibody identifies a cell surface receptor for reovirus. *J. Immunol.* **131,** 2533–2538.

Olmsted, J. B. (1981) Affinity purification of antibodies from diazotized paper blots of heterogeneous protein samples. *J. Biol. Chem.* **256,** 11955–11957.

Owen, F. L., Ju, S. T., and Nisonoff, A. (1977) Presence on idiotype-specific suppressor T cells of receptors that interact with molecules bearing the idiotype. *J. Exp. Med.* **145,** 1559–1566.

Patrick, J. and Lindstrom, J. (1973) Autoimmune response to acetylcholine receptor. *Science* **180,** 871–872.

Pedersen, S. E. and Ross, E. M. (1985) Functional activation of β-adrenergic receptors by thiols in the presence or absence of agonists. *J. Biol. Chem.* **260,** 14150–14157.

Rapiejko, P. J., George, S. T., and Malbon, C. C. (1988) Primary structure of a human protein which bears structural similarities to members of the rhodopsin/beta-adrenergic receptor family. *Nucleic Acids Res.* **16,** 8721–8722.

Relyveld, E. H. and Ben-Efraim, S. (1981) Preparation of highly immunogenic protein conjugates by direct coupling to glutaraldehyde-treated cells: Comparison with commonly used preparations. *J. Immunol. Methods* **40,** 209– 217.

Rockson, S. G., Homcy, C. J., and Haber, E. (1980) Anti-alprenolol antibodies in the rabbit. *Circ. Res.* **46,** 808–813.

Roof, D. J., Applebury, M. L., and Sternweis, P. C. (1985) Relationships within the family of GTP-binding proteins isolated from bovine central nervous system. *J. Biol. Chem.* **260,** 16242–16249.

Ros, M., Northup, J. K., and Malbon, C. C. (1988) Steady-state levels of G-proteins and β-adrenergic receptors in rat fat cells. *J. Biol. Chem.* **263,** 4362–4368.

Roth, R. A., Cassell, D. J., Wong, K. Y., Maddux, B. A., and Goldfine, I. D. (1982) Monoclonal antibodies to the human insulin receptor block insulin binding and inhibit insulin action. *Proc. Natl. Acad. Sci. USA* **79,** 7312–7316.

Rubenstein, R. C., Wong, S. K.-F., and Ross, E. M. (1987) The hydrophobic tryptic core of the β-adrenergic receptor retains G_s regulatory activity in response to agonists and thiols. *J. Biol. Chem.* **262,** 16655–16662.

Sato, M., Kubota, Y., Malbon, C. C., and Tohyama, M. (1989) Immunohistochemical evidence that most rat corticotrophs contain β-adrenergic receptors. *Neuroendocrinology* **50,** 577–583.

Sawutz, D. G., Bassel-Duby, R., and Homcy, C. J. (1987) High-affinity binding of reovirus type 3 to cells that lack β-adrenergic receptor activity. *Life Sci.* **40,** 399–406.

Schreiber, A. B., Couraud, P. O., Andre, C., Vray, B., and Strosberg, A. D. (1980) Anti-alprenolol antiidiotypic antibodies bind to β-adrenergic receptors and modulate catecholamine-sensitive adenylate cyclase. *Proc. Natl. Acad. Sci. USA* **77,** 7385–7389.

Schreiber, A. B., Lax, I., Yarden, Y., Eshhar, Z., and Schlessinger, J. (1981) Monoclonal antibodies against receptor for epidermal growth factor induce early and delayed effects of epidermal growth factor. *Proc. Natl. Acad. Sci. USA* **78,** 7535–7539.

Schreiber, A. B., Liberman, T. A., Lax, I., Yarden, Y., and Schlessinger, J. (1983) Biological role of epidermal growth factor-receptor clustering. *J. Biol. Chem.* **258,** 846–853.

Sege, K. and Peterson, P. A. (1978) Use of antiidiotypic antibodies as cell-surface receptor probes. *Proc. Natl. Acad. Sci. USA* **75,** 2443–2447.

Shechter, Y., Maron, R., Elias, D., and Cohen, I. R. (1982) Autoantibodies to insulin receptor spontaneously develop as antiidiotypes in mice immunized with insulin. *Science* **216,** 542–545.

Shorr, R. G. L., Strohsacker, M. W., Lavin, T. N., Lefkowitz, R. J., and Caron, M. G. (1982) The adrenergic receptor of the turkey erythrocyte: Molecular heterogeneity revealed by purification and photoaffinity labeling. *J. Biol. Chem.* **257,** 12341–12350.

Sigel, M. B., Sinha, Y. N., and Vanderlaan, W. P. (1983) Production of antibodies by inoculation into lymph nodes. *Methods Enzymol.* **93,** 3–12.

Smith, D. E. and Fisher, P. A. (1984) Identification, developmental regulation, and response to heat shock of two antigenically related forms of a major nuclear envelope protein in *Drosoohila* embryos: Application of an improved method for affinity purification of antibodies using polypeptides immobilized on nitrocellulose blots. *J. Cell Biol.* **99,** 20–28.

Strader, C. D., Pickel, V. M., Joh, T. H., Strohsacker, M. W., Shorr, R. G. L., Lefkowitz, R. G., and Caron, M. G. (1983) Antibodies to the β-adrenergic receptor: Attenuation of catecholamine-sensitive adenylate cyclase and demonstration of postsynaptic receptor localization in brain. *Proc. Natl. Acad. Sci. USA* **80,** 1840–1844.

Strader, C. D., Sigal, I. S., Blake, A. D., Cheung, A. H., Register, R. B., Rands, E., Zemcik, B. A., Candelore, M. R., and Dixon, R. A. F. (1987) The carboxyl terminus of the hamster β-adrenergic receptor expressed in mouse L cells is not required for receptor sequestration. *Cell* **49,** 855–863.

Strosberg, A. D. (1984) Antiidiotypic antibodies as immunological internal images of hormones, in *Idiotypy in Biology and Medicine* (Academic), pp. 365–383.

Sutcliffe, J. G., Shinnick, T. M., Green, N., and Lerner, R. A. (1983) Antibodies that react with predetermined sites on proteins. *Science* **219,** 660–666.

Takano, T., Kubota, Y., Malbon, C. C., and Tohyama, M. (1989) β-adrenergic receptors in the vasopressin-containing neurons in the paraventricular and supraoptic nucleis of the rat. *Brain Research* **499,** 174-179.

Temeles, G. L., Gibbs, J. B., D'Alonzo, J. S., Sigal, I. S., and Scolnick, E. M. (1985) Yeast and mammalian ras proteins have conserved biochemical properties. *Nature* **313,** 700–703.

Towbin, H., Staehelin, T., and Gordon, J. (1979) Electrophoretic transfer of proteins from polyacrylamide gels to nitrocellulose sheets: Procedure and some applications. *Proc. Natl. Acad. Sci. USA* **76,** 4350–4354.

Tzartos, S. J. and Lindstrom, J. M. (1980) Monoclonal antibodies used to probe acetylcholine receptor structure: Localization of the main immunogenic region and detection of similarities between subunits. *Proc. Natl. Acad. Sci. USA* **77,** 755–759.

Van Obberghen, E., Kasuga, M., Le Cam, A., Hedo, J. A., Itin, A., and Harrison, L. C. (1981) Biosynthetic labeling of insulin receptor: Studies of subunits in cultured human IM-9 lymphocytes. *Proc. Natl. Acad. Sci. USA* **78,** 1052–1056.

Venter, J. C., Fraser, C. M., and Harrison, L. C. (1980) Autoantibodies to β_2-adrenergic receptors: A possible cause of adrenergic hyporesponsiveness in allergic rhinitis and asthma. *Science* **207,** 1361–1362.

Venter, J. C. and Fraser, C. M. (1981) The development of monoclonal antibodies to β-adrenergic receptors and their use in receptor purification and characterization, in *Monoclonal Antibodies in Endocrine Research* (Fellows, R. and Eisenbarth, G., eds.), Raven, New York, pp. 119–134.

Venter, J. C. and Fraser, C. M. (1983) The structure of alpha- and beta-adrenergic receptors. *Trends Pharmacol. Sci.* **4,** 256–258.

Ventimiglia, R., Greene, M. I., and Geller, H. M. (1987) Localization of β-adrenergic receptors on differentiated cells of the central nervous system in culture. *Proc. Natl. Acad. Sci. USA* **84,** 5073–5077.

Wanaka, A., Kiyama, H., Murakami, T., Matsumoto, M., Kamada, T., Malbon, C. C., and Tohyama, M. (1989) Immunocytochemical localization of β-adrenergic receptors in the rat brain. *Brain Res.* **485,** 125-140.

Wanaka, A., Malbon, C. C., Matsumoto, M., and Tohyama, M. (1989) Presence of catecholamine axon-terminals which contain β-adrenergic receptor in the periventricular zone of the rat hypothalamus. *Brain Res.* **479,** 190–193.

Wang, H.-S., Berrios, M., and Malbon, C. C. (1988a) Indirect immunofluorescence localization of β-adrenergic receptors and G-proteins in human epidermoid carcinoma A431 cells. *Biochem. J.* **263,** 519–533.

Wang, H. -S., Berrios, M., and Malbon, C. C. (1989b) Localization of β-adrenergic receptors in A431 cell *in situ*: Effect of chronic exposure to agonist. *Biochem. J.* **263,** 533–538.

Wang, H. Y., Lipfert, L., Malbon, C. C., and Bahouth, S. (1989c) Site-directed anti-peptide antibodies define the topography of the β-adrenergic receptor. *J. Biol. Chem.* **264,** 14424–14431.

Wasserman, N. H., Penn, A. S., Freimuth, P. I., Treptow, N., Wentzel, S., Cleveland, W. L., and Erlanger, B. F. (1982) Antiidiotypic route to antiacetylcholine receptor antibodies and experimental myasthenia gravis. *Proc. Natl. Acad. Sci. USA* **79,** 4810–4814.

Weiss, E., Hadcock, J., Johnson, G. L., and Malbon, C. C. (1987) Antipeptide antibodies directed against cytoplasmic rhodopsin sequences recognize the β-adrenergic receptor. *J. Biol. Chem.* **262,** 4319–4323.

Weiss, E. R., Kelleher, D. J., and Johnson, G. L. (1988) Mapping sites of interaction between rhodopsin and transducin using rhodopsin antipeptide antibodies. *J. Biol. Chem.* **263,** 6150–6154.

Wrenn, S. and Haber, E. (1979) An antibody specific for the propranolol binding site of cardiac muscle. *J. Biol. Chem.* **254,** 6577–6582.

Yarden, Y., Rodriguez, H., Wong, S. K.-F., Brandt, D. R., May, D. C., Burnier, J., Harkins, R. N., Chen, E. Y., Ramachandran, J., Ullrich, A., and Ross, E. M. (1986) The avian β-adrenergic receptor: Primary structure and membrane topology. *Proc. Natl. Acad. Sci. USA* **83,** 6795–6799.

Young, R. A. and Davis, R. W. (1983) Efficient isolation of genes by using antibody probes. *Proc. Natl. Acad. Sci. USA* **80,** 1194–1198.Zucker, C. S., Cowman, A. F., and Rubin, G. M. (1985) Isolation and structure of a rhodopsin gene from *D. melanogaster. Cell* **40,** 851–858.

Zucker, C. S., Cowman, A. F., and Rubin, G. M. (1985) Isolation and structure of a rhodopsin gene from D. melanogaster. *Cell* **40,** 851–858.

CHAPTER 6

Autoradiographic Studies of Beta-Adrenergic Receptors

Barry B. Wolfe

1. Introduction

A great deal of information concerning the properties, functions, and regulation of beta-adrenergic receptors has come from studies utilizing either whole organs or homogenized or soluble preparations from organs (Wolfe et al., 1977; Lefkowitz et al., 1984; Wolfe and Molinoff, 1988). These types of experiments have led to our current understanding of how these receptors are functionally coupled to adenylate cyclase and how tissues respond to either increases or decreases in the level of receptor stimulation (often by changing the number or properties of the receptors). These types of studies, however, do not yield information regarding the localization of receptors, except on a relatively large anatomical scale where, commonly, entire tissues (e.g., cerebral cortex, heart, and so on) are used to study receptors. Additionally, studies utilizing radioligand binding to homogenized preparations of tissues require relatively large amounts of tissue. In response to these types of problems, the methodology of receptor localization using autoradiographic techniques has been developed. Autoradiography is a method, described in detail below, of localizing and quantifying neurotransmitter receptors in thin sections of tissues. Briefly, autoradiography involves the incubation of slide-

The β-Adrenergic Receptors Ed.: J. P. Perkins ©1991 The Humana Press Inc.

mounted tissue sections with a radioligand that, hopefully, binds only to the receptor of interest, in this case beta-adrenergic receptors. Incubation is followed by exposure of the labeled sections to a film emulsion that is sensitive to the bound radioactivity. The exposed film emulsion is then processed, resulting in silver grains being developed where the radioligand was apposed. Correlations can be found that relate the density of silver grains in the film emulsion to the exact amount of radioactivity bound to a given amount of tissue. Thus, using such techniques, receptors can be localized to areas as small as 5–10 μm in diameter. This can be done in tissue sections as thin as 6–10 μm, allowing receptor measurements to be performed on samples that weigh as little as 1 ng. This is several orders of magnitude more sensitive than methods utilizing radioligand binding to homogenized preparations. The price that is usually paid for such increases in sensitivity is a tremendous increase in the difficulty in performing and analyzing experiments. Additionally, data generated using autoradiographic methods are often associated with a greater degree of assay variability than, for example, data generated using homogenized tissue preparations. Thus, in general, only experiments requiring increased anatomical resolution and/or sensitivity should utilize this technique.

2. Methodology

2.1. Preparation of Tissue

Commonly, organs such as brain, kidney, or heart are collected immediately following sacrifice or, alternatively, are first perfused *in situ* with either saline or low concentrations of fixative (Lew and Summers, 1985). Sucrose (5.5% w/v) added to the perfusion fluid for cryoprotection has been recommended (Kuhar, 1985). Organs are then rapidly frozen onto cryostat chucks using liquid nitrogen or dry ice. Tissues can be stored at –70°C for several weeks or months. The tissues are then cut into thin sections using a cryostat/microtome at a thickness of 6–32 μm. The thinner the section, the smaller the block of tissue being cut must be. Thus, for 6 μm thick sections, the block of tissue should be less than 0.5 cm in diameter for optimal results whereas for 24 μm thick sections entire rat brains or kidneys can be sectioned. Sections are picked up from the knife with a gelatin-coated (subbed) microscope slide and thaw-mounted by warming the underside of the slide with a finger. This tissue must be desiccated prior to refreezing, or ice crystals that destroy morphology will form (Herkenham, 1984). Additionally, desiccation will prevent the tissue sections from floating off the slides during the incubation with the radioligand. A good method for this involves keeping the freshly cut, thaw-mounted sections on ice until a full box is collected, and then placing the box of slides into a desiccator under vacuum at 4°C overnight. These sections can then be stored at −70°C for weeks prior to labeling.

2.2. Preliminary Experiments

2.2.1. Choice of Radioligand

Several excellent radioligands have been developed and tested for use in studying beta-adrenergic receptors. Each ligand has its own set of advantages and disadvantages. Some properties of ligands that have been utilized in autoradiographic studies of beta-adrenergic receptors are outlined in Table 1. As a general rule, because of lower specific activity (S/A in Table 1), ligands labeled with ^{3}H need to expose film for a much longer time than do ligands labeled with ^{125}I in order to produce an appropriately dark autoradiogram. The exposure times are often 20–50 d for tritiated ligands, compared to 0.2–1 d for iodinated ligands. The percentage of specific binding for each of these radioligands (Table 1) is acceptable; the iodinated ligands tend to have 80–95% specific binding, and [^{3}H]DHA typically will have 65–90% specific binding, depending on ligand concentration and choice of tissue.

2.2.2. The "Wipe" Method

Two specific methods are used to obtain data utilizing the general methodology outlined in this chapter. One method, discussed above and below, involves the production and subsequent analysis of autoradiograms. Another commonly employed technique, which is much more rapid in yielding results is called the "wipe" method. This latter method eliminates the production and analysis of autoradiograms, since, following the incubation with radioligand and subsequent washing steps to remove unbound and nonspecifically bound ligand (discussed below),

Table 1
Ligands Utilized for Visualization of Beta-Adrenergic Receptors

Ligand	S/A, Ci/mmol	K_d, nM	Reference
³H]DHA	60–120	1–2	Palacios and Kuhar, 1980
[¹²⁵I]IPIN	2200	0.04–0.06	Rainbow et al., 1984
[¹²⁵I]ICYP	2200	0.05–0.10	Summers et al., 1985

the slide-mounted tissue sections are simply wiped off the slide with a glass fiber filter, and the section with its associated radioactivity is placed into a scintillation vial or γ tube for determination of radioactive content. This method, although it allows rapid determination of the amount of radioligand associated with the entire tissue section, loses the anatomical resolution of the autoradiographic technique. Thus, the "wipe" method is utilized only in preliminary experiments, described in this section, that are designed to establish conditions appropriate for labeling the slide-mounted tissue sections. Alternatively, if rapid data acquisition is not an important issue, all of the experiments in this section can be performed by generating and analyzing autoradiograms.

2.2.3. Wash Conditions

Usually, the initial experiment performed to establish overall label-ing conditions is optimization of the time of washing the slides following incubation with radioligand. Since this experiment does not depend on the ligand having reached equilibrium, the time of incubation with radioligand employed in this experiment is important only to the extent that sufficient counts are bound to ensure a reasonable signal. Washing of the sections is designed to eliminate unbound ligand and maximally decrease nonspecific binding without affecting specifically bound ligand. If one is establishing conditions using a new tissue or a new radioligand, both total and nonspecific binding must be examined in these and subsequent experiments. Since one cannot know *a priori* the "correct" definition of nonspecific binding, a reasonable guess, based on values reported in the literature, must be chosen. A "reasonable" choice for beta-adrenergic receptors seems to be the binding that occurs in the presence of 100 μM (–) isoproterenol, although this choice, or any other, needs to be validated in later experiments as described below. Typically, washing is carried out at low temperatures (0–4°C) to minimize loss of specifically bound ligand, although with a ligand that has a very slow rate of dissociation, such as [^{125}I]ICYP, higher temperatures (e.g., 22°C) have been used (Zarbin et al., 1983). Often the wash buffer is changed periodically during washing. In our experience, however, periodically raising and lowering the slides in the wash buffer accomplishes the same goal of eliminating the unstirred, locally high concentrations of ligand that build up when ligand is being removed from the sections by washing. If autoradiograms are to be produced, a quick dip into distilled water following the buffer wash removes buffer salts. Sections are then dried either by placing them on a hot (56°C) plate for a minute or by desiccating them under vacuum overnight at 4°C. This latter method works well only on thin (<10 µm) sections.

2.2.4. Time to Equilibrium

Once conditions are established for proper washing, experiments are carried out to establish the time required for the ligand to reach steady state. Typically, the incubation with radioligand is performed at room temperature for convenience. Total and nonspecific binding are determined at several times of incubation. The data should yield information regarding the proper length of time to incubate and the stability of the ligand–receptor complex. It should be noted that if saturation curves are to be performed in later experiments, the concentration used to establish the time to steady state should be the lowest concentration to be used in the saturation experiments, since higher concentrations will equilibrate more rapidly, whereas lower concentrations will equilibrate more slowly. It should be noted that, although [^{125}I]ICYP has become a popular ligand with which to perform autoradiographic experiments, the very slow rate of dissociation of this ligand means that at concentrations around the K_d value (5–10 pM) the ligand will equilibrate very slowly (Weiland and Molinoff, 1981). Thus, complete saturation curves with this ligand are essentially impossible to perform, since equilibrium times are in the range of 10–20 h at concentrations of 5–10 pM. Typically, the ligand is used at concentrations (50–100 pM) well above its K_d value to minimize this problem. However, this introduces a second problem, namely, at such concentrations, a relatively high percentage of the labeling is to serotonin-1B receptors (Pazos et al., 1985b). This problem is discussed in detail below. [^{125}I]ICYP is in fact an excellent radioligand with which to study these receptors, if the binding is performed in the presence of an appropriate concentration of isoproterenol to mask the binding to beta-adrenergic receptors (Pazos et al., 1985b).

2.2.5. Pharmacological Specificity

To determine that the radioligand is binding to beta-adrenergic receptors and not to other receptors, such as described above for [^{125}I]ICYP, inhibition curves for several competing drugs are generated. In practice, the tissue sections are incubated with radioligand in the absence and presence of several concentrations of drugs with known affinity for beta-adrenergic receptors. The amount of radioligand bound is then determined, either by the "wipe" method or by densitometry, and inhibition curves are constructed. K_d values determined by this method are in excellent agreement with those determined by radioligand binding to homogenized preparations (Palacios and Kuhar, 1980; Rainbow et al., 1984). Stereoselectivity, an important characteristic of receptors, also can be demonstrated by these experiments. Additionally, such experiments yield information regarding a reasonable definition of nonspecific binding. For example, Rainbow et al. (1984) found that, although the K_d values of (–)-propranolol and (–)-isoproterenol for competing for [^{125}I]IPIN binding to slide-mounted brain sections were of the appropriate magnitude, high concentrations (>1 μM) of (–)-propranolol inhibited more binding than did high concentrations (>10 μM) of (–)-isoproterenol. Additionally, concentrations of (–)-isoproterenol between 10 and 200 μM all inhibited the same amount (approximately 85%) of the binding of [^{125}I]IPIN forming a "plateau." Thus, on the basis of these data the authors chose 100 μM (–)-isoproterenol to define nonspecific binding. Since [^{125}I]ICYP binds to serotonin receptors, [^{125}I]IPIN is structurally similar to [^{125}I]ICYP, and propranolol has a relatively high affinity for serotonin receptors (Pazos et al., 1985b), it is likely that much of the nonspecific binding of [^{125}I]IPIN that is inhibited by propranolol is caused by the ligand binding to serotonin receptors.

2.2.6. Saturability

As with studies of radioligand binding to homogenized preparations, the binding of ligands to slide-mounted tissue sections can be shown to be a saturable process. Incubating sections with increasing concentrations of ligand and plotting the resulting data according to the method of Scatchard (1949) (or more correctly, Rosenthal, 1967) yields estimates of the K_d value and the density of binding sites. For some receptors, the K_d value determined for radioligands by autoradiographic methods has been reported to be 10–20-fold higher than that determined using homogenized preparations (Dohanich et al., 1986). In general, this does not appear to be the case for the ligands utilized for visualizing beta-adrenergic receptors. Thus, Palacios and Kuhar (1980) reported the K_d value (1–2 nM) for [^{3}H]DHA determined in autoradiographic studies to be in close agreement with literature values from studies using homogenized preparations. Similar agreement for [^{125}I]IPIN (K_d = 50 pM) was reported by Rainbow et al. (1984). On the other hand, K_d values reported for [^{125}I]ICYP using autoradiographic methods (50–100 pM; Lew and Summers, 1985; Summers et al., 1985; Goldie et al., 1986) are higher than those reported using homogenized preparations (13–25 pM; Wolfe and Harden, 1981; Neve et al., 1986). This may result from the binding being carried out under nonequilibrium conditions, as discussed above.

2.3. Production of Autoradiograms

2.3.1. LKB Ultrofilm

Once the sections have been labeled, rinsed, and dried, they are tightly apposed to LKB Ultrofilm™ for a period of time determined by the specific activity of the radioligand, the density of receptors in the tissue, and the concentration of radioligand. This film is an X-ray film that lacks the typical plastic scratch coat over the emulsion, thus allowing low energy β particles, such as those emitted by tritium, to expose the film. Additionally, the lack of a scratch coat allows the emulsion and the radioactivity to come into close proximity, which reduces the blurring of the image. Following exposure, the film is developed by standard methods to produce an autoradiogram. The tissue section from which the autoradiogram is produced can then be histologically stained with thionin or cresyl violet (Paxinos and Watson, 1982) to allow for anatomical identification. For analysis, the autoradiogram and the stained section can be aligned as described below, allowing the assignment of densities to various anatomical structures. The practical limit of resolution of this technique is (optimistically) 100 μm, although at this resolution the ability to properly align tissue sections with autoradiograms is probably the limiting factor. Recently, Amersham Corp., has introduced a film that appears to be identical to LKB Ultrofilm. In our hands, preliminary experiments have detected no differences in the response of the two films to isotopes or in the degree of film background.

2.3.2. Emulsion-Coated Coverslips

To increase the resolution of autoradiography, one can utilize the methods developed by Young and Kuhar (1979) in which the slides, prepared in the manner described above, are physically attached to emulsion-coated coverslips. Thus, coverslips the same size as microscope slides are dipped into Kodak™, NTB2, or NTB3 liquid emulsion. When the emulsion is dry, the emulsion-coated coverslips are glued at one end to the slides on which the labeled tissue sections are mounted. *(See* Kuhar [1985] for excellent pictures of this technique.) The slide and coverslip are clipped together to ensure tight apposition. Following an appropriate exposure time, the clip is removed and a spacer placed between the slide and coverslip to allow for sequential development of the emulsion followed by histological staining of the tissue section. Removing the spacer allows for the developed autoradiogram and the stained section to return to their original juxtaposition, and the number of silver grains in the emulsion can be associated with the underlying morphology at a resolution of approx 5–10 μm. This higher resolution method allows one to distinguish, at least in some tissues, the types of cells on which beta-adrenergic receptors are located (e.g., McCarthy, 1983; Healy et al., 1985).

2.4. Quantitation of Autoradiograms

2.4.1. Densitometric Analysis of Autoradiograms

One of the major advantages of LKB Ultrofilm is the ability to quantify the amount of radioactivity bound to a specific area of tissue by measuring the absorbance of light passing through a darkened section of the film corresponding to the area of tissue. Currently there are dozens of commercially available video-based densitometers that will accomplish these measurements. Typically, these systems utilize a video camera that captures the image of an autoradiogram through which a standardized light source is shining. The histologically stained image of the section is also captured and aligned with the autoradiographic image to allow for identification of anatomically-defined structures on the autoradiogram. The autoradiographic image is then analyzed by a microcomputer to determine the amount of light absorbed by a given region of the autoradiogram. These absorbance values are converted into the amount of radioactivity bound to the section by the use of a standard curve (Rainbow et al., 1982; Geary et al., 1985). A standard curve must be used, since the exposure of film by radioactivity is not a linear process (i.e., the film is saturable) either with time of exposure or with amount of radioactivity apposed to it. A standard curve is constructed by creating a series of "tissue-mash standards." These standards are made by homogenizing a sample of tissue in a Teflon™ glass homogenizer (with no liquid added) and then thoroughly mixing with several different concentrations of the radioisotope of interest. These radioactive "tissue mashes" are lightly (400 × g) centrifuged to remove air bubbles and then frozen in small microfuge tubes. The end of the tube is cut off, the frozen mash extruded, and a section of the resulting cylinder attached to a cryostat chuck. Sequential sections of the cylinder are cut at the same thickness as the tissue to be analyzed, and alternate sections are either thaw-mounted onto subbed slides or saved for analysis of protein and radioactivity content. The sections of "tissue mash" mounted onto slides are then apposed to LKB Ultrofilm, using the same conditions as used for the tissue to be analyzed. The resulting autoradiograms are then subjected to video densitometry, and a standard curve is constructed by plotting the absorbance of the film vs the amount of radioactivity/mg of protein (nCi/mg). Knowing the specific activity of the radioligand used in a given experiment allows conversion into units (fmol/mg) comparable to those used in experiments with homogenized tissue preparations.

2.4.2. Grain Counting

If autoradiograms are produced using emulsion-coated coverslips, the amount of radioactivity associated with a given structure can still be quantified, although, typically, not in units that are as meaningful as is possible with LKB Ultrofilm. Thus, the number of silver grains/μm^2 can be counted, either manually or automatically by a video-based image analysis system attached to a microscope. The data from such analyses usually are presented as raw data (i.e., grains/μm^2) and, if no standard curve is generated, it is not possible to convert such data into units such as fmol/mg. The basis for constructing standard curves for such analyses has been established in studies of other receptors (Unnerstall et al., 1981), and in a few studies of beta-adrenergic receptors this methodology has been implemented (e.g., Vandermolen et al., 1986).

2.5. Problems Involved with Autoradiography
2.5.1. Tritiated Ligands

When tritiated ligands such as [³H]DHA are used for producing autoradiograms, a problem arises from the fact that the emissions from tritium are weak and easily quenched. Thus, for example, when ³H-amino acids are incubated with brain sections to nonspecifically label cells to an equal degree, the resulting autoradiogram is not homogeneously dark as expected, but rather shows that radioactivity associated with gray matter is more efficient at producing an image than is radioactivity associated with white matter (Alexander et al., 1981). It appears that white matter is more effective in quenching the radiation from tritium than is gray matter. This gives rise to potentially serious problems in quantitatively interpreting autoradiograms from tissues containing different amounts of gray and white matter. One possible solution to this problem has been suggested by Herkenham (1984): When sections were treated with organic solvents to remove lipids prior to exposure of film, this differential quench was eliminated. Although this approach theoretically has great benefit, it must be noted that it is likely that the treatment, in addition to removing lipids, will remove some or all of the radioligand bound to the receptor (Kuhar and Unnerstall, 1982). Thus, to avoid the added artifact of unknown loss of ligand, this approach must be validated for each ligand and each tissue.

2.5.2. Iodinated Ligands

Unlike tritium, the emissions from ¹²⁵I are of high energy and are not completely quenched by tissue even as thick as 32 μm. Although this avoids the problem of differential quench described above, it can lead to a different problem. If tissue sections are cut with some amount of "chatter," or unevenness in thickness, this difference in thickness will lead to differences in the degree of exposure of the film. Similarly, if sections differ in thickness from one another, the resulting autoradiograms will differ in darkness even though the concentration of receptors is the same in the sections. Since the beta particles from tritium do not penetrate tissue significantly deeper than 5 μm or so, this is not a problem with tritium.

2.5.3. Assigning Functional Significance to Receptors

When receptors are studied at a more gross anatomical level, the functionality of the receptors usually can be assayed. This can be done by examining the effects of receptor stimulation by agonists on physiological processes, such as muscle contraction, or on biochemical processes, such as glucose production or synthesis of cyclic AMP. Using autoradiographic techniques, however, the ability to directly assay functionality of the receptors is lost. Thus, the question of whether or not receptors in a given area of a tissue are physiologically significant can be studied only indirectly by autoradiography. In publications utilizing autoradiographic techniques, there is a great deal of speculation regarding the potential importance of certain receptors localized to specific regions, but evidence in this regard is difficult to obtain. One approach to this problem, discussed below, has been to examine the regulation of receptor density following physiological or pharmacological manipulations that alter the degree of stimulation of the receptors under study. The assumption of such studies is that if receptor density is altered, for example, by chronic blockade of the receptor or by denervation, this implies that the receptors are normally receiving a tonic innervation and thus are likely to subserve some function.

3. Studies Utilizing Autoradiography

3.1. Brain

3.1.1. Mapping Receptor Localization

The earliest experiments in this field were those performed by Palacios and Kuhar (1980) in which they examined [^{3}H]DHA binding to slide-mounted sections of rat brain. They showed that the characteristics of the ligand binding to brain sections were nearly identical to those reported for binding to homogenized preparations of brain (Alexander et al., 1975; Bylund and Snyder, 1976). Using a method involving emulsion-coated coverslips (Young and Kuhar, 1979), it was shown that beta-adrenergic receptors are most highly concentrated in the outer layers of the cortex (I–III), the caudate putamen, and the molecular layer of the cerebellum. These results disagreed with those in earlier reports of beta-adrenergic receptor localization in brain using a fluorescent derivative of propranolol (Atlas and Levitzki, 1976; Melamed et al., 1976). The advent of receptor autoradiography was instrumental in the demonstration that the previous data obtained using the fluorescent ligand, were incorrect.

More complete maps of the distribution of beta-adrenergic receptors in the brain of several species have been published subsequently (Wamsley et al., 1981; Rainbow et al., 1984; Shaw et al., 1984; Goffinet and Rockland, 1985; Goffinet and DeVolder, 1985; Pazos et al., 1985a; Goffinet and Caviness, 1986; Reznikoff et al., 1986). In general, it seems that for all mammals, the distribution of receptors is similar: High levels are found in the cerebral cortex, in certain portions of the hippocampus, such as CA1; in limbic structures, such as caudate putamen, nucleus accumbens, and the olfactory tubercule; as well as in the molecular layer of the cerebellum. Conversely, low levels of beta-adrenergic receptors are reported in such areas as the granular layer of the cerebellum, most of the brain stem, the hypothalamus, and areas of the hippocampus, such as CA3.

3.1.2. Localizing Subtypes of Beta-Adrenergic Receptors

Subtypes of beta-adrenergic receptors have been shown to exist in several tissues including the brain (Lands et al., 1967; Minneman et al., 1979b). Palacios and Kuhar (1982) used [^{3}H]DHA coupled with the somewhat selective (20-fold) agonist of beta-2 receptors, zinterol, to visualize beta-1 receptors and to calculate by subtraction the density of beta-2 receptors. These authors concluded that beta-1 receptor distribution showed a marked regional heterogeneity, whereas beta-2 receptors were more or less homogeneously distributed throughout the rat forebrain. Using [^{125}I]iodopindolol and much more selective (1000-fold) antagonists for both beta-1 (ICI 89,406) and beta-2 (ICI 118,551) receptors, Rainbow et al. (1984) were able to clearly visualize and quantify both subtypes of beta-adrenergic receptors in rat brain. In these studies it was shown that not only are beta-1 receptors heterogeneously distributed, but beta-2 receptors have a heterogeneous distribution as well. Thus, such structures as layers I–III, V, and VI of the cerebral cortex, area CA1 of the hippocampus, the dentate gyrus, the islands of calleja, and certain thalamic nuclei (such as the gelatinosis, ventroposterior, and dorsal lateral geniculate) are all specifically enriched in beta-1 receptors. In contrast, such areas as layer IV of the cerebral cortex, the molecular layer of the cerebellum, the olfactory tubercule, and certain thalamic areas (such as the lateral posterior, paraventricular, and reticular nuclei) are specifically enriched in beta-2 receptors. Typical autoradiograms generated by these types of experiments are shown in Fig. 1.

Zarbin et al. (1986) examined the distribution of beta-adrenergic receptors in the retina of several species, including human. They found that the receptors were most highly concentrated in the inner plexiform layer and found less dense labeling in the outer nuclear layer, the outer plexiform layer, and the inner nuclear layer.

Elena et al. (1987) examined the distribution of the subtypes of beta-adrenergic receptors in rabbit eye. They reported that most of the receptors were of the beta-2 subtype. A high density of receptors was found on conjunctival, corneal, and ciliary process epithelium.

3.1.3. Regulation of Receptor Density in Brain

The study of the regulation of receptor density at high anatomical resolution is one major goal of autoradiographic experiments. Experiments designed to study receptor regulation have been performed using homogenized preparations under a great variety of experimental paradigms. The effects of denervation (Sporn et al., 1977), chronic receptor blockade (Wolfe et al., 1978), antidepressant administration (Wolfe et al., 1978; Kellar et al., 1981), and hormonal alterations (Wolfe et al., 1976; Minneman et al., 1982) have all been studied. These experiments have, in general, examined the effects of in vivo treatment of rats on the density of beta-adrenergic receptors in homogenized preparations of the cerebral cortex.

For example, Kellar et al. (1981) demonstrated that chronic administration of electroconvulsive shock (ECS) to rats results in a decrease in the density of beta-adrenergic receptors in homogenates of the cerebral cortex and hippocampus. Using quantitative autoradiography, Kellar et al. (1985) showed not only that chronic (12 d) ECS caused decreases in the cortex and hippocampus, but also that these decreases were regionally localized and confined exclusively to the beta-1 subtype. Thus, these authors reported that ECS caused a 25–50% decrease in beta-1 receptors

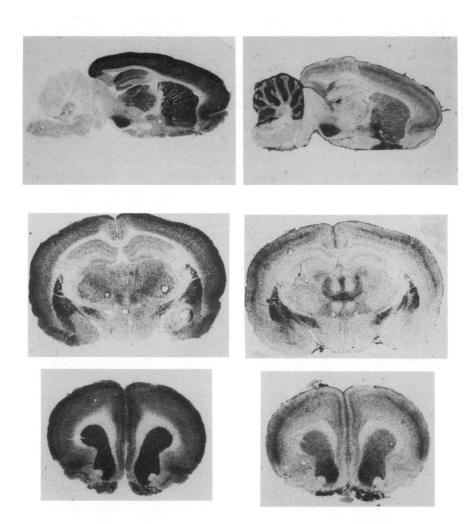

Fig. 1. Distribution of subtypes of beta-adrenergic receptors in rat brain. Cryostat-cut sections of rat brain were labeled with [^{125}I]iodopindolol as described in the text. The left column shows incubations in the presence of a beta-2 blocker (ICI 118,551); in the right column a beta-1 blocker (ICI 89,406) was included to show beta-1 and beta-2 receptors, respectively. The upper row displays a sagittal section of a rat brain; the middle and lower rows display coronal sections. The sagittal section is approx 2.5 mm lateral of bregma, the middle row is approx 6 mm rostral to the interaural line, and the lower row is approx 11 mm rostral to the interaural line.

in all layers of the cerebral cortex, the cingulate cortex, the amygdala, certain nuclei of the thalamus, CA1 of the hippocampus, and the dentate gyrus. No effects were observed in either receptor subtype in the caudate putamen, CA3 of the hippocampus, many thalamic nuclei, the olfactory tubercule, or the molecular layer of the cerebellum. In a similar study, Biegon and Israeli (1986) reported that 10 d of ECS resulted in a decrease in the density of beta-adrenergic receptors in the frontal cortex and hippocampus, including CA1 and dentate gyrus, but not CA3. Although receptor subtypes were not studied in these latter experiments, the data reported agreed with those of Kellar et al. (1985) in that no changes were observed in the caudate putamen, the cerebellum, globus pallidus, substantia nigra, or in several thalamic nuclei.

In another study examining alterations in beta-adrenergic receptors following a paradigm that mimics a clinical treatment of depression, Biegon (1986) reported that chronic administration of desipramine to rats resulted in decreases in the density of beta-adrenergic receptors in selective regions of the rat brain. Using autoradiographic techniques she reported decreases throughout the cortex and hippocampus, but not in other areas of the brain rich in these receptors, such as the caudate putamen, olfactory tubercule, superior colliculus, dorsomedial thalamus, substantia nigra, or pineal.

Other studies to determine the effects of increased stimulation on the density and distribution of beta-adrenergic receptors have been performed. Thus, Davenport et al. (1986) and Vos et al. (1987) examined the effects of the chronic administration of the lipophilic beta-adrenergic receptor agonist, clenbuterol, on the subtypes of beta-adrenergic receptors in rat brain. These authors reported that clenbuterol selectively decreased (by 50–75%) the density of beta-2 receptors in several areas of rat brain. This included areas high in beta-2 receptors (e.g., molecular layer of the cerebellum, layer IV of cerebral cortex) as well as areas high in beta-1 receptors and low in beta-2 receptors (e.g., layer I–III of cerebral cortex, area CA1 of the hippocampus). These selective changes occurred even though the drug was shown to have equal affinity for both receptor subtypes. The most likely explanation for the observation was suggested to be that clenbuterol is an agonist at beta-2 receptors and an antagonist at beta-1 receptors. This, in fact, has recently been shown to be the case by Ordway et al. (1987).

In experiments of opposite design, Johnson et al. (1985,1989) examined the effects of chronic denervation on the density of beta-adrenergic receptors in rat brain. Quantitative autoradiography was used to localize and quantify changes in the subtypes of beta-adrenergic receptors following intracerebroventricular administration of 6-hydroxydopamine to rats. These authors reported that denervation caused a 30–50% increase in the density of beta-1 receptors in several areas of rat cerebral cortex, certain thalamic nuclei, the amygdala, and several areas of the hippocampus. In contrast with previous reports (Minneman et al., 1979a), these authors reported that the density of beta-2 receptors also was altered by 6-hydroxydopamine in specific areas of the brain. For example, beta-2 receptor density was increased in the cingulate and motor cortex, but not in somatosensory or entorhinal cortex. Beta-2 receptors were also increased in the striatum, several thalamic nuclei, the cerebellum, and specific areas of the hippocampus.

The development of beta-adrenergic receptors in the brain has been examined using autoradiographic techniques. Thus, Shaw et al. (1984) have shown that not only do beta-adrenergic receptors increase in density during the first few weeks following birth (as has been shown using homogenized preparations of CNS tissue [Harden et al., 1977]), but in cat cerebral cortex the pattern changes with age as well. Thus, in the visual cortex of a three-day-old kitten beta-adrenergic receptors are concentrated in layers II and IV, whereas in the adult cat the receptors are localized mainly in layers I, II, and III. Similarly, Goffinet et al. (1986) have examined normal development of beta-adrenergic receptors in mouse forebrain using autoradiography. They reported that receptor density is concentrated in the inner layers of the cerebral cortex at E17, but that by postnatal day 4, the receptors are more concentrated in the outer layers of the cortex (the pattern that is seen in the adult mouse brain).

Three studies have examined the relationship of beta-adrenergic receptors with specific areas of sensory cortex. Vos et al. (1985) showed that beta-2 adrenergic receptors were specifically colocalized with structures known as "whisker barrels" in layer IV of the somatosensory cortex. There was, however, no pattern of beta-1 receptor labeling in this area of the brain. Additionally, when specific whiskers on the rat's face were deafferented at birth, the change in the "whisker barrel" pattern visualized by histochemistry was mirrored by changes in the pattern of labeling of beta-2 receptors in the somatosensory cortex. In a somewhat related study, Aoki et al. (1986) reported that beta-adrenergic receptors were specifically associated with the visual area of the cat cortex. In particular, in areas 17, 18, and 19, beta-1 receptors were concentrated in layers I–III, whereas beta-2 receptors were more diffusely localized. Dark-rearing the cats, a paradigm that can cause profound developmental changes in visual cortex, did not affect the density or distribution of either subtype of beta-adrenergic receptor. Similarly, Goffinet and Rockland (1985) examined the distribution of beta-adrenergic receptors in ferret visual cortex and reported that the highest densities occurred in layers I–III of area 17.

3.2. Peripheral Tissues

3.2.1. Respiratory Tract

Barnes et al. (1982, 1983) originally reported the autoradiographic localization of beta-adrenergic receptors in tissues from mammalian airways. These authors, in addition to several others reporting similar results (Barnes and Basbaum, 1983; Xue et al., 1983; Conner and Reid, 1984; Carstairs et al., 1984, 1985; Gatto et al., 1984; Smith and Sidhu, 1984; Goldie et al., 1986) have shown that beta-adrenergic receptors are highly concentrated on alveolar walls, surface epithelium, and submucosal glands, with much lower densities found on airway and vascular smooth muscle. Not surprisingly, most (75%) of the receptors have been found to be of the beta-2 subtype; smooth muscle contains 100% beta-2 receptors, whereas alveoli, and submucosal glands contain a mixture of subtypes (Carstairs et al., 1985).

Only a few studies have utilized autoradiographic techniques to examine the regulation of beta-adrenergic receptors in respiratory tissue. Thus, Barnes et al. (1984) have shown that administration of glucocorticoids to fetal rabbits resulted in a specific increase in the density of beta-adrenergic receptors on alveoli. These authors reported that 24 h after an injection of β-methasone to pregnant rabbits, the density of beta-adrenergic receptors associated with fetal alveoli was doubled, but no changes were observed in other parts of the airways or in fetal myocardium. Alternatively, Conner and Reid (1984) showed that the density of beta-adrenergic receptors is decreased in rat lung following 6 d of daily subcutaneous administration of isoproterenol. Silver grains associated with several structures including alveoli, smooth muscle, and epithelium, were decreased by 50–60% in treated animals, and Rosenthal plots of ligand binding to homogenized preparations of the lungs from these animals indicated that the density of beta-adrenergic receptors was decreased approximately 60–70% in treated animals, which agrees with the autoradiographic data. Gatto et al. (1987) reported that beta-adrenergic receptor density in guinea-pig alveolar and conducting-airway epithelium, as well as in bronchiolar and vascular smooth muscle, was significantly decreased when guinea pigs were exposed to ovalbumin. This implies that epinephrine released from the adrenal in response to the antigenic challenge regulates the density of beta-adrenergic receptors in pulmonary tissues.

3.2.2. Kidney

The distribution of beta-adrenergic receptors in rat kidney was originally described by Summers and Kuhar (1983), who utilized $[^{125}I]ICYP$. They reported that beta-receptors were most highly concentrated on glomeruli and on distal and cortical collecting tubules, but were present at low density on proximal tubules and vascular elements. These data were confirmed and amplified by Munzel et al. (1984). Several investigators subsequently reported the localization of beta-adrenergic receptor subtypes in rat kidney by autoradiographic techniques (Engel et al., 1985; Healy et al., 1985; Summers et al., 1985). Each of these latter studies noted that beta-1 receptors were almost exclusively associated with juxtaglomerular granule cells, glomeruli, and distal tubules, whereas beta-2 receptors were diffusely localized and found mainly associated with medullary tubules at the cortico–medullary junction. A similar distribution was reported in dog kidney (Lew and Summers, 1987). Engel et al. (1985) and Lew and Summers (1985) also examined the distribution of beta-adrenergic receptor subtypes in guinea-pig kidney and reported that, although beta-1 receptors were localized in a manner similar to that found in rat kidney, beta-2 receptors were not. Thus, beta-2 receptors were not found at the cortico–medullary junction, but rather were located primarily in the straight portion of the proximal tubule. In addition, there was no evidence of binding to distal or cortical collecting tubules. Some caution should be used in interpreting each of these studies, since they all utilized [^{125}I]ICYP as a radiolabel and, in addition to noting the problems with this ligand discussed above, Lew and Summers (1986) reported that the labeling of slices of mouse kidney by [^{125}I]ICYP was not uniformly inhibited by (–)-isoproterenol. They noted, however, that the labeling was inhibited completely and stereospecifically by the isomers of propranolol and pindolol, that such compounds as cinanserin, haloperidol, or phentolamine could inhibit selectively the medullary binding, and that (–)-isoproterenol selectively inhibited the cortical binding sites. These authors concluded that, at least in mouse kidney, much of the labeling by [^{125}I]ICYP in renal medulla was unrelated to beta-adrenergic receptors and more likely represented nonspecific labeling of lipophilic sites in the tissue.

Robinson and Wolfe (1984), utilizing [^{125}I]iodopindolol as a radioligand and using (–)-isoproterenol as a definition of nonspecific binding, reported that the densities of beta-adrenergic receptor subtypes in rat kidney were altered differentially when either epinephrine or norepinephrine was chronically administered to rats. For example, chronic subcutaneous administration of norepinephrine resulted in a 61% decrease in beta-1 receptors in glomeruli, but only a 25% decrease in beta-2 receptors. Conversely, administration of epinephrine resulted in only a 42% decrease in beta-1 receptors and a 66% decrease in beta-2 receptors. These differential effects are likely to result from the differences in affinity of these agonists for the subtypes.

3.2.3. Cardiovascular Tissues

Muntz et al. (1984) originally demonstrated the distribution of beta-adrenergic receptors in cardiovascular tissues in a study utilizing [^{3}H] DHA and emulsion-coated coverslips. These authors reported that specific [^{3}H]DHA labeling was found over myocardial arterioles, arteries, and myocytes, and that the grain density associated with each cardiovascular element was 1047, 219, and 231 grains/10^{-2} mm^2, respectively. This study showed that the selective (20-fold) beta-1 receptor antagonist, metoprolol, was a more potent inhibitor of [^{3}H]DHA binding to myocytes than it was of [^{3}H]DHA binding to arterioles. It suggested that most of the labeling of myocytes represented beta-1 receptors, whereas the labeling of arterioles represented mainly beta-2 receptors. In this study, the effects of myocardial ischemia on the regional density of beta-adrenergic receptors was examined. Occlusion of the left anterior descending coronary artery for 1 h resulted in a selective increase (18%) in the density of silver grains associated with cardiac myocytes; no changes were observed over cardiac blood vessels. It should be noted that these experiments did not attempt to convert grain densities into receptor densities. In subsequent studies, however, these authors (Muntz et al., 1986; Vandermolen et al., 1986) examined the relationship between the density of silver grains in the emulsion and the amount of ligand bound to the tissue. They showed that in the range of exposure used, there was a linear relationship between the density of grains and the amount of radioactivity bound to tissue. Using this relationship, they measured the affinity of [^{3}H]DHA for beta-adrenergic receptors on myocytes and arterioles. The dissociation constant for myocytes was found to be approx 1.6 nM, in good agreement with K_d values reported for many other tissues. The dissociation constant found for [^{3}H]DHA binding to arterioles was 0.26 nM, however. It is not clear why this value should be so different from that for myocytes and from all other values reported in the literature. Part of the difference may result from a reported selectivity of [^{3}H]DHA for beta-2 receptors (Neve et al., 1986), although this selectivity was reported to be only twofold. Muntz et al. (1986) noted that the increased number of silver grains associated with arterioles compared to myocytes reported earlier (Muntz et al., 1984) is mainly the result of differences in affinity, rather than differences in density of receptors. Molenaar et al. (1987) have reported that guinea-pig atrium contains approx 85% beta-1 adrenergic receptors and that there is an even distribution of both receptor subtypes in this tissue.

Saito et al. (1988) reported that the rat atrioventricular (av) node contains a higher concentration of beta-adrenergic receptors than the adjacent interventricular septum. A high percentage of beta-2 adrenergic receptors (44%) was found in the av node.

Amenta et al. (1985) reported that most of the beta-adrenergic receptors in rabbit aoreta were associated with the media. The intima and adventitia contained much lower levels of receptors. In the media, the receptors appeared to be found in two distinct layers. Thus, the outermost portion of the media and the portion closest to the intima were more heavily labeled than the more intermediate layer. These authors suggested, based on anatomical localization, that these subgroups of receptors may be innervated separately and subserve separate functions. Lipe and Summers (1986) examined the receptors on the splenic vasculature of the dog and reported that almost all of the receptors were of the beta-2 subtype. The receptors in the veins were over cells adjacent to the lumen; in the arteries, most beta-adrenergic receptors were associated with the medial layer, with fewer receptors toward the intima or adventitia.

Buxton et al. (1987) have examined the distribution of beta-receptor subtypes in human cardiac tissue. They reported that receptor subtypes were evenly distributed over the myocardium of the right atrial appendage, left ventricular papillary muscle, and left atrial free wall. On the other hand, the pericardium and the intimal surface of the coronary arteries possessed mainly beta-2 adrenergic receptors. In both the right and left atrial appendages, approx 40% of the receptors were of the beta-2 subtype. The authors note that the surprisingly high percentage of beta-2 adrenergic receptors could have clinical relevance.

3.2.4. Miscellaneous Tissues

The methodology described in this chapter has been used to examine the localization of beta-adrenergic receptors in several other tissues. Thus, for example, DeSouza (1985) and Schimchowitsch et al. (1986) showed, using $[^{125}I]ICYP$ and emulsion-coated coverslips, that the beta-adrenergic receptors found in bovine, rat, and human pituitary are mainly of the beta-2 subtype. Additionally, in rat pituitary, receptors are concentrated most densely in the intermediate lobe, with progressively lower densities in the posterior and anterior lobes, respectively. On the other hand, the human pituitary appears to contain the highest density of beta-2 receptors in the posterior portions. DeSouza (1985) suggests that these observations support a role for epinephrine and norepinephrine in modulating pituitary function. This hypothesis is supported by another study by DeSouza (1987), in which beta-adrenergic receptors were significantly increased in chronically adrenalectomized rats. This increase was not reversed by dexamethasone treatment.

Johnston and Summers (1985) reported the distribution of beta-adrenergic receptors in rabbit ear. They noted that beta-2 receptors were most highly concentrated in the central ear artery, the hyline cartilage, nerve trunks, epithelium, and sebaceous glands.

Dube et al. (1986) showed that beta-adrenergic receptors in rat ventral prostate are exclusively associated with epithelial cells, with no receptors present on stromal cells. The silver grains were mostly associated with the apical pole of the glandular cells and were much less concentrated in the basal nuclear region of the epithelium. Very low concentrations of grains were found in the lumen of the acini. Castration caused a decrease in the receptor concentration, and treatment with dihydrotestosterone reversed the effects of castration. These authors suggest that the specific localization of beta-adrenergic receptors and their regulation by androgens is consistent with the idea that these receptors play a physiological role in androgen action in prostatic tissue.

Marchetti et al. (1987) examined the development and distribution of beta-adrenergic receptors in rat ovary and its associated nerves. Receptor concentration increased progressively between ages 12 and 27 d, and reached a peak at 37 d. At 27 d, receptors were mostly localized to the interstitial cells, whereas at 37 d the corpora lutea were strongly labeled. When the nerves (superior ovarian nerve and the plexus nerve) are cut, the density of beta-adrenergic receptors in the corpora leutea decrease dramatically (two to threefold), suggesting either that the receptors are located presynaptically on nerve terminals or that the decreased input of some transmitter results in a downregulation of the receptors.

Ek et al. (1987) have examined the distribution of beta-adrenergic receptors in cat colon. Overall, the majority (60%) of the receptors were found to be of the beta-2 subtype. In the circular smooth muscle, on the other hand, essentially all of the receptors were of the beta-2 subtype, whereas the longitudinal muscle contains a significant proportion of beta-1 receptors.

3.3. Cell Culture

One of the goals of receptor autoradiography is to localize specific receptor types on identifiable cells. Because of the small size of brain cells and the fact that receptors are often found on dendrites and axons, to accomplish this goal using sections of brain tissue is technically beyond the limits of the methods available. Thus, "On which brain cells do beta-adrenergic receptors mainly reside?" is still an open question. To attempt to address this question, researchers have utilized primary cultures from rat brain to examine the receptor types on each cell that can be identified.

Thus, Hosli and Hosli (1982) used dissociated cell cultures from rat cerebellum, brain stem, and spinal cord to examine the receptors on specific
cells. They reported specific binding of [³H]DHA to Purkinje cells in
cerebellar cultures. In cultures of spinal cord and brain stem, labeling of
many types of large neurons was observed, and in all cultures, glial cells
were found to possess binding sites for [³H]DHA. Hosli et al. (1983)
followed up the latter observation by showing that stimulation of beta-
adrenergic receptors on glia that were maintained in primary culture
resulted in a hyperpolarization, thus demonstrating a functional consequence of stimulating these receptors.

McCarthy and his coworkers (McCarthy, 1983; Burgess and
McCarthy, 1985; Burgess et al., 1985) have examined the distribution of
beta-adrenergic receptors on immunocytochemically defined cells,
derived from neonatal rat cerebral cortex, that were maintained for several
days in primary culture. These authors have established an automated,
computer-assisted method of counting grains on single cells; using these
methods, they determined that, at least in culture, the majority of labeling
of beta-adrenergic receptors was associated with polygonal astroglia.
Surprisingly, no labeling was found to be associated with immunocytochemically defined neurons. Similarly, oligodendroglia did not exhibit
any specific labeling. It thus seems possible that the majority of the beta-
adrenergic receptors in the cerebral cortex were associated with glia, not
neurons. Alternatively, it is possible that, as time in culture increased, the
density of beta-adrenergic receptors on neurons dramatically decreased,
and that the density measured in culture was not a reflection of what
occurs naturally. In a later paper, Trimmer and McCarthy (1986) demonstrated that beta-adrenergic receptors present on polygonal astroglia
develop mainly between prenatal day 16 and postnatal day 1. This
increase in receptors was observed whether the cells were cultured
for 1–5 d or 8–22 d. However, the authors noted that the level of receptor
expression in these cells was a function of the culture methodology.

3.4. Axonal Transport of Beta-Adrenergic Receptors

In an elegant set of experiments using autoradiographic techniques,
Zarbin et al. (1983) have examined the axonal transport of beta-adrenergic
receptors in rat sciatic nerve. In these experiments [¹²⁵I]ICYP was used
to label beta-adrenergic receptors in axons following a double ligation of

sciatic nerve. Receptors were found to accumulate both proximally and distally to the ligatures. Receptors were also transported in the isolated segment between the ligatures, which was interpreted to indicate that receptor movement occurs via fast transport. Pharmacological analysis of the receptors demonstrated that they were of the beta-2 subtype, and that although antagonists were equipotent at receptors found proximal and distal to a ligature, agonists were more potent (10–30 x) at receptors accumulating proximally to a ligature. Additionally, receptors proximal to a ligature were much more sensitive to the effects of guanine nucleotides that caused a decrease in the potency of agonists. Thus, the anterogradely transported receptors, but not retrogradely transported receptors, appear to be in func-tional association with guanine nucleotide binding regulatory proteins.

4. Conclusion

It has been only a decade since the first experiments were performed that allowed one to visualize and quantify beta-adrenergic receptors in tissues at unprecedented anatomical resolution. During those years, the methodology has been refined and tested such that these are now routine techniques available in dozens of laboratories. Using these techniques it has been shown that beta-adrenergic receptors and their subtypes are extremely heterogeneously distributed throughout most organs. Although there are several potential explanations for these observations (some of which are artifactual, as noted above), it is usually hoped that knowing the location of these and other receptors will provide a clue as to the functional innervation of the tissue and perhaps to the function of the receptors themselves. It seems that one of the challenges of the future lies in perfecting assays of receptor function that are as sensitive in that area as quantitative autoradiography currently is in detecting the presence of receptors.

Acknowledgments

During the course of preparation of this manuscript the author was supported by the following grants from the National Institutes of Health (NS22040, GM31155, GM34781) and an Established Investigator Award from the American Heart Association.

References

Alexander, G. M., Swartzman, R. J., Bell, R. D., Yu, J., and Renthal, A. (1981) Quantitative measurements of local cerebral metabolic rate for glucose utilizing tritiated 2-deoxyglucose. *Brain Res.* **223**, 59–67.

Alexander, R. W., Williams, L. T., and Lefkowitz, R. J. (1975) Identification of cardiac beta-adrenergic receptors by (–)³H-alprenolol binding. *Proc. Natl. Acad. Sci. USA* **72**, 1564–1568.

Amenta, F., Cavallotti, C., and DeRossi, M. (1985) Histoautoradiographic localization of (–)-³H-dihydroalprenolol in rabbit aorta. *Pharmacology* **30**, 160–167.

Aoki, C., Kaufman, D., and Rainbow, T. C. (1986) The ontogeny of the laminar distribution of beta-adrenergic receptors in the visual cortex of cats, normally reared and dark-reared. *Brain Res.* **392**, 109–116.

Atlas, D. and Levitzki, A. (1976) An irreversible blocker for the beta-adrenergic receptor. *Biochem. Biophys. Res. Commun.* **69**, 397–403.

Barnes, P. and Basbaum, C. B. (1983) Mapping of adrenergic receptors in the trachea by autoradiography. *Exp. Lung. Res.* **5**, 183–192.

Barnes, P., Basbaum, C. B., and Nadel, J. A. (1983) Autoradiographic localization of autonomic receptors in airway smooth muscle. Marked differences between large and small airways. *Am. Rev. Respir. Dis.* **127**, 758–762.

Barnes, P., Basbaum, C. B., Nadel, J. A., and Roberts, J. M. (1982) Localization of beta-adrenoceptors in mammalian lung by light microscope autoradiography. *Nature* **299**, 444–447.

Barnes, P., Jacobs, M., and Roberts, J. M. (1984) Glucocorticoids preferentially increase fetal alveolar beta-adrenoreceptors: Autoradiographic evidence. *Pediatric Res.* **18**, 1191–1194.

Biegon, A. and Israeli, M. (1986) Localization of the effects of electroconvulsive shock on beta-adrenergic receptors in the rat brain. *Eur. J. Pharmacol.* **123**, 329–334.

Biegon, A. (1986) Effect of chronic desipramine treatment on dihydroalprenolol, imipramine, and desipramine binding sites: A quantitative autoradiographic study in the rat brain. *J. Neurochem.* **47**, 77-80.

Burgess, S. K. and McCarthy, K. D. (1985) Autoradiographic quantitation of beta-adrenergic receptors on neural cells in primary cultures. I. Pharmacological studies of [¹²⁵I]iodopindolol binding of individual astroglial cells. *Brain Res.* **335**, 1–9.

Burgess, S. K., Trimmer, P. A., and McCarthy, K. D. (1985) Autoradiographic quantitation of beta-adrenergic receptors on neural cells in primary cultures. II. Comparison of receptors on various types of immunocytochemically identified cells. *Brain Res.* **335**, 11–19.

Buxton, B. F., Jones, C. R., Molenaar, P., and Summers, R. J. (1987) Characterization and autoradiographic localization of beta-adrenoceptor subtypes in human cardiac tissues. *Br. J. Pharmacol.* **92**, 299–310.

Bylund, D. B. and Snyder, S. H. (1976) *Beta*-adrenergic receptor binding in membrane preparations from mammalian brain. *Mol. Pharmacol.* **12**, 568–580.

Carstairs, J. R., Nimmo, A. J., and Barnes, P. J. (1985) Autoradiographic visualization of beta-adrenoceptor subtypes in human lung. *Am. Rev. Respir. Dis.* **132,** 541–547.

Carstairs, J. R., Nimmo, A. J., and Barnes, P. J. (1984) Autoradiographic localization of beta-adrenoceptors in human lung. *Eur. J. Pharmacol.* **103,** 198–190.

Conner, M. W. and Reid, L. M. (1984) Mapping of beta-adrenergic receptors in rat lung: Effect of isoproterenol. *Exp. Lung. Res.* **6,** 91–101.

Davenport, P. A., Artymyshyn, R. P., Vos, P., Frazer, A., and Wolfe, B. B. (1986) Chronic administration of clenbuterol to rats causes a specific decrease in the density of beta-2 adrenergic receptors: An autoradiographic study. *Soc. Neurosci. Abstr.* **12,** 416.

DeSouza, E. B. (1985) Beta-2-adrenergic receptors in pituitary. Identification, characterization, and autoradiographic localization. *Neuroendocrinology* **41,** 289–296.

DeSouza, E. B. (1987) Modulation of beta-adrenergic receptors in the pituitary gland following adrenalectomy in rats. *Neurosci. Lett.* **73,** 281–286.

Dohanich, G. P., Halpain, S., Lambdin, L. T., and McEwen, B. S. (1986) Features of ligand binding in homogenate and section preparations. *Brain Res.* **366,** 338–342.

Dube, D. Poyet, P., Pelletier, G, and Labrie, F. (1986) Radioautographic localization of beta-adrenergic receptors in the rat ventral prostate. *J. Androl.* **7,** 169–174.

Ek, B., Jodal, M., and Lundgren, O. (1987) Autoradiographic location of beta- adrenoceptor subtypes in cat colon smooth muscle. *Acta Physiol. Scand.* **129,** 353–360.

Elena, P. P., Dosina-Boix, M., Moulin, G., and Lapalus, P. (1987) Autoradiographic localization of beta-adrenergic receptors in rabbit eye. *Invest. Ophthalmol. Vis. Sci.* **28,** 1436–1441.

Engel, G., Maurer, R., Perrot, K., and Richardson, B. P. (1985) Beta-adrenoceptor subtypes in sections of rat and guinea-pig kidney. *Nauyn Schmiedebergs Arch. Pharmacol.* **328,** 354–357.

Gatto, C., Green, T. P., Johnson, M. G., Marchessault, R. P., Seybold, V., and Johnson, D. E. (1987) Localization of quantitative changes in pulmonary beta- receptors in ovalbumin-sensitized guinea pigs. *Am. Rev. Respir. Dis.* **136,** 150–154.

Gatto, C., Johnson, M. G., Seybold, V., Kulik, T. J., Lock, J. E., and Johnson, D. E. (1984) Distribution and quantitative developmental changes in guinea pig pulmonary beta-receptors. *J. Appl. Physiol.* **57,** 1901–1907.

Geary, W. A., Toga, A. W., and Wooten, G. F. (1985) Quantitative film autoradiography for tritium: Methodological considerations. *Brain Res.* **337,** 99– 108.

Goffinet, A. M. and Caviness, V. S. (1986) Autoradiographic localization of beta-1 and alpha-1 adrenoceptors in the midbrain and forebrain of normal and reeler mutant mice. *Brain Res.* **366,** 193–202.

Goffinet, A. M. and DeVolder, A. (1985) Autoradiographic analysis of adrenergic receptors in the mammalian brain. *Acta. Neurol. Belg.* **85,** 82–109.

Goffinet, A. M., Hemmendinger, L. M., Caviness, V. S. (1986) Autoradiographic study of beta-1 adrenergic receptor development in the mouse forebrain. *Brain Res.* **389,** 187–191.

Goffinet, A. M. and Rockland, K. S. (1985) Laminar distribution of alpha-1 and beta-1 adrenoceptors in ferret visual cortex. *Brain Res.* **333,** 11–17.

Goldie, R. G., Papadimitriou, J. M., Patterson, J. W., Rigby, P. J., and Spina, D. (1986) Autoradiographic localization of beta-adrenoceptors in pig lung using [^{125}I]iodocyanopindolol. *Br. J. Pharmacol.* **88,** 621–628.

Harden, T. K., Wolfe, B. B., Sporn, J. R., Perkins, J. P., and Molinoff, P. B. (1977) Ontogeny of beta-adrenergic receptors in rat cerebral cortex. *Brain Res.* **125,** 99–108.

Healy, D. P., Munzel, P. A., and Insel, P. A. (1985) Localization of beta-1 and beta-2 adrenergic receptors in rat kidney by autoradiography. *Circ. Res.* **57,** 278–284.

Herkenham, M. (1984) Autoradiographic demonstration of receptor distributions, in *Brain Receptor Methodologies, Part A* (Marangos, P. J., Campbell, I. C., and Cohen, R. M., eds.), Academic Press, pp. 127–152.

Hosli, E. and Hosli, L. (1982) Evidence for the existence of alpha- and beta-adrenoceptors on neurones and glial cells of cultured rat central nervous system. An autoradiographic study. *Neuroscience* **7,** 2873–2881.

Hosli, L., Hosli, E., Zehnter, C. Lehmann, R., and Lutz, T. W. (1983) Alpha- and beta-adrenoceptors on cultured glial cells. *Adv. Biochem. Psychopharmacol.* **37,** 417–420.

Johnson, E. W., Rainbow, T. C., and Wolfe, B. B. (1985) Quantitative autoradiography of rat cortical beta-1 and beta-2 adrenergic receptors after 6-hydroxydopamine treatment. *Soc. Neurosci. Abstr.* **11,** 1046.

Johnson, E. W., Wolfe, B. B., and Molinoff, P. B. (1989) Regulation of subtypes of beta-adrenergic receptors in the rat brain following treatment with 6-hydroxydopamine. *J. Neurosci.* **9,** 2297–2305.

Johnston, H. G. and Summers, R. J. (1985) Localization of beta-adrenoceptors in the rabbit ear by light microscopic autoradiography. *Eur. J. Pharmacol.* **115,** 97–101.

Kellar, K. J., Cascio, C. S., Bergstrom, D. A., Butler, J. A., and Iadarola, P. (1981) Electroconvulsive shock and reserpine: Effects on beta-adrenergic receptors in rat brain. *J. Neurochem.* **37,** 830–836.

Kellar, K. J., Stockmeier, C. A., Rainbow, T. C., and Wolfe, B. B. (1985) Electroconvulsive shock selectively down-regulates beta-1 adrenergic receptors in specific areas of the rat brain. *Soc. Neurosci. Abstr.* **11,** 812.

Kuhar, M. J. (1985) Receptor localization with a microscope, in *Neurotransmitter Receptor Binding,* 2nd Ed. (Yamamura, H. I., ed.), Raven Press, New York, pp. 153–176.

Kuhar, M. J. and Unnerstall, J. R. (1982) In vitro labeling receptor autoradiography: Loss of label during ethanol dehydration and preparative procedures. *Brain Res.* **244,** 178–181.

Lands, A. M., Arnold, A., McAuliff, J. P., Luduena, F. P., and Brown, T. G. (1967) Differentiation of receptor systems activated by sympathomimetic amines. *Nature* **214,** 597–598.

Lefkowitz, R. J., Caron, M. G., and Stiles, G. L. (1984) Mechanisms of membrane receptor regulation. *New Engl. J. Med.* **310,** 1570–1579.

Lew, R. and Summers, R. J. (1985) Autoradiographic localization of beta-adrenoceptor subtypes in guinea-pig kidney. *Br. J. Pharmacol.* **85,** 341–348.

Lew, R. and Summers, R. J. (1986) Autoradiographic analysis of (–)-[^{125}I]-CYP bindinginmousekidney. *Clin. Exp. Pharmacol. Physiol.* **13,** 211–221.

Lew, R. and Summers, R. J. (1987) The distribution of beta-adrenoceptors in dog kidney: An autoradiographic analysis. *Eur. J. Pharmacol.* **140,** 1–11.

Lipe, S. and Summers, R. J. (1986) Autoradiographic analysis of the distribution of beta-adrenoceptors in the dog splenic vasculature. *Br. J. Pharmacol.* **87,** 603– 609.

Marchetti, B., Cioni, M., Badr, M., Follea, N., and Pelletier, G. (1987) Ovarian adrenergic nerves directly participate in the control of luteinizing hormone-releasing hormone and beta-adrenergic receptors during puberty: A biochemical and autoradiographic study. *Endocrinology* **121,** 219–226.

McCarthy, K. D. (1983) An autoradiographic analysis of beta-adrenergic receptors on immunocytochemically defined astroglia. *J. Pharmacol. Exp. Ther.* **226,** 282–290.

Melamed, E., Lahav, M., and Atlas, D. (1976) Direct localization of beta-adrenoceptor sites in rat cerebellum by a new fluorescent analogue of propranolol. *Nature* **261,** 420–421.

Minneman, K. P., Dibner, M. D., Wolfe, B. B., and Molinoff, P. B. (1979a) Beta-1 and beta-2 adrenergic receptors in rat cerebral cortex are independently regulated. *Science* **204,** 866–868.

Minneman, K. P., Hegstrand, L. R., and Molinoff, P. B. (1979b) Simultaneous determination of beta-1 and beta-2 adrenergic receptors in tissues containing both receptor subtypes. *Mol. Pharmacol.* **18,** 34–46.

Minneman, K. P., Wolfe, B. B., and Molinoff, P. B. (1982) Selective changes in the density of beta-1 adrenergic receptors in rat striatum following chronic drug treatment and adrenalectomy. *Brain Res.* **252,** 309–314.

Molenaar, P., Canale, E., and Summers, R. J. (1987) Autoradiographic localization of beta-1 and beta-2 adrenoceptors in guinea pig atrium and regions of the conducting system. *J. Pharmacol. Exp. Ther.* **241,** 1048–1064.

Muntz, K. H., Calianos, T. A., Vandermolen, D. T., Willerson, J. T., and Buja, L. M. (1986) Differences in affinity of cardiac beta-adrenergic receptors for (^{3}H)dihydroalprenolol. *Am. J. Physiol.* **250,** H490–H497.

Muntz, K. H., Olson, E. G., Lariviere, G. R., DeSouza, S., Mukherjee, A., Willerson, J. T., and Buja, L. M. (1984) Autoradiographic characterization of beta- adrenergic receptors in coronary blood vessels and myocytes in normal and ischemic myocardium of the canine heart. *J. Clin. Invest.* **73,** 349–357.

Munzel, P. A., Healy, D. P., and Insel, P. A. (1984) Autoradiographic localization of beta-adrenergic receptors in rat kidney slices using [^{125}I]iodocyanopindolol. *Am. J. Physiol.* **246,** F240-F245.

Neve, K. A., McGonigle, P. M., and Molinoff, P. B. (1986) Quantitative analysis of the selectivity of radioligands for subtypes of beta-adrenergic receptors. *J. Pharmacol. Exp. Ther.* **238,** 46–53.

Ordway, G., O'Donnell, J., and Frazer, A. (1987) Effects of clenbuterol on central beta-1 and beta-2 adrenergic receptors of the rat. *J. Pharmacol. Exp. Ther.* **241,** 187–195.

Palacios, J. M. and Kuhar, M. J. (1980) Beta-adrenergic receptor localization by light microscopic autoradiography. *Science* **208,** 1378–1380.

Palacios, J. M. and Kuhar, M. J. (1982) Beta-adrenergic receptor localization in the rat brain by light microscopic autoradiography. *Neurochem. Int.* **4,** 473–490.

Paxinos, G. and Watson, C. (1982) *The Rat Brain in Stereotaxic Coordinates* (Academic Press, New York), p. 2.

Pazos, A., Engel, G., and Palacios, J. M. (1985b) Beta-adrenoceptor blocking agents recognize a subpopulation of serotonin receptors in brain. *Brain Res.* **343,** 403–408.

Pazos, A., Probst, A., and Palacios, J. M. (1985a) Beta-adrenergic receptor subtypes in the human brain: Autoradiographic localization. *Brain Res.* **358,** 324–328.

Rainbow, T. C., Bleisch, W. V., Biegon, A. and McEwen, B. S. (1982) Quantitative densitometry of neurotransmitter receptors. *J. Neurosci. Methods* **5,** 127–138.

Rainbow, T. C., Parsons, B., and Wolfe, B. B. (1984) Quantitative autoradiography of beta-1 and beta-2 adrenergic receptors in rat brain. *Proc. Natl. Acad. Sci. USA* **81,** 1585–1589.

Reznikoff, G. A., Manaker, S., Rhodes, C. H., Winokur, A., and Rainbow, T. C. (1986) Localization and quantification of beta-adrenergic receptors in human brain. *Neurology* **36,** 1067–1073.

Robinson, D. M. and Wolfe, B. B. (1984) Regulation of subtypes of renal beta-adrenergic receptors. *Soc. Neurosci. Abstr.* **10,** 671.

Rosenthal, H. E. (1967) Graphical method for the determination and presentation of binding parameters in a complex system. *Anal. Biochem.* **20,** 525–532.

Saito, K., Kurihara, M., Cruciani, R., Potter, W. Z., and Saavedra, J. M. (1988) Characterization of beta-1 and beta-2 adrenoceptors subtypes in the rat atrioventricular node by quantitative autoradiography. *Circ. Res.* **62,** 173–177.

Scatchard, G. R. (1949) The attraction of proteins for small molecules and ions. *Ann. NY Acad. Sci.* **51,** 660–672.

Schimchowitsch, S., Williams, L. M., and Pelletier, G. (1986) Autoradiographic localization of beta-adrenergic receptors in the rabbit pituitary. *Brain Res. Bull.* **7,** 705–710.

Shaw, C., Needler, M. C., Wilkinson, M., Aoki, C., and Cynader, M. (1984) Alterations in receptor number, affinity and laminar distribution in cat visual cortex during the critical period. *Prog. Neuropsychopharmacol. Biol. Psychiatry* **8,** 627–634.

Smith, D. M. and Sidhu, N. K. (1984) In vivo autoradiographic demonstration of beta-adrenergic binding sites in adult rat type II alveolar epithelial cells. *Life Sci.* **34,** 519–527.

Sporn, J. R., Harden, T. K., Wolfe, B. B., and Molinoff, P. B. (1977) Supersensitivity in rat cerebral cortex: Pre- and postsynaptic effects of 6-hydroxydopamine at noradrenergic synapses. *Mol. Pharmacol.* **13,** 1170–1180.

Summers, R. J. and Kuhar, M. J. (1983) Autoradiographic localization of beta- adrenoceptors in rat kidney. *Eur. J. Pharmacol.* **91,** 305–310.

Summers, R. J., Stephenson, J. A., and Kuhar, M. J. (1985) Localization of beta-adrenoceptor subtypes in rat kidney by light microscopic autoradiography. *J. Pharmacol. Exp. Ther.* **232,** 561–569.

Trimmer P. A. and McCarthy, K. D. (1986) Immunocytochemically defined astroglia from fetal, newborn and young adult rats express beta-adrenergic receptors in vitro. *Brain Res.* **392,** 151–165.

Unnerstall, J. R., Kuhar, M. J., Niehoff, D. L., and Palacios, J. M. (1981) Benzodiazepine receptors are coupled to a subpopulation of GABA receptors: Evidence from a quantitative autoradiographic study. *J. Pharmacol. Exp. Ther.* **218,** 797–804.

Vandermolen, D. T., Muntz, K. H., and Buja, L. M. (1986) Quantification of beta-adrenergic receptors in canine cardiac myocytes using autoradiography and an internal standard. *Lab. Invest.* **54,** 353–359.

Vos, P., Davenport, P. A., Artymyshyn, R. P., Frazer, A., and Wolfe, B. B. (1987) Selective regulation of beta-2 adrenergic receptors by the chronic administration of the lipophilic beta-adrenergic receptor agonist clenbuterol: An autoradiographic study. *J. Pharmacol. Exp. Ther.* **242,** 707–712.

Vos, P., Kaufman, D., Mansfield, E., Wolfe, B. B., and Hand, P. J. (1985) Beta-2 adrenergic receptors are specifically associated with the "whisker barrels" in the somatosensory cortex of the rat. *Soc. Neurosci. Abstr.* **11,** 666.

Wamsley, J. K., Palacios, J. M., Young, W. S., and Kuhar, M. J. (1981) Autoradiographic determination of neurotransmitter receptor distributions in the cerebral and cerebellar cortices. *J. Histochem. Cytochem.* **29,** 125–135.

Weiland, G. A. and Molinoff, P. B. (1981) Quantitative analysis of drug-receptor interactions: I. Determination of kinetic and equilibrium properties. *Life Sci.* **29,** 313–330.

Wolfe, B. B. and Harden, T. K. (1981) Guanine nucleotides modulate the affinity of antagonists at beta-adrenergic receptors. *J. Cyclic Nucl. Res.* **7,** 303–312.

Wolfe, B. B., Harden, T. K., and Molinoff, P. B. (1976) Beta-adrenergic receptors in rat liver: Effects of adrenalectomy. *Proc. Acad. Natl. Sci.* **73,** 1343–1347.

Wolfe, B. B., Harden, T. K., and Molinoff, P. B. (1977) In vitro study of beta adrenergic receptors. *Ann. Rev. Pharmacol. Toxicol.* **17,** 575–604.

Wolfe, B. B., Harden, T. K., Sporn, J. R., and Molinoff, P. B. (1978) Presynaptic modulation of beta-adrenergic receptors in rat cerebral cortex after treatment with antidepressants. *J. Pharmacol. Exp. Ther.* **207,** 446-457.

Wolfe, B. B. and Molinoff, P. B. (1988) Catecholamine receptors, in *Handbook of Experimental Pharmacology* Vol 90/I (Trendelenburg, U. and Weiner, N., eds.), Springer Verlag, Berlin, pp. 321–417.

Xue, Q. F., Maurer, R., and Engel, G. (1983) Selective distribution of beta- and alpha-1 adrenoceptors in rat lung visualized by autoradiography. *Arch. Int. Pharmacodyn. Ther.* **266,** 308–314.

Young, W. S. and Kuhar, M. J. (1979) A new method for receptor autoradiography. ³H-Opioid receptor labeling in mounted tissue sections. *Brain Res.* **179,** 255–270.

Zarbin, M. A., Palacios, J. M., Wamsley, J. K., and Kuhar, M. J. (1983) Axonal transport of beta-adrenergic receptors. Antero- and retrogradely transported receptors differ in agonist affinity and nucleotide sensitivity. *Mol. Pharmacol.* **24,** 341–348.

Zarbin, M. A., Wamsley, J. K., Palacios, J. M., and Kuhar, M. J. (1986) Autoradiographic localization of high affinity GABA, benzodiazepine, dopaminergic, adrenergic and muscarinic cholinergic receptors in the rat, monkey and human retina. *Brain Res.* **374,** 75–92.

CHAPTER 7

Beta-Adrenergic Receptors in Pathophysiologic States and in Clinical Medicine

Paul A. Insel

1. Introduction

As is well documented in other chapters of this book, beta-adrenergic receptors play an important role in modulating a variety of target cell responses to catecholamines. Accordingly, there has been much speculation as to whether alterations in these receptors contribute to disease states characterized by alterations in catecholamine response. Within the past decade, radioligand binding techniques have begun to allow a direct test of such speculation. The focus of this chapter is on the application of those techniques, and available evidence regarding alterations in beta-adrenergic receptors in various disease states is reviewed (Table 1). The principal emphasis is on information available in human subjects. The use of peripheral blood cells as a means to assess beta-adrenergic receptors in humans is discussed initially, and then the various disease states listed in Table 1 are reviewed. In general, material published since a previous review on this topic by Motulsky and Insel (1982) is emphasized.

The β-Adrenergic Receptors Ed.: J. P. Perkins © 1991 The Humana Press Inc.

Table 1
Disease States with Possible Alterations
in Beta-Adrenergic Receptors

Atopic disorders
Cardiac disorders
Hypertension
Diabetes mellitus
Thyroid disorders
Pheochromocytoma
Pseudohypoparathyroidism
Neurologic disorders
Psychiatric disorders

2. Beta-Adrenergic Receptors in Blood Cells

2.1. Identification
of Beta-Adrenergic Receptors in Blood Cells

The application of radioligand binding techniques to the study of beta-adrenergic receptors in humans presents several problems that must be considered in evaluating data on this topic. The first problem is the relative inaccessibility of the key target cells that may preferentially express disease-related alterations in beta-adrenergic receptors. At the present time it is difficult, if not impossible, to assay beta-adrenergic-receptor expression on organs, such as heart, liver, lung, brain, and kidney, from living subjects, although these may be tissues in which alterations in receptors are most clinically pertinent. As a result, the most common approach in humans has been to use peripheral blood cells as accessible "marker cells" that may demonstrate changes that occur on other tissues (Motulsky and Insel, 1982; Insel, 1985; Insel and Motulsky, 1987). Beta-adrenergic receptors have been detected on leukocytes (including lymphocytes, monocytes, and polymorphonuclear [PMN]cells), erythrocytes, and platelets (Williams et al., 1976; Galant et al., 1978; Brodde et al., 1981; Steer and Atlas, 1982; Sager, 1983; Marinetti et al., 1983; Halper et al., 1984; Kerry et al., 1984; Cook et al., 1985,1987; Motulsky et al., 1986; DeBlasi et al., 1986a; Szefler et al., 1987; Liggett et al., 1988; reviewed in Insel, 1985) (Table 2).

Table 2
Beta-Adrenergic Receptors in Human Blood Cells

Cell type	Beta-adrenergic subtypes	Number of receptors/cell	Functional effects	Selected references
Erythrocytes	Beta-2	~900	Change in deformability	Rasmussen et al., 1975 Sager, 1982,1983
Lymphocytes	Beta-2	~1000	Inhibit immune responses	Bourne et al., 1974 Williams et al., 1976 Brodde et al., 1981 Kahn et al., 1986 Landmann et al., 1984; DeBlasi et al., 1986a Motulsky et al., 1986 Szefler et al., 1987
Polymorpho-nuclear leukocytes	Beta-2	~1000	Inhibit oxidant production, chemiluminescence, and release of lysosomal enzymes	Bourne et al., 1974 Busse, 1977 Galant et al., 1978b Marinetti et al., 1983 Schopf and Lemmel, 1983 Busse and Sosman, 1984 Tecoma et al., 1986 Fantozzi et al., 1984 Mack et al., 1986 Davies et al., 1987 Nielson, 1987
Platelets	Beta-2	~5–20	Block aggregation and secretion	Steer and Atlas, 1982 Kerry et al., 1984 Cook et al., 1985 Wang and Brodde, 1985 Winther et al., 1985 Cook et al., 1987

Several kinds of information regarding beta-adrenergic receptors can be obtained from studies conducted with peripheral blood cells. One can "count receptors," i.e., determine the number of receptors on either intact cells or membrane preparations by performing binding isotherms (assessment of binding at varying concentrations) of labeled antagonists, such as [^{125}I]iodopindolol, [^{125}I]iodocyanopindolol, or [^{3}H]dihydroalprenolol. The radioiodinated probes are generally preferred, because their higher specific activity allows use of a smaller amount of receptor protein (i.e., a smaller number of cells and therefore less blood). One also can assess the affinity of receptors for radioligands and for agonist and antagonist drugs that compete for radioligand binding sites. Moreover, studies with washed membrane preparations allow estimates of receptor interaction with the guanine nucleotide binding protein, G_s; these interactions are assayed by the ability of guanine nucleotides to modify the apparent affinity of agonists in competitive binding studies. Functional response of beta-adrenergic receptors and G_s can be assessed as the stimulation of cyclic AMP formation in studies of cyclic AMP accumulation in intact cells or assay of adenylate cyclase activity in membrane preparations, in particular with leukocytes. The distribution of beta-adrenergic receptors between cell surface and a sequestered compartment can be assessed by binding of the hydrophilic antagonist [^{3}H]CGP-12177, or by competitive binding studies between unlabeled CGP-12177 and hydrophobic radioligands (e.g., [^{125}I]iodocyanopindolol, [^{125}I]iodopindolol, and [^{3}H]dihydroalprenolol) (DeBlasi et al., 1985,1986a; Motulsky et al., 1986; Cook et al., 1987). In addition, photoaffinity labeling can be used to identify mononuclear leukocyte (MNL) beta-adrenergic receptors; the use of *p*-azido-*m*-[^{125}I]iodobenzylcarazolol with MNL membranes facilitates identification of two specifically labeled peptides of molecular mass ~68,000 and ~55,000 dalton (as assessed by sodium dodecyl sulfate polyacrylamide gel electrophoresis) (Feldman and Lai, 1986).

2.2. Properties
of Beta-Adrenergic Receptors on Blood Cells

Properties of beta-adrenergic receptors on blood cells are different from those on other tissues in several respects:

1. Receptors on these cells are primarily exposed to circulating catecholamines (norepinephrine and epinephrine) and not to catecholamines present in the central nervous system or peripheral tissue at postganglionic sympathetic neuronal sites (almost exclusively norepinephrine).
2. Beta-receptors on blood cells are largely or exclusively of the beta-2 adrenergic subtype. Thus, such receptors may not adequately reflect changes in beta-1 adrenergic receptors.
3. Blood cells, such as erythrocytes and platelets, lack nuclei, erythrocytes are unable and platelets have a limited capacity to synthesize protein. In addition, the smaller complement of cellular organelles in platelets and erythrocytes provides a more limited repertoire of mechanisms for "processing" of receptors.
4. Leukocytes—in particular, lymphocytes—comprised several subpopulations that are dynamically regulated (Maisel et al., 1990) and that express different numbers of beta-adrenergic receptors (Sheppard et al., 1977; Pochet et al., 1979; Krawietz et al., 1982; Crary et al., 1983; Landmann et al., 1984; Griese et al., 1988; Khan et al., 1986; Maisel et al., 1989). Thus, redistribution among various subpopulations can lead to changes in beta-receptor expression in a mixed population of peripheral blood MNL, the cells that are typically assayed. This problem of cellular heterogeneity is not limited to MNL. For example, less mature forms of erythrocytes (e.g., reticulocytes) may express beta-adrenergic receptors with properties different those of mature erythrocytes (Larner and Ross, 1981). The limited number of beta-adrenergic receptors on platelets (<20/cell) also suggests that only a small number or proportion of platelets preferentially express beta-receptors (Cook et al., 1985; Wang and Brodde, 1985).

5. The life span of blood cells is generally less than that of most other cell types (erythrocytes—approx 120 d; leukocytes—several days/several wks; platelets—8–10 d).
6. Normal "physiological" changes in beta-receptor expression in blood cells may be difficult to discern from disease-related changes. Thus, receptor number, affinity, or other properties (e.g., sensitivity of binding of agonists to guanine nucleotides, perhaps secondary to altered interaction of receptor and G_s) may occur with ontogeny (e.g., Roan and Galant, 1982; Whitsett et al., 1982; Reinhardt et al., 1984), aging (Halper et al., 1984; Feldman et al., 1984a; Insel and Motulsky, 1987), exercise training (Butler et al., 1982; Mäki et al., 1987; Ohman et al., 1987), and as a function of circadian rhythm (e.g., Titinchi et al., 1984; Frohler et al., 1985). Other data suggest that lymphocyte beta-receptor number also may correlate with various measures of type A behavior, as assessed by interviewing techniques (Kahn et al., 1987).

In spite of these problems, many studies have been published regarding alterations in expression of beta-adrenergic receptors in peripheral blood cells from human subjects. Whether expression of beta-receptors in these cells accurately "mirrors" properties of receptors on other tissues is still an unresolved question. Studies that carefully compare biochemical and pharmacological properties of beta-receptors on peripheral blood cells and tissues from animals, or, if possible, from humans, will ultimately be required for resolution of this issue. Recently published data indicate a good correlation between beta-adrenergic receptor number in MNL and right atrial membrane preparations (in particular, the beta-2 adrenergic receptors in those preparations [Michel et al., 1986; Brodde, 1988; Michel et al., 1988b]), between beta-adrenergic receptor number in MNL and maximal contractile response of atrial muscle to the beta-adrenergic agonist isoproterenol (Brodde et al., 1986b; Michel et al., 1986), and between beta-adrenergic receptor number in MNL and lung (Liggett et al., 1988) and skeletal muscle (Liggett et al., 1989a) membranes.

An underlying assumption of much of the work conducted on beta-adrenergic receptors using model systems, such as tissue-culture cells and experimental animals is that the systems provide clinically meaningful information. This assumption has not yet been rigorously tested. However, homologous regulation of beta-adrenergic receptors by beta-adrenergic agonists and antagonists, a phenomenon that has been demonstrated in many experimental systems, has also been shown on blood cells from humans. Thus, beta-adrenergic receptors on MNL and PMN leukocytes are susceptible to an agonist-mediated decrease in receptor number (downregulation), agonist-induced "uncoupling" of receptors from stimulation of adenylate cyclase, and decrease in beta-adrenergic-mediated generation of cyclic AMP synthesis (desensitization) (e.g., Galant et al., 1978a,1980; Tohmeh and Cryer, 1980; Krall et al., 1980; Greenacre and Conolly, 1978; Tashkin et al., 1982; Sano et al., 1983; Aarons et al., 1983; Feldman et al., 1983; Davies and Lefkowitz, 1983; Galant and Britt, 1984; Samuelson and Davies, 1984; Brodde et al., 1985a; Tecoma et al., 1986; Neve and Molinoff, 1986; Martinsson et al., 1987; Maisel et al., 1989a). Conversely, administration of beta-adrenergic antagonists can increase (upregulate) beta-adrenergic receptors on blood cells (e.g., Aarons et al., 1980; Fraser et al., 1981; Brodde et al., 1985b; Neve and Molinoff, 1986; Michel et al., 1988b). In addition, as with many other cell types, beta-adrenergic receptors on human MNL can undergo a rapid redistribution (internalization) of receptors and homologous desensitization of cyclic AMP synthesis in response to beta-agonists (DeBlasi et al., 1985,1986a,b; Feldman and Lai, 1986; Motulsky et al., 1986; Sandnes et al., 1987). Platelet, beta-adrenergic receptors may also be internalized by beta-agonists (Cook et al., 1987). Results from these types of studies emphasize that beta-adrenergic receptors of human blood cells are capable of modulation by agonists in a manner that resembles that of other cell types. The results show that changes in circulating (or perhaps tissue) catecholamines have the potential to alter expression of beta-adrenergic receptors in blood cells, although the extent to which this occurs in MNL by endogenous catecholamines in vivo is not clearly defined (Liggett et al., 1988; Maisel et al., 1988). Thus, changes in receptor number (and perhaps affinity) in disease states may be secondary to changes in catecholamines rather than to disease-induced alterations in beta-adrenergic receptors themselves.

3. Beta-Adrenergic Receptors in Clinical Disorders

3.1. Atopic Disorders

The notion that a defective, beta-adrenergic receptor system or imbalance between alpha- and beta-adrenergic receptors might be responsible for asthma and other atopic ("allergic") disorders was first proposed by Szentivanyi (1968) 20 y ago. This concept was based primarily on a variety of clinical observations, and it has provided a heuristic framework for subsequent studies. The hypothesis implies that individuals with atopic disorders develop (or are genetically endowed) with target cells whose beta-adrenergic receptors are unable to maintain normal function of the tissues responsible for expressing atopic symptoms (e.g., respiratory tract, nasal mucosa, and so on). Certain data seem to support this hypothesis. The findings include results obtained in physiologic assays of receptor function (e.g., Busse, 1977; Kaliner et al., 1982; Barnes and Pride, 1983; Davis et al., 1986; Goldie et al., 1986), assays of beta-adrenergic receptor-mediated cyclic AMP generation (e.g., Parker and Smith, 1973; Busse, 1977), and beta-adrenergic receptor binding studies (e.g., Galant et al., 1979; Kariman, 1980; Brooks et al.; 1979; Sano et al., 1983). Other studies in patients with atopic disorders have failed to find changes in beta-adrenergic receptor number and have raised important methodologic questions about the studies that observed such changes (e.g., Galant et al., 1980; Ruoho et al., 1980; Tashkin et al., 1982; Reinhardt et al., 1984; Davies et al., 1986).

Additional work has emphasized the difficulty in distinguishing disease-related changes from the effect of beta-adrenergic agonist therapy or changes induced by severe disease in association with increased levels of circulatory catecholamines (Conolly and Greenacre, 1976; Galant et al., 1980; Bruynzeel et al., 1979; Berg et al., 1982; Tashkin et al., 1982; Sano et al., 1983). Other work has shown that glucocorticoids, which are commonly used in treatment of atopic disease, can block or reverse these agonist-induced changes (Hui et al., 1982a; Davies and Lefkowitz, 1983; Samuelson and Davies, 1984; Brodde et al., 1985a).

The possibility that only certain tissues may express beta-adrenergic receptor abnormalities has been inferred from the results of Sano et al. (1983). Those workers performed parallel studies on lymphocytes and PMN granulocytes and observed a decreased number of beta receptors in lymphocytes, but not in PMN, of asthmatic subjects untreated with adrenergic agonists. However, beta-adrenergic receptor number was decreased in both cell types in agonist-treated patients. The possibility that such changes might occur only in certain subpopulations of lymphocytes (Landmann et al., 1984; Khan et al., 1986; Maisel et al., 1989a) has not been explored, although abnormal circadian variation in expression of lymphocyte beta-2 adrenergic receptors of asthmatic subjects has been noted (Titinchi et al., 1984).

In other recent studies, Meurs et al. have observed that challenge with allergen causes a variety of changes in the beta-2 adrenergic receptor-adenylate cyclase system of lymphocyte membranes of allergic asthmatic patients. These include uncoupling and downregulation of beta-adrenergic receptors as well as nonspecific refractoriness of adenylate cyclase to hormonal agonists (beta-adrenergic-agonists and histamine) and nonhormonal stimuli (Gpp(NH)p and NaF) (Meurs et al., 1982,1986,1987). These results could not be replicated when cells were treated with the beta-agonist isoproterenol in vitro. Instead, the changes were similar to those produced by treatment of cells with a phorbol ester, thus suggesting that allergen-induced changes might result from an activation of protein kinase C, the presumed cellular receptor for phorbol esters.

One mechanism accounting for beta-adrenergic receptor subsensitivity in atopic disease might be antibodies directed against beta-adrenergic receptors. Such antibodies might block binding or function of the receptor and conceivably could arise secondary to infections or genetic predisposition. Data documenting that such antibodies are present in atopic subjects has been provided in several studies (Venter et al., 1980; Fraser et al., 1981; Blecher et al., 1984). However, subjects without atopic diseases can also show a detectable titer of antireceptor antibodies, and the frequency and titer of the antibodies in atopic individuals is generally low. In one study of 376 pediatric asthmatic patients, only about 5% of the asthmatic patients and about 9% of what were termed "high-risk" patients had probable evidence of autoantibodies to beta-adrenergic receptors (Blecher et al., 1984). Thus, at the present time, it seems unlikely that antireceptor antibodies play a major role in pathogenesis of atopic disease in most patients with these disorders.

In summary, many patients with atopic diseases show evidence of subsensitivity at beta-adrenergic receptors, but this may relate in part to desensitization/downregulation produced by increased levels of endogenous catecholamines or by adrenergic agonist therapy. Evidence that allergen exposure can induce a refractoriness of adenylate cyclase, perhaps secondary to an activation of protein kinase C, provides an additional mechanism to account for beta-adrenergic subsensitivity. The extent to which alterations in properties of beta-adrenergic receptors *per se* contribute to the pathogenesis of asthma and other atopic diseases remains an incompletely tested issue. However, the effectiveness of beta-adrenergic-agonist-mediated bronchodilation in treatment of asthmatic patients argues against a major defect in beta receptors in asthma (Nadel and Barnes, 1984). The questionable suitability of blood cells as appropriate models to study expression of receptors in the airway and other target sites involved in atopy complicates one's ability to draw firm conclusions. Finally, antibeta-adrenergic receptor antibodies appear not to play a major role in pathogenesis of atopic disorders, but may contribute to the abnormal adrenergic response in some patients with these disorders.

3.2. Cardiac Disorders

Beta-adrenergic receptors are widely recognized as playing an important role in enhancement of cardiac function—in particular, of rate (chronotropy) and force (inotropy) of cardiac contraction and in contributing to the genesis of cardiac arrhythmias. Although adrenergic agonists are sometimes employed to elicit such effects, more commonly beta-adrenergic antagonists are used to block one or more of those actions in patients with cardiac disorders. Beta-adrenergic receptors, aside from their usefulness as therapeutic "targets," are involved in contributing to the pathophysiology of cardiac disease. Studies with experimental animals and, to a lesser extent, with human subjects have explored this question in two principal settings, myocardial failure and myocardial ischemia, which will be discussed below. In addition, less-detailed information is available that suggests subsets of women patients with mitral valve prolapse and symptoms of beta-adrenergic hypersensitivity may have enhanced coupling of beta-adrenergic receptors to G proteins, as assessed by studies of radioligand binding in neutrophil membranes (Davies et al., 1987).

3.2.1. Myocardial Failure

Myocardial failure (congestive heart failure) represents a decompensation of homeostatic mechanisms that are used by the cardiovascular system to maintain cardiac output and blood pressure. Patients suffering from heart failure utilize adrenergic stimulation to support cardiac function; these patients thus often appear hyperadrenergic (e.g., tachycardic, diaphoretic, and so on) and can be shown to have higher plasma and urinary catecholamine levels (in particular, norepinephrine) as well as enhanced sympathetic nerve traffic (Thomas and Marks, 1978; Cohn et al., 1984; Bristow, 1984b; Viquerat et al., 1985; Leimbach et al., 1986). It has been hypothesized that excessive sympathetic tone may contribute to the pathophysiology of decreased cardiac function in the setting of congestive heart failure (as reviewed by Francis and Cohn, 1986).

The work of Bristow and colleagues has provided direct evidence for alteration in beta-adrenergic receptors of human myocardium from patients with cardiac failure (Bristow et al., 1982; Ginsburg et al., 1983; Bristow, 1984a,b). This finding has been confirmed by other investigators (e.g., Brodde et al., 1986a; Bohm et al., 1988; Limas et al., 1989b; Murphee and Saffitz, 1989).

Bristow and colleagues initially showed that the number of ventricular myocardial beta-adrenergic receptors was decreased about 50% in such patients, and that this decrease in receptor number was associated with a decrease in sensitivity of myocardium to adrenergic agonist-mediated contractile response and to beta-adrenergic agonist-mediated stimulation of adenylate cyclase (Bristow et al., 1982). Responses to another class of receptor (e.g., histamine) and to non-receptor stimuli (NaF, forskolin) were not decreased in ventricular preparations from patients with heart failure (Bristow, 1984a). Moreover, the loss in beta-adrenergic receptor and response appeared to be relatively specific for the ventricular chamber that was in failure (Bristow, 1984a; Bristow et al., 1986). The decrease in receptor number seems to selectively involve cardiac beta-1 adrenergic receptors rather than beta-2 adrenergic receptors, which are also present in the heart and may mediate inotropic and chronotropic responses to catecholamines, perhaps preferentially to circulating catecholamines (Bristow et al., 1986,1989; Brodde et al., 1986a). Recent studies indicate that endomyocardial biopsies can be used to assess beta-adrenergic receptors in patients with heart failure; in these patients, the reduction in myocardial beta-adrenergic receptor density was related to the degree of heart failure (Fowler et al., 1986). Other recent studies that have involved use of quantitative autoradiography indicate a selective reduction of beta-adrenergic receptors in subendocardial myocytes in patients with severe congestive heart failure (Murphee and Saffitz, 1989).

What is the molecular mechanism for this selective loss in the number of beta-adrenergic (even more so, of beta-1 adrenergic) receptors in heart failure? One possibility, for which there is as yet limited direct evidence is that patients with heart failure have disease-induced alterations in beta-adrenergic receptors that lead to their enhanced degradation, perhaps secondary to antibodies directed against the receptors (Limas et al., 1989a). A more likely possibility is that the decrease in receptor number and response is secondary to a catecholamine-mediated downregulation of myocardial beta-adrenergic receptors (Bristow 1984b; Limas et al., 1989b). Thus, compensatory increases in plasma and, perhaps more transiently, in myocardial levels of norepinephrine that are associated with cardiac failure may induce homologous desensitization of the heart to beta-adrenergic agonist stimulation. This type of desensitization has been produced in several experimental systems involving the heart, including infusion of animals with beta-adrenergic agonists, transplantation of catecholamine-producing pheochromocytoma, and exposure of isolated cardiac cells to high levels of catecholamines (Tse et al., 1979; Marsh et al., 1980; Chang et al., 1982; Snavely et al., 1983; Linden et al., 1984; Tsujimoto et al., 1984; Hayes et al., 1984; Karliner et al., 1986; Reithmann and Werden, 1989). In addition, other clinical data indicate that administration of beta-adrenergic agonists, such as dobutamine and pirbuterol, to patients with congestive heart failure produces only shortlived enhancement in hemodynamic parameters (Colucci et al., 1981; Unverferth et al., 1980). Since beta-2 adrenergic receptors appear not to be down-regulated in heart failure, use of beta-2 adrenergic agonists may potentially provide beneficial effects in myocardial function in patients with severe heart failure. However, postreceptor alterations, perhaps in G proteins, in patients with heart failure (Bristow et al., 1989; Insel and Ransnäs, 1988) may limit efficacy of beta-2 adrenergic agonists.

Assessment of beta-adrenergic receptors in heart failure has also involved studies of receptor expression in peripheral blood MNL. Several groups of investigators (but not all [Maisel et al., 1989b]) have reported that MNL from patients with heart failure express a decreased number of beta-adrenergic receptors as well as a decrease in beta-adrenergic-stimulated cyclic AMP generation (Colucci et al., 1981; Minakuchi et al., 1981; Brodde et al., 1986b). This is intriguing, since MNL possess only beta-2 adrenergic receptors, and cardiac tissue has predominantly beta-1 adrenergic receptors. If one were to hypothesize that down-regulation of cardiac beta-1 adrenergic receptors in heart failure were to result from enhanced levels of plasma norepinephrine (which shows a much higher affinity for beta-1 than for beta-2 adrenergic receptors), then it would be difficult to explain the downregulation of beta-2 adrenergic receptors on MNL. Thus, other factors, such as elevated levels of epinephrine or catecholamine-induced changes in lymphocyte subpopulations (which would express different numbers of beta-adrenergic receptors [Crary et al., 1983; Landmann et al., 1984; Khan et al., 1986; Maisel et al., 1989a]), may also contribute to the down-regulation of lymphocyte beta-adrenergic receptors produced in heart failure.

3.2.2. Myocardial Ischemia

Myocardial ischemia is another cardiac disorder in which beta-adrenergic receptor number may change. Although no data are available in humans, results obtained in experimental animals in which coronary arteries are ligated indicate that beta-adrenergic receptor number increases and beta-adrenergic-agonist stimulated cyclic AMP formation and metabolic responses are enhanced within several minutes after the onset of ischemia (Mukherjee et al., 1979, 1982; Maisel et al., 1985; Dominiak and Türck, 1986). The mechanism for this "upregulation" in beta-receptor number is not yet fully defined, but recent studies in guinea pigs indicate that an "externalization" of receptors may occur with ischemia. Thus, ischemia is associated with a decrease in the number of cardiac beta-adrenergic receptors identifiable in a "light vesicle" membrane fraction (one that sediments at high speeds and lacks G_s and catalytic adenylate cyclase activities) in parallel with an increased number of cell surface (sarcolemmal) receptors (Maisel et al., 1985,1987a). This "externalization" may represent a net decrease in receptors that are tonically internalized in response to agonist or other factors (Maisel et al., 1987b). In addition to changes in beta-adrenergic receptor number, myocardial ischemia may be associated with changes in the coupling of receptors to the G_s protein and to the stimulation of adenylate cyclase (Devos et al., 1985; Freissmuth et al., 1987).

The apparent "externalization" of beta-adrenergic receptors from intracellular sites, where the receptors are nonfunctional, to sarcolemmal sites where the receptors are active, is potentially reversible with reversal of ischemia and does not occur in guinea pigs that are pretreated with beta-adrenergic antagonists, such as propranolol (Maisel et al., 1987a,b). Propranolol treatment also promotes "externalization" of receptor from "light vesicle" to sarcolemmal sites, suggesting that the beta-antagonist and ischemia may involve externalization of (or loss of internalization to) the same pool of light vesicle receptors. From a clinical perspective, it is possible that the increase in sarcolemmal beta-adrenergic receptor number with ischemia is related to ischemic-related cardiac arrhythmias or to myocardial necrosis. Moreover, treatment of patients with beta-adrenergic antagonists has been shown to improve mortality in the setting of acute myocardial infarction as well as for many months to years after myocardial infarction (as reviewed by Vedin and Wilhelmsson, 1983 and Sleight, 1986). It is tempting to speculate that this clinical effect may relate to the action of beta-adrenergic antagonists in preventing or altering ischemic-induced increases in the number of sarcolemmal beta-adrenergic receptors.

3.2.3. Beta-Adrenergic-Blocker Withdrawal Syndrome

The ability of propranolol to externalize beta-adrenergic receptors may represent a blockade of ongoing internalization of beta-adrenergic receptors in response to ambient concentrations of agonist. This effect of propranolol may also explain why treatment with certain beta-adrenergic blockers can upregulate beta-adrenergic receptors in heart as well as other tissues (Aarons et al., 1980; Fraser et al., 1981; Brodde et al., 1985b; Hedberg et al., 1985; Golf and Hansson, 1986). This antagonist-induced increase in beta-adrenergic receptor number, in association with the rapid disappearance of plasma and tissue levels of propranolol after abrupt discontinuation of the drug has been proposed as the basis for the propranolol (or beta-adrenergic blocker) withdrawal syndrome (Prichard et al., 1983; Wood, 1983). This clinical syndrome is characterized by hyperadrenergic signs and symptoms, which include myocardial ischemia and arrhythmias, and is sometimes observed in patients who abruptly terminate therapy with propranolol or other beta-adrenergic antagonists.

Therapy with "full" antagonists other than propranolol can also upregulate beta-adrenergic receptor in animals and people, but partial agonists that have intrinsic sympathomimetic activity may not produce beta-adrenergic receptor upregulation, and also may not promote "externalization" of cardiac beta-adrenergic receptors (Brodde et al., 1985b; Neve and Molinoff, 1986; Golf and Hansson, 1986; Hedberg et al., 1987; Maisel et al., 1987a,b; Michel et al., 1988b). Moreover, recent data obtained with human right atrial and lymphocyte membranes suggest that antagonist-induced upregulation of beta-adrenergic receptors may be subtype-selective such that beta-1 selective blockers (e.g., metroprolol and atenolol) increase cardiac beta-1 receptors but not cardiac or lymphocyte beta-2 adrenergic receptors (Michel et al., 1988); others have not found evidence for those types of subtype-selective effects (Golf and Hansson, 1986). It is also possible that patients treated with other beta-adrenergic blockers, typified by the drug tertatolol, may actually have a drug-induced decrease in beta-adrenergic receptor number (DeBlasi et al., 1986c,1988). *(See also,* chapter 8).

3.2.4. Summary

In summary, patients with congestive heart failure show a prominent decrease in beta-adrenergic receptor number in failing ventricles and perhaps also in MNL. This decrease may be caused by a catecholamine-mediated downregulation and desensitization of receptors. Myocardial ischemia in experimental animals is associated with a rapid increase in beta-adrenergic receptor number, perhaps as a consequence of net externalization of receptor from intracellular to sarcolemmal sites. Therapy with certain beta-adrenergic antagonists can upregulate receptors in MNL and in the heart, perhaps also as a consequence of externalization of receptors. The discrepancy between a rapid loss in plasma and tissue levels of a beta-adrenergic blocker and the slower reversal of antagonist-induced upregulation of receptors may produce a beta-adrenergic blocker withdrawal syndrome.

3.3. Hypertension

Considerable evidence has suggested that alterations in the sympathetic nervous system contribute to the development and/or maintenance of systemic hypertension *(see* reviews by Abboud, 1982; Folkow, 1982; Kuchel, 1983; Insel and Motulsky, 1984; Rosendorff et al., 1985; Westfall and Meldrum, 1985; Gavras, 1986; Feldman, 1987; de Champlain et al., 1989; Michel et al., 1990). In addition, sympatholytic drugs, in particular beta-adrenergic blocking drugs, are important in the management of hypertensive patients. The precise mechanism(s) by which beta-adrenergic blockers lower blood pressure has not yet been resolved, but beta-adrenergic receptors are known to contribute to the regulation of blood pressure through effects at several target sites, including the central nervous system, adrenergic nerve terminals, blood vessels, heart, and kidney (Frishman, 1987).

I believe that an elevation in blood pressure is a physical sign produced by what will ultimately be recognized as several different "diseases." Our current lack of molecular understanding regarding pathogenesis of "essential" or "primary" hypertension leads many investigators to group patients together, even though differences in certain biochemical and clinical features can be discerned. For example, some subsets (typically about 30%) of hypertensive patients have moderate (typically <50%) increases in plasma catecholamine levels, at least as reported in some studies (Kuchel, 1983; Goldstein, 1983). Increases in plasma norepinephrine in hypertensive patients seem to be most readily observed in patients younger than age 40 (Goldstein et al., 1983). It is attractive to imagine that changes in beta-adrenergic receptors may occur as primary or as early alterations in the pathogenesis of hypertension, at least in some forms of the disease. Since agonists can act at beta-adrenergic receptors in vessels to produce vasodilation and lower blood pressure, one hypothesis is that hypertension is associated with a decrease in beta-adrenergic receptor number and/or affinity and, in turn, of beta-2 adrenergic-mediated vasodilation.

Results in support of such a hypothesis have been reported in certain forms of hypertension in experimental animals, although the data are not entirely consistent (Hamet, 1983; Limas and Limas, 1985; Vatner et al.,1985a; Rosendorff et al., 1985; Brodde, 1988; Feldman, 1987). Moreover, cardiac hypertrophy that results from hypertension may also be associated with a decrease in beta-adrenergic receptor number and function (Limas and Limas, 1978,1985; Kumano et al., 1983). Hypertensive animals may also have a reduced ability to exhibit agonist-induced changes in subcellular compartmentation of beta-adrenergic receptors (Limas and Limas, 1984). In genetic forms of hypertension, data from some studies indicate that changes in beta-adrenergic receptors in the heart and vessels can be detected prior to the development of hypertension (e.g., Limas and Limas, 1978). This would suggest that the changes in the receptors may not merely be the result of hypertension *per se*, although other data raise questions about such a conclusion (Feldman, 1987; Michel et al., 1989,1990).

Studies in human subjects have shown that certain patients with labile hypertension seem to have enhanced cardiac and vascular response to isoproterenol, whereas patients with chronic, established hypertension can have blunted or enhanced response to beta-adrenergic agonists (Kuchel, 1983; Alexander, 1987; Feldman, 1987; de Champlain et al., 1989). Changes in beta-adrenergic receptors have been assessed on peripheral blood cells. The results generally indicate that receptor number either is not altered in patients with essential hypertension or is modestly increased, perhaps in parallel with the increased level of mean arterial blood pressure (Feldman et al., 1984b; Brodde et al., 1985c; *see* reviews by Alexander, 1987; Feldman, 1987; Brodde, 1988). Other preliminary data suggest an increase in beta-adrenergic receptor number and in beta-adrenergic-mediated adenylate cyclase activity in the right atria of patients with essential hypertension (Hill et al., 1986). It is difficult to reconcile this increase in beta-adrenergic receptor number with the reported decrease in receptor number often observed in studies with experimental animals. It is also paradoxical that patients have increases in beta-adrenergic receptors even though increases in circulating catecholamines are often found in hypertensive patients, since an increase in catecholamines would be expected to decrease receptor number. This paradox has suggested that hypertension may be associated with an abnormal "regulation" of beta-adrenergic receptors (Limas and Limas, 1984,1987; Brodde, 1988).

One example of altered regulation of beta-adrenergic receptors in hypertension has been provided by the work of Feldman and coworkers (1984b,1987b). These investigators demonstrated that subjects with essential hypertension have an altered ability of beta-adrenergic receptors in MNL membranes to exhibit high-affinity binding of the agonist isoproterenol in competition binding studies (Feldman et al., 1984). This inability to form a high-affinity binding state (thought to represent a ternary complex between agonist-receptor and G_s protein [*see* review by Lefkowitz et al., 1983 and chapter by Stadel and Lefkowitz, this vol.]) was associated with a decrease in beta-adrenergic-modulated adenylate cyclase activity in blood cells from hypertensive subjects. Similar findings with respect to loss of high-affinity binding of agonists and decrease in adenylate cyclase activity are associated with upright posture and increases in plasma catecholamines (Feldman et al., 1983). It is thus conceivable that chronic small increases in catecholamines, especially epinephrine, may produce an uncoupling of beta-adrenergic receptors in hypertensive subjects. In recent studies, Feldman and colleagues have shown that a low-salt diet (10 meq/d) can ameliorate the hypertension-associated changes in lymphocyte beta-adrenergic receptor binding and function, thus indicating that these changes are reversible and may be related to dietary sodium intake (Feldman et al., 1987b). Since low-salt diets typically enhance plasma catecholamine levels (Fraser et al., 1981), the mechanism by which the changes in beta-adrenergic receptor binding are reversed with low-salt diet is not clear.

To summarize, hypertension is likely to be a heterogeneous disease that is "caused" or influenced by several different factors. These may include alterations in beta-adrenergic receptor number or in receptor affinity for agonists, as shown in peripheral blood cells of some hypertensive patients. Mechanisms for these changes are not yet well defined, although catecholamine-induced effects or abnormal regulation of beta-adrenergic receptors may be involved.

3.4. Diabetes Mellitus

Changes in beta-adrenergic receptors can be associated with the absolute or relative deficiency of insulin that characterizes the diabetic state. In experimental animals in which diabetes is produced by administration of an agent that destroys pancreatic beta cells (the insulin-producing cells), the number of myocardial and adipocyte beta adrenergic receptors (presumably beta-1 adrenergic receptors) is decreased (Savarese and Berkowitz, 1979; Ingebretsen et al., 1983) and skeletal muscle beta-adrenergic receptor number (presumably beta-2 adrenergic receptors) is slightly increased (Olson et al., 1981). The role that these changes play in the altered metabolic state of diabetes mellitus is not clear.

In human subjects, limited data indicate that subjects with type 1, or insulin-dependent, diabetes mellitus (IDDM) have a normal number of lymphocyte beta-2 adrenergic receptors, but an increased glycemic (beta-adrenergic) response to catecholamines (Shamoon et al., 1980; Serusclat et al., 1983). Other data indicate that children with IDDM have a decrease in MNL membrane beta-adrenergic receptor number and that children with less well controlled disease have lower receptor number than those whose disease is under better control (Noji et al., 1986).

The status of beta-adrenergic receptors in other target cells in human subjects with diabetes mellitus has not been carefully defined. Several investigators have shown, however, that subjects with IDDM have a defective increment in plasma epinephrine in response to hypoglycemia (Bolli et al., 1983; Cryer and Gerich, 1985). This increase in plasma epinephrine mediates an increase in blood sugar by beta-adrenergic effects that lead to both a decrease in glucose utilization and an increase in glucose production (Cryer and Gerich, 1985). Thus, patients with IDDM may have offsetting changes in the adrenergic system with lower epinephrine levels, but enhanced beta-adrenergic responses in target cells. Since the liver is the principal site of glucose production, a testable hypothesis (but one not yet tested) would be that diabetic subjects have an enhancement in hepatic beta-2 adrenergic receptor number or in receptor affinity for catecholamines (Kawai et al., 1986).

3.5. Thyroid Hormone Disorders

In this book, Stiles has discussed beta-adrenergic receptor regulation by thyroid hormone. Other reviews have also discussed this topic (Bilezikian and Loeb, 1983; Malbon et al., 1988).

In human subjects, some data suggest that basal levels of thyroid hormone are not important for adrenergic receptor expression on peripheral blood cells (Rosen et al., 1984); other studies indicate that administration of triiodothyronine may (Ginsberg et al., 1981; Andersson et al., 1983) or may not (Hui et al., 1982b) increase leukocyte beta-adrenergic receptor number. Some workers have failed to find an increase in MNL beta-adrenergic receptor number in thyrotoxic patients compared to controls (Williams et al., 1979), other groups report that treatment of thyrotoxicosis will lower beta-adrenergic receptor number (Andersson et al., 1983; Cognini et al., 1983). Data obtained with gluteal adipocytes indicate that patients with hyperthyroidism show very slight increases in beta-adrenergic receptor number but prominent increases in beta-adrenergic-mediated cyclic AMP generation and lipolysis, as well as in agonist affinity for the receptor (Richelsen and Sørensen, 1987). Other data confirm the increase in beta-adrenergic receptor number in fat, as well as in skeletal muscle membrane in subjects treated with triiodothyronine (Liggett et al., 1989b). Patients with hyperthyroidism show numerous clinical features of a hyperadrenergic state (tachycardia, tremor, diaphoresis, thermogenesis, and so on), and beta-adrenergic antagonists play a useful role in clinical management of such patients. Thus, alteration in beta-adrenergic receptor binding and/or function is likely to contribute to clinical thyrotoxicosis; this conclusion, however, is not well-supported by recent experimental data (Liggett et al., 1989b). Since thyroid hormone is known to effect changes in gene transcription (Samuels et al., 1988), it is possible that the hormone enhances the rate of transcription of the gene encoding beta-adrenergic receptors, at least in some tissues, of such patients (Gross and Lues, 1985). In addition, thyroid hormone might effect an increased transcription of G_s (Malbon et al., 1988; Ransnäs et al., 1988).

Conversely, one would predict lower rates of transcription of beta-adrenergic receptors in cells of patients with thyroid hormone deficiency. Consistent with this idea are the findings that thyroidectomy can decrease MNL beta-adrenergic receptor number (Cognini et al., 1983) and that adipocytes from patients with hypothyroidism show a decrease in beta-adrenergic receptor number and in beta-adrenergic-mediated cyclic AMP generation and lipolysis (Richelsen and Sørensen, 1987). Little direct data are as yet available regarding the question of whether people with hypothyroidism, like hypothyroid animals, have a decreased ability of beta-adrenergic receptors to couple productively to G_s (e.g., Malbon et al., 1984; Richelsen and Sørensen, 1987; Malbon et al., 1988). Since thyroid disorders produce metabolic changes in many different tissues, it should prove useful to learn more about alteration in beta-adrenergic receptor number and coupling to G_s in additional tissues from hyperthyroid and hypothyroid individuals.

3.6. Pheochromocytoma

Pheochromocytoma is a tumor of chromaffin tissue that results in increased levels of plasma catecholamines. The plasma level of catecholamines in patients with pheochromocytoma increases three- to tenfold over that found in normal individuals, although there is considerable interindividual and intraindividual variation (Bravo et al., 1979; Bravo and Gifford, 1984). In addition, the relative proportion of epinephrine and norepinephrine are quite variable, presumably reflecting tumor content and activity of phenylethanolamine *N*-methyltransferase (the enzyme responsible for converting norepinephrine to epinephrine).

One might imagine that the effect of pheochromocytoma on beta-adrenergic receptors would be directly predicted from the impact of chronically elevated levels of plasma catecholamines on tissue receptors. Thus, desensitization and downregulation of beta-adrenergic receptors would be an expected consequence in this setting. Evidence of such changes has been reported in animals with transplanted pheochromocytoma as well as in some patients with these tumors (Greenacre and Conolly, 1978; Smith et al., 1977; Snavely et al., 1983; Tsujimoto et al., 1984; Ratje and Wisser, 1986). In addition, other factors seem to contribute to alteration in beta-adrenergic receptors and receptor-mediated responses in pheochromocytoma. Thus, rats infused with norepinephrine seem to show a somewhat different pattern of downregulation of beta-adrenergic subtypes than do animals with transplantable pheochromocytoma that produce norepinephrine (Snavely et al., 1985). This may result from the release of other factors from the tumors, perhaps chromaffin storage granule proteins, neuropeptide Y, or opioid peptides.

In addition, the high circulatory levels of catecholamines in pheochromocytoma can produce tissue damage, in particular in the heart (Van Vliet et al., 1966), and this too may contribute to cardiovascular complications of this disease. A further complication is that hypertension, a virtual *sine qua non* of pheochromocytoma, may itself have effects on tissue beta-adrenergic receptors, as discussed above.

Beta-adrenergic receptors have been assessed in peripheral blood cells from relatively small numbers of patients with pheochromocytoma (Chobanian et al., 1982; Ratje and Wisser, 1986). The data suggest prominent decreases in beta-adrenergic receptor number and beta-agonist-stimulated adenylate cyclase in such cells. Removal of the tumor restores beta-adrenergic receptor number and function to values observed in normal controls (Ratje and Wisser, 1986).

Desensitization of beta-adrenergic response of cells from patients with pheochromocytoma has been demonstrated in multiple tissues (Smith et al., 1977; Greenacre and Conolly, 1978; Ratje and Wisser, 1986). This desensitization probably represents an important negative feedback mechanism for maintaining hemostasis in the face of what may be lethal levels of catecholamines in patients with these tumors.

3.7. Pseudohypoparathyroidism

Pseudohypoparathyroidism (PHP) is a rare genetic disorder characterized by abnormal responsiveness of tissues to parathyroid hormone as well as to several other hormones that work through stimulation of adenylate cyclase (Levine et al., 1983; Van Dop and Bourne, 1983). In a portion of these patients (those with PHP-Ia), the activity of G_s is deficient (as assessed by cholera toxin-stimulated [^{32}P]ADP ribosylation or by reconstitution of G_s-deficient S49 lymphoma cells), although in other patients (those with PHP-Ib), G_s activity is apparently normal. Beta-adrenergic receptor-mediated stimulation of adenylate cyclase also is decreased in cells from patients with PHP-Ia (Carlson and Brickman, 1983). Radioligand binding studies conducted with beta-adrenergic receptors in erythrocytes indicate that this decrease in response seems to represent a decreased ability of the receptors to form high-affinity complexes between the agonist-bound receptor and G_s protein (Heinsimer et al., 1984). Such results provide evidence supporting the view that the ability of receptors to form such complexes is important for beta-adrenergic receptor-mediated stimulation of adenylate cyclase in normal human cells.

3.8. Neurologic Disorders

Neurologic disorders that are characterized by dysfunction of the autonomic nervous system may be associated with alterations in beta-adrenergic receptors. These disorders include primary neurologic diseases as well as diseases, such as diabetes mellitus, that may secondarily manifest abnormalities in autonomic nervous system activity. Most commonly, the clinical picture associated with these disorders includes orthostatic (upright posture) hypotension, often together with supine hypertension, as well as abnormalities of bladder and bowel control, pupillary changes, and abnormal sweating (*see* review by Robertson and Hollister, 1987). These clinical findings are often associated with relatively low levels of circulating catecholamines. Patients with autonomic dysfunction can show marked hypersensitivity to beta-adrenergic (as well as alpha-adrenergic) agonists and thus demonstrate enhanced chronotropic and vasodilatory responses to beta-adrenergic agonists (Robertson et al., 1984; Robertson and Hollister, 1987). Lymphocyte beta-adrenergic receptor number can be increased in patients showing such hypersensitivity (e.g., Bannister et al., 1981; Hui and Conolly, 1981; Chobanian et al., 1982). Thus, enhanced numbers of receptors, perhaps resulting in part from the withdrawal of tonic "downregulation" by ambient catecholamines, may contribute to adrenergic hypersensitivity in autonomic dysfunctional syndromes.

Denervation of particular organs is also associated with beta-adrenergic supersensitivity, in part attributable to an increased number of postsynaptic receptors, but also, perhaps more importantly, to a lack of norepinephrine reuptake (Vatner et al., 1985b). Similarly, many studies have indicated that treatment of animals with drugs that deplete tissue catecholamines (e.g., guanethidine, reserpine, 6-hydroxydopamine) will also produce an upregulation of beta-adrenergic receptors (e.g., Glaubiger et al., 1978; Tenner et al., 1982; Kajiyama et al., 1982); adrenal demedullation can produce a similar effect (Sundaresan et al., 1987). I am not aware of direct evidence for these types of changes in humans with drug- or surgically induced denervation, or with catecholamine depletion.

Patients with dementia of the Alzheimer's type have been tested for beta-adrenergic receptors or receptor-mediated cyclic AMP formation in MNL, cultured fibroblasts and various brain regions; in general, no statistically significant ($p > .05$) changes have been noted compared to age-matched controls (Oppenheim et al., 1984; Ebstein et al., 1984; Jenni-Eiermann et al., 1984; Volicier et al., 1985), although some region- and receptor-subtype-specific changes have been noted (Shimohama et al., 1987). Patients with Parkinson's disease also fail to show prominent differences in beta-adrenergic receptors in postmortem brain or cerebral microvessels compared to age-matched controls (Cash et al., 1984,1985).

3.9. Psychiatric Disorders

An extensive literature has been published in which attempts have been made to relate alterations in catecholamines and catecholamine receptors to several different psychiatric conditions (*see* review by Insel and Cohen, 1987). Norepinephrine in the brain has been implicated as an important regulator of several functions, including mood, memory, neuroendocrine control, and stimulation of the autonomic nervous system. Several of these functional responses are thought to be controlled by beta-adrenergic receptors in the central nervous system. Studies in humans have suggested that beta-1 adrenergic receptors may be localized to neural cells, whereas beta-2 adrenergic receptors may be predominantly associated with either glial or vascular elements (Cash et al., 1986). However, quantitative autoradiographic studies, such as those indicating selective regional localization of beta-1 and beta-2 adrenergic receptors in rat brain (Rainbow et al., 1984), are not yet available for human brain. In addition, beta-adrenergic receptors, perhaps predominantly beta-1 adrenergic receptors, are downregulated by treatment of animals with tricylic antidepressants or other maneuvers (e.g., monoamine oxidase inhibitors or electroconvulsive therapy) that increase tissue levels of catecholamines (e.g., Banerjee et al., 1977; Pandey et al., 1979; Minneman et al., 1981; Cohen et al., 1982; Lerer, 1984).

Data in human subjects has been limited by the paucity of studies available on central nervous system tissue from patients who have psychiatric disorders. Some information has been obtained from postmortem studies, but most results have been provided by studies obtained using peripheral blood cells.

Much effort has been directed at patients with affective (mood) disorders, since the "catecholamine hypothesis"—which states that depressed patients have a decreased level of brain catecholaminergic activity and manic patients an excess of such activity—has provided a framework for the possible pathogenesis of these disorders (Schildkraut, 1965; Insel and Cohen, 1987). Several groups of investigators have reported somewhat lower (~20%) numbers of beta-adrenergic receptors and decreases in beta-adrenergic agonist-stimulated cyclic AMP levels in peripheral blood cells from patients with affective disorders (e.g., Pandey et al., 1987 and references therein; Wood et al., 1986) although other investigators have failed to find such changes (e.g., Healy et al., 1983; Zohar et al., 1983). In one study, patients with endogenous depression and psychomotor agitation (but not all depressed subjects) had a decrease in beta-adrenergic agonist-stimulated cyclic AMP accumulation in MNL, but a similar number of beta-receptors on those cells (Mann et al., 1985). A careful evaluation of the affinity of these beta-receptors for agonists was not undertaken, although subsequent data obtained by the same investigators indicate a blunting in isoproterenol stimulated cyclic AMP generation in MNL from depressed patients over a several-log concentration range of isoproterenol (Halper et al., 1988). Of interest, one group has observed a decreased number of beta-adrenergic receptors in lymphoblasts (derived by transformation with Epstein-Barr virus of lymphocytes) from patients with affective disorders in three out of five families (Wright et al., 1984). The binding studies were conducted after cells had been passed in culture for more than 6 mo, thus suggesting a possible genetic basis for the changes in beta-receptor binding. It is disappointing that another group of investigators could not replicate the findings of Wright et al. (Berrettini et al., 1987).

Studies of schizophrenia have emphasized the possibility of changes in dopamine or the dopamine receptor system, rather than changes in beta-adrenergic receptors. Limited studies on postmortem brains have failed to detect changes in [^{3}H]dihydroalprenolol binding in frontal cortex and caudate from schizophrenia patients; however, regions such as subcortical limbic structures, which would be hypothesized to be more likely to have lesions in schizophrenia, have not been studied (Insel and Cohen, 1987). Studies on peripheral blood leukocytes fail to show a change in beta-adrenergic receptor number or affinity in schizophrenic patients (e.g., Pandey et al., 1987).

Limited studies on patients with anorexia nervosa or bulimia suggest that the former group have a lower number of beta-adrenergic receptors on MNL and that this value returns to control levels with weight recovery (Lonati-Galligani and Pirke, 1986).

3.10. Other Disorders

The role of abnormalities in beta-adrenergic receptors has been assessed in a variety of other diseases, although in general, less detailed information is available. These include:

1. *Cystic Fibrosis:* Abnormal beta-adrenergic receptor sensitivity and decrease in beta-adrenergic binding in neutrophils has been reported in patients with cystic fibrosis (Davis et al., 1978,1980; Galant et al., 1981). Subsequent studies have failed to replicate the decrease in beta-adrenergic receptor number in such patients (Davis et al., 1983), although lymphocytes prepared from decompensated, hospitalized patients with cystic fibrosis show decreased beta-adrenergic receptor-stimulated adenylate cyclase activity (Feldman et al., 1987a). These and other data indicate that beta-receptor changes are unlikely to be of primary pathogenetic importance in this disease (Frizell et al., 1986; Widdicombe, 1986).

2. *Chronic Renal Insufficiency.* Patients with chronic renal failure sometimes have a blunting of adrenergic responsivity, which improves after dialysis therapy (Zucchelli et al., 1985). Chronic renal failure appears not to be associated with a reduction in number of beta-adrenergic receptors on MNL (in fact, an increase in receptor number may occur) (Souchet et al., 1986), but instead may result from a postreceptor defect in the coupling of beta-adrenergic receptors to adenylate cyclase (Brodde and Daul, 1984) or, perhaps, from an alteration in receptors caused by one or more factors present in uremic plasma ultrafiltrates (Bree et al., 1987).

3. *Chronic Lymphocytic Leukemia:* Several groups have reported that lymphocytes prepared from patients with chronic lymphocytic leukemia have a decreased number of beta-adrenergic receptors (Sheppard et al., 1977; Bidart et al., 1981; Paietta and Schwarzmeier, 1983). These findings might be related to the

evidence that different T cell subpopulations express different numbers of beta-adrenergic receptors (Khan et al., 1986). Thus, the clonal expansion of lymphocytic proliferation in leukemia may preferentially involve a population of cells endowed with a smaller number of receptors.

4. *Lactic Acidosis:* Davies has reported that incubation of neutrophils, together with both lactate (16 m*M*) and acidosis (pH 7.1) but neither alone, will produce an uncoupling of beta-adrenergic receptors (Davies, 1984). Receptor number was not altered, but receptor affinity for agonist and isoproterenol-stimulated cyclic AMP accumulation were markedly decreased by lactic acidosis. Subsequent studies have confirmed these findings and have indicated that redox couplets and pH can independently regulate the ability of beta-adrenergic receptors to bind agonist (Davies, 1986). These effects may relate to impaired beta-adrenergic receptor responsiveness observed in lactic acidosis and other types of metabolic acidosis, including, for example, chronic renal insufficiency, as described above.

4. Conclusions and Perspective

As reviewed above, alterations in beta-adrenergic receptors have been hypothesized to contribute to disease processes in several different organ systems. Suggestive data have been found for disease-related changes in beta-adrenergic receptors in human subjects with atopic disorders and certain cardiovascular, neurologic, and psychiatric diseases. It is somewhat disappointing to realize that the availability of radioligand binding and other biochemical assays of beta-adrenergic-receptor function have not yet provided clear, unequivocal evidence of a "beta-adrenergic receptor disease." This is in spite of the obvious importance of beta-adrenergic receptors for mediating numerous physiologic responses and a rather vigorous effort on the part of investigators to test for alterations in beta-adrenergic receptors in many different disease states.

Why has it been so difficult to find clearer evidence of changes in beta-adrenergic receptors in disease? One problem is the feedback control of receptor number by catecholamines. Since catecholamine levels increase as part of the body's stress response, it becomes difficult to separate disease-related changes from superimposed changes in receptors that are secondary to increases in catecholamines.

A second problem relates to the inability to assess beta-adrenergic receptors in internal organs *in situ*. The unresolved issue of the suitability of using accessible blood cells as mirrors of other tissues was discussed at the outset of this review. It is possible to examine beta-adrenergic receptors in other human cell types, such as fat cells (e.g., *see* Berlan and Lafonton, 1985), and in myocardial samples (Brodde et al., 1986b; Bristow et al., 1982,1986), but these studies generally require surgical intervention. The use of endomyocardial biopsies for studying cardiac beta-adrenergic-receptors provides a clinical approach that may prove important (Fowler et al., 1986). Work continues toward the development of radioisotopic scans for beta-adrenergic receptors in vivo, but progress has been slow (Hughes et al., 1986, and references therein).

A further problem in clinical studies is the excessive attention that has been devoted to disease-related changes in receptor number without adequate consideration of the possible changes in beta-adrenergic receptor affinity for agonists and interaction of receptors with G proteins. Some efforts directed at testing these important interactions have begun, but state-of-the-art techniques have not yet been widely applied. It is likely that disease settings associated with altered beta-adrenergic receptor-mediated functional responses will have altered expression of G proteins or of agonist promoted "coupling" of receptors to these proteins.

In sum, in spite of almost 10 years of effort, clinical studies of adrenergic receptors remain a major challenge. New techniques are probably going to be required to develop more definitive insights. The application to human tissues of methods to assess compartmentation of beta-adrenergic receptors (DeBlasi et al., 1985,1986a; Motulsky et al., 1986; Sandnes et al., 1987; Cook et al., 1987), of using photoaffinity labeling of receptors (Feldman and Lai, 1986), of preparing antibodies to receptors (Dixon et al., 1986) and of cloning human beta-adrenergic receptors (Kobilka et al., 1987; Chung, 1987; Emorine et al., 1987; Schofield et al., 1987; Frielle et al., 1987) provide useful new approaches that should help in defining disorders of beta-adrenergic receptors in human subjects.

Acknowledgments

I thank Pamela Bucher and Patti Kiecolt for typing of the manuscript, Otto-Erich Brodde for providing an unpublished manuscript, and Harvey Motulsky and Alan Maisel for helpful comments. Work in the author's laboratory is supported by grants from NIH (GM 31987, GM 40781, and HL 35018 and HL 35847) and NSF (DCB 85-02168).

References

Aarons, R. D., Nies, A. S., Gerber, J. G., and Molinoff, P. B. (1983) Decreased beta-adrenergic receptor density on human lymphocytes after chronic treatment with agonists. *J. Pharmacol Exp. Ther.* **224**, 1–6.

Aarons, R. D., Nies, A. S., Gal, J., Hegstrand, L. R., and Molinoff, P. B. (1980) Elevation of beta-adrenergic receptor density in human lymphocytes after propranolol administration. *J. Clin. Invest.* **65**, 949–957.

Abboud, F. (1982) The sympathetic system in hypertension. Hypertension **4** (Suppl. II), II-208– 225.

Alexander, R. W. (1987) Adrenergic receptors in cardiovascular disease in *Adrenergic Receptors in Man* (Insel, P. A., ed.), Marcel Dekker, New York, pp. 237–258.

Andersson, R. G. G., Milsson, O. R., and Kuo, J. F. (1983) Beta-adrenoceptor-adenosine 3'5'-monophosphate system in human leukocytes before and after treatment for hyperthyroidism. *J. Clin. Endocrinol. Metab.* **51**, 42–45.

Banerjee, S. P., Kung, L. S., Riggi, S. J., and Chanda, S. K. (1977) Development of beta-adrenergic receptor subsensitivity by antidepressants. *Nature* **268**, 455,456.

Bannister, R., Boylston, A. W., Davies, I. B., Mathias, C. J., Sever, P. S., and Sudera, D. (1981) Beta-receptor numbers and thermodynamics in denervation supersensitivity. *J. Physiol.* **319**, 369–377.

Barnes, P. J. and Pride, M. B. (1983) Dose-response curves to inhaled beta-adrenoceptor agonists in normal and asthmatic subjects. *Br. J. Clin. Pharmacol.* **15**, 677–682.

Berlan, M. and Lafontan, M. (1985) Evidence that epinephrine acts preferentially as an antipolytic agent in abdominal human subcutaneous fat cells: Assessment by analysis of beta and alpha$_2$-adrenoceptor properties. *Europ. J. Clin. Invest.* **15**, 341–348.

Berrettini, W. H., Capellari, C. B., Nurnberger, J. I., and Gershon, E. S. (1987) Beta adrenergic receptors on lymphoblasts. A study of manic-depressive illness. *Neuropsychobiology* **17**, 15–18.

Bidart, J. M., Motte, P., Bohuon, C., and Bellet, D. (1981) Lymphocyte aminergic binding changes in chronic lymphocytic leukemia. *Leukemia Res.* **5**, 443–446.

Bilezikian, J. P. and Loeb, J. N. (1983) The influence of hyperthyroidism and hypothyroidism on alpha- and beta-adrenergic receptor systems and adrenergic responsiveness. *Endocr Rev.* **4**, 378–388.

Blecher, M., Lewis, S., Hicks, J. M., and Josephs, S. (1984) Beta-blocking autoantibodies in pediatric bronchial asthma. *J. Allergy Clin. Immunol.* **74**, 246–251.

Bohm, M., Diet, F., Feiler, G., Keunkes, B., Kreuzer, E., Weinhold, C., and Erdmann, E. (1988) Subsensitivity of the failing human heart to isoprenaline and milrinone is related to beta-adrenoceptor downregulation. *J. Cardiovasc. Pharmacol.* **12**, 726–732.

Bolli, G., Feo, P. D., Compagnucci, P., Cartechini, M. G., Angeletti, G., Sonteusanio, F., Brunett, P., and Gerich, J. E. (1983) Abnormal glucose counter regulation in insulin-dependent diabetes mellitus. Interaction of anti-insulin antibodies and impaired glucagon and epinephrine secretion. *Diabetes* **32**, 134–141.

Bourne, H. R., Lichtenstein, L. M., Melmon, K. L., Henney, C. S., Weinstein, Y., and Shearer, G. (1974) Modulation of inflammation and immunity by cyclic AMP. *Science* **189**, 19–28.

Bravo, E. L. and Gifford, R. W., Jr. (1984) Pheochromocytoma: Diagnosis, localization and management. *N. Engl. J. Med.* **311**, 1298–1303.

Bravo, E. L., Tarazi, R. C., Gifford, R. W., and Stewart, B. H. (1979) Circulating and urinary catecholamines in pheochromocytoma. Diagnostic and pathophysiologic implications. *N. Engl. J. Med.* **301**, 682–686.

Bree, F., Souchet, T., Baatard, R., Fontenaille, C., Lhoste, F., and Tillement, J. P. (1987) Inhibition of (-)[^{125}I]iodocyanopindolol binding to rat lung-beta-adrenoceptors by uremic plasma ultrafiltrates. *Biochem. Pharmacol.* **36**, 3121–3125.

Bristow, M. R. (1984a) Myocardial beta-adrenergic receptor downregulation in heart failure. *Int. J. Cardiol.* **5**, 648–652.

Bristow, M. R. (1984b) The adrenergic nervous system in heart failure. *N. Eng. J. Med.* **311**, 850, 851.

Bristow, M. R., Hershberger, R. E., Port, J. D., Minobe, W., and Rasmussen, R. (1989) Beta-1 and beta-2 adrenergic receptor-mediated adenylate cyclase stimulation in nonfailing and failing human ventricular myocardium. *Mol. Pharmacol.* **35**, 295–303.

Bristow, M. R., Ginsberg, R., Minobe, W., Cubicciotti, R. S., Sageman, W. S., Lurie, K., Billingham, M. E., Harrison, D. C., and Stinson, E. D. (1982) Decreased catecholamine sensitivity and beta-adrenergic receptor density in failing human hearts. *N. Eng. J. Med.* **307**, 205–211.

Bristow, M. R., Ginsberg, R., Umans, V., Fowler, M., Minobe, W., Rasmussen, R., Zera, P., Merlove, R., Shah, P., Jamieson, S., and Stinson, E. B. (1986) Beta-1 and beta-2 adrenergic subpopulations in nonfailing and failing human ventricular myocardium. Coupling of both receptor subtypes to muscle contraction and selective beta-1 receptor downregulation in heart failure. *Circ. Res.* **59**, 297–309.

Brodde, O.-E. (1988) Beta-adrenoceptors, in *Receptor Pharmacology and Function* (Williams, M., Glennon, R. A., and Timmermans, P. B. W. M., eds.), Marcel Dekker, New York, pp. 207–255.

Brodde, O.-E. and Daul, A. (1984) Alpha- and beta-adrenoceptor changes in patients on maintenance hemodialysis. *Contrib. Nephrol.* **41**, 99–107.

Brodde, O.-E., Brinkmann, M., Schemuth, R., O'Hara, N., and Dave, A. (1985) Terbutaline-induced desensitization of human lymphocyte beta-2 adrenoceptors. Accelerated restoration of beta-adrenoceptor responsiveness by prednisone and ketotifen. *J. Clin. Invest.* **76**, 1096–1101.

Brodde, O.-E., Daul, A., Stuka, N., O'Hara, N., and Borchard, U. (1985b) Effects of beta-adrenoceptor antagonist administration on beta-2 adrenoceptor density in human lymphocytes. The role of "intrinsic sympathomimetic activity." *Nauyn Schmiedebergs Arch. Pharmacol.* **328**, 417–422.

Brodde, O.-E., Engel, G., Hoyer, D., Bock, K. D., and Weber, F. (1981) The beta-adrenergic receptor in human lymphocytes. Subclassification by the use of a new radioligand (+)-[^{125}I]iodocyanopindolol. *Life Sci.* **29**, 2189–2198.

Brodde, O.-E., Schüler, S., Kretsch, R., Brinkman, M., Borst, H. G., Hetzer, R., Reidemeister, J. C., Warnecke, H., and Zerkowski, H. R. (1986a) Regional distribution of beta-adrenoceptors in the human heart: Coexistence of functional beta–1 and beta–2 adrenoceptors in both atria and ventricles in severe congestive cardiomyopathy. *J. Cardiovasc. Pharmacol.* **8**, 1235–1242.

Brodde, O.-E., Stuka, N., Demuth, V., Fesel, R., Bergerhausen, J., Daul, A., and Bock, K. D. (1985c) Alpha- and beta-adrenoceptors in circulating blood cells of essential hypertensive patients: Increased receptor density and responsiveness. *Clin. Exp. Hyperten. (A)* **7**(8), 1135–1150.

Brodde, O.-E., Kretsch, R., Ikezono, K., Zerkowski, H. R., and Reidemeister, J. C. (1986b) Human beta-adrenoceptors: Relation of myocardial and lymphocyte beta-adrenoceptor density. *Science* **231**, 1584–1585.

Brooks, S. M., McGowan, K., Bernstein, L., Altenau, P., and Peagler, J. (1979) Relationship between number of beta-adrenergic receptors in lymphocytes and disease severity in asthma. *J. Allergy Clin. Immunol.* **63**, 401–406.

Bruynzeel, P. L., Van den Berg, W., Hamelink, M. L., Bogaard, W. van den Houben, L. A. M. J., and Kreukniet, J. (1979) Desensitization of the beta-adrenergic receptor on leukocytes after long term oral use of a beta sympathicomimetic: Its effect on the beta-adrenergic blockade hypothesis of Szentivanyi. *Ann. Allergy* **43**, 105–109.

Busse, W. W. (1977) Decreased granulocyte response to isoproterenol in asthma during upper respiratory tract infection. *Am. Rev. Respir. Dis.* **115,** 783–792.

Busse, W. W. and Sosman, J. M. (1984) Isoproterenol inhibition of isolated human neutrophil function. *J. Allergy Clin. Immunol.* **73,** 404–410.

Butler, J., O'Brien, M., O'Malley, K., and Kelly, J. G. (1982) Relationship of beta-adrenoceptor density to fitness in athletes. *Nature* **298,** 60–63.

Carlson, H. E. and Brickman, A. S. C. (1983) Blunted plasma cyclic adenosine monophosphate response to isoproterenol in pseudohypoparathyroidism. *J. Clin. Endocrinol. and Metab.* **56,** 1323–1326.

Cash, R., Lasbennes, F., Sercombe, R., Seylaz, J., and Agid, Y. (1985) Adrenergic receptors in cerebral microvessels in control and Parkinsonian subjects. *Life Sci.* **37,** 531–536.

Cash, R., Ruberg, M., Raisman, R., and Agid, Y. (1984) Adrenergic receptors in Parkinson's disease. *Brain Res.* **322,** 269–275.

Cash, R., Raisman, R., Lanfumey, L., Ploska, A., and Agid, Y. (1986) Cellular localization of adrenergic receptors in rat and human brain. *Brain Res.* **3701,** 127–135.

Chang, H. Y., Klein, R. M., and Kunos, G. (1982). Selective desensitization of cardiac beta-receptors by in vivo infusion of catecholamines in rats. *J. Pharmacol. Exp. Ther.* **221,** 784–789.

Chobanian, A. V., Tifft, C. P., Sackel, H., and Pitzuzella, A. (1982) Alpha- and beta-adrenergic receptor activity in circulating blood cells of patients with idiopathic orthostatic hypotension and pheochromocytoma. *Clin. Exp. Hypertens. (A)* **4,** 793–806.

Chung, S. Z., Lentes, K. U., Gocayne, J., Fitzgerald, M., Robinson, D., Kerlavage, A. R., Fraser, C. M., and Venter, J. C. (1987) Cloning and sequence analysis of the human brain beta-adrenergic receptor. *FEBS Lett.* **211,** 200–206.

Cognini, G., Piantanell, L., Paotinell, E., Orlandoni, P., Pelligrini, A., and Masera, N. (1983) Decreased beta-adrenergic receptor density in mononuclear leukocytes from thyroidectomized patients. *Acta Endocrinol.* **103,** 1–5.

Cohen, R. M., Campbell, I. C., Dauphin, M., Tallman, J. F., and Murphy, D. L. (1982) Changes in alpha- and beta-receptors with monoamine oxidase inhibiting antidepressants. *Neuropharmacology* **21,** 293–298.

Cohn, J., Levine, T. B., Olivari, M. T., Garberg, V., Lura, D., Francis, G. S., Simon, A. B., and Rector, T. (1984) Plasma norepinephrine as a guide to prognosis in patients with chronic congestive heart failure. *New Engl. J. Med.* **311,** 819–823.

Colucci, W. S., Alexander, R. W., Williams, G. H., Rude, R. F., Holman, B. L., Konstam, M. A., Wynne, J., Mudge, G. H. J., and Braunwald, E. (1981) Decreased lymphocyte beta-adrenergic receptor density in patients with heart failure and tolerance to the beta-adrenergic agonist pirbuterol. *New Engl. J. Med.* **305,** 185–190.

Conolly, M. E. and Greenacre, J. K. (1976) The lymphocyte beta-adrenoceptor in normal subjects and patients with bronchial asthma. *J. Clin. Invest.* **58,** 1307–1316.

Cook, N., Nahorski, S. R., and Barnett, D. (1985) (-)[^{125}I]pindolol binding to the human platelet beta-adrenoceptor: Characterization and agonist interactions. *Eur. J. Pharmacol.* **113,** 247–254.

Cook, N., Nahorski, S. R., and Barnett, D. B. (1987) Human platelet β_2-adrenoceptors: agonist-induced internalisation and downregulation in intact cells. *Br. J. Pharmacol.* **92,** 587–596.

Crary, B., Hansen, S. L., Borysenko, M., Kutz, I., Hoban, C., Ault, K. A., Weiner, H. L., and Benson, H. (1983) Epinephrine-induced changes in the distribution of lymphocyte subsets in peripheral blood of humans. *J. Immunol.* **131,** 1178-1181.

Cryer, P. E. and Gerich, J. E. (1985) Glucose counterregulation, hypoglycemia and intensive insulin therapy in diabetes mellitus. *New Engl. J. Med.* **313,** 232-241.

Davies, A. O. (1984) Rapid desensitization and uncoupling of human beta-adrenergic receptors in an *in vitro* model of lactic acidosis. *J. Clin. Endocrinol. Metab.* **59,** 398-405.

Davies, A. O. (1986) Effects of endogenous redoxactive compounds on coupling of human beta-2 adrenergic receptors. *Am. J. Med. Sci.* **292,** 257–263.

Davies, A. O. and Lefkowitz, R. J. (1983) In vitro desensitization of beta- adrenergic receptors in human neutrophils. Attenuation by corticosteroids. *J. Clin. Invest.* **71,** 565–571.

Davies, A. O., Mares, A., Pool, J. L., and Taylor, A. A. (1987) Mitral valve prolapse with symptoms of beta-adrenergic hypersensitivity. *Am. J. Med.* **82,** 193–201.

Davis, P. B., Braunstein, M., and Jay, C. (1978) Decreased adenosine 3'5'monophosphate response to isoproterenol in cystic fibrosis leukocytes. *Pediatr. Res.* **12,** 703–707.

Davis, P. B., Dieckman, L., Boat, T. F., Stern, R. C., and Doetshuk, C. F. (1983) Beta-adrenergic receptors in lymphocytes and granulocytes from patients with cystic fibrosis. *J. Clin. Invest.* **71,** 1787–1795.

Davis, P. B., Shelhamer, J. R., and Kaliner, M. (1980) Abnormal adrenergic and cholinergic sensitivity in cystic fibrosis. *New Eng. J. Med.* **302,** 1453–1456.

Davis, P. B., Simpson, D. M., Paget, G. L., and Turi, V. (1986) Beta-adrenergic responses in drug-free subjects with asthma. *J. Allergy Clin. Immunol.* **77,** 871-879.

DeBlasi, A., Fratelli, M., and Marasco, O. (1988) Certain beta-blockers can decrease beta-adrenergic receptor number I. Acute reduction in receptor number by tertatolol and bopindolol. *Circ. Res.* **63,** 273–278.

DeBlasi, A., Lipartiti, M., Motulsky, H. J., Insel, P. A., and Fratelli, M. (1985) Agonist-induced redistribution of beta-adrenergic receptors on intact human mononuclear leukocytes. Redistributed receptors are nonfunctional. *J. Clin. Endocrinol. Metab.* **61,** 1081–1088.

DeBlasi, A., Cotecchia, S., Fratelli, M., and Lipartiti, M. (1986a) Agonist-induced beta-adrenergic receptor internalization in intact human mononuclear leukocytes: Effects of temperature of mononuclear leukocyte separation. *J. Lab Clin. Med.* **107,** 86–94.

DeBlasi, A., Maisel, A., Feldman, R. D., Ziegler, M. G., Fratelli, M., Di Lallo, M., Smith, D., Lai, C. C., and Motulsky, H. J. (1986b) In vivo regulation of beta adrenergic receptors on human mononuclear leukocytes: Assessment of receptor number, function, and location following posture change, exercise, and infusion of isoproterenol. *J. Clin. Endocrinol. Metab.* **63,** 847–855.

DeBlasi, A., Lipartiti, M., Pirone, F., Rochat, C., Prost, J. F., and Garattini, S. (1986c) Reduction of beta-adrenergic receptors by tertatolol: An additional mechanism for beta-adrenergic blockade. *Clin. Pharmacol. Ther.* **39,** 245–253.

de Champlain, J., Gonzalez, M., Lebeau, R., Eid, H., Petrovitch, M., and Nadeau, R. A. (1989) The sympatho-adrenal tone and reactivity in human hypertension. *Clin. Exp. Hypertension (A)* 11, Supp. 1, 159–171.

Devos, C., Robberecht, P., Nokin, P., Wallbroeck, M., Clinet, M., Camus, J. C., Beaufort, P., Schoenfeld, P., and Christophe, J. (1985) Uncoupling between beta-adrenoceptors and adenylate cyclase in ischemic dog myocardium. *Nauyn. Schmiedebergs Arch. Pharmacol.* **331,** 71–75.

Dixon, R. A. F., Kobilka, B. K., Strader, D. J., Benovic, J. L., Dahlman, H. G., Frielle, T., Bolanowski, M. A., Bennett, C. D., Rands, E., Diehl, R. E., Mumford, R. A., Slater, E. E., Sigal, I. S., Caron, M. G., Lefkowitz, R. J., and Strader, C. D. (1986) Cloning of the gene and cDNA for mammalian beta-adrenergic receptors and homology with rhodopsin. *Nature* **321,** 75–79.

Dominiak, P. and Türck, D. (1986) Alterations of beta-adrenoceptors subsequent to myocardial infarction. *Basic Res. Cardiol.* 81 (Suppl. 1), 243–251.

Ebstein, R. P., Oppenheim, G., and Stessman, J. (1984) Alzheimer's disease: Isoproterenol and prostaglandin E_1-stimulated cyclic AMP accumulation in lymphocytes. *Life Sci.* **34,** 2239–2243.

Emorine, L. J., Marutto, S., Delavier-Klutchko, Kaveri, S. V., Durien-Trautmann, O., and Strosberg, A. D. (1987) Structure of the gene for the human beta-2 adrenergic receptor: Expression and promoter characterization. *Proc. Natl. Acad. Sci. USA* **84,** 695–699.

Fantozzi, R., Brunelleschi, E., Cambri, S., Blandina, P., Masin, B., and Mannaioy, P. E. (1984) Autocoid and beta-adrenergic agonist modulation of *N*-formyl methionyl leucyl-phenylalanine evoked lysosomal enzyme release from human neutrophils. *Agents Actions* **14,** 441–450.

Feldman, R. D. (1987) Beta-adrenergic receptor alterations in hypertension-physiological and molecular correlates. *Can. J. Physiol. Pharmacol.* **65,** 1666–1672.

Feldman, R. D. and Lai, C-Y. C. (1986) Characterization of the human lymphocyte beta-adrenergic receptor by photoaffinity labeling: Alteration with desensitization. *Circ. Res.* **58,** 384–388.

Feldman, R. D., Fick, R. B., McArdle, W., and Lai, C.-Y. C. (1987a) Are lymphocyte beta-adrenoceptors altered in patients with cystic fibrosis? *Clin. Sci.* **73,** 407–410.

Feldman, R. D., Lawton, W. J., and McArdle, W. L. (1987b) Low sodium diet corrects the defect in lymphocyte beta-adrenergic responsiveness in hypertensive subjects. *J. Clin. Invest.* **79,** 290–294.

Feldman, R. D., Limbird, L. E., Nadeau, J., Robertson, D., and Wood, A. J. J. (1984a) Alteration in leukocyte beta-receptor affinity with aging. A potential explanation for altered beta-adrenergic receptors in the elderly. *New Engl. J. Med.* **310,** 815–819.

Feldman, R. D., Limbird, L. E., Nadeau, J., Robertson, D., and Wood, A. J. J. (1984b) Leukocyte beta-receptor alteration in hypertensive subjects. *J. Clin. Invest.* **73,** 648–653.

Feldman, R. D., Limbird, L. E., Nadeau, J., Fitzgerald, G. A., Robertson, D., and Wood, A. J. J. (1983) Dynamic regulation of leukocyte beta-adrenergic receptor-agonist interactions by physiological changes in circulating catecholamines. *J. Clin. Invest.* **72,** 164–170.

Folkow, B. (1982) Physiological aspects of primary hypertension. *Physiol. Rev.* **62,** 347–504.

Fowler, M. B., Laser, J. A., Hopkins, G. L., Minobe, W., and Bristow, M. R. (1986) Assessment of beta-adrenergic receptor pathway in the intact failing human heart: Progressive receptor downregulation and subsensitivity to agonist response. *Circulation* **74,** 1290–1302.

Francis, G. S. and Cohn, J. N. (1986) The autonomic nervous system in congestive heart failure. *Annu. Rev. Med.* **37,** 235–247.

Fraser, C. M., Venter, J. C., and Kaliner, M. (1981) Autonomic abnormalities and autoantibodies to beta-adrenergic receptors. *N. Engl. J.* **305,** 1165–1169.

Fraser, J., Nadeau, J., Robertson, D., and Wood, A. J. J. (1981) Regulation of human leukocyte beta-receptors by endogenous catecholamine. Relationship of leukocyte beta-receptor density to the cardiac sensitivity to isoproterenol. *J. Clin. Invest.* **67,** 1777–1784.

Freissmuth, M., Schutz, W., Weindlmayer-Göttel, M., Zimpfer, M., and Spiss, C. K. (1987) Effects of ischemia on the canine myocardial beta-adrenoceptor-linked adenylate cyclase system. *J. Cardiovasc. Pharmacol.* **10,** 568–574.

Frielle, T., Collins, S., Daniel, K. W., Caron, M. G., Lefkowitz, R. J., and Kobilka, B. K. (1987) Cloning of the cDNA for the human beta-1 adrenergic receptor. *Proc. Natl. Acad. Sci. USA* **84,** 7920–7924.

Frishman, W. (1987) Adrenergic receptors as pharmacological targets: The beta-adrenergic blocking drugs, in *Adrenergic Receptors in Man* (Insel, P.A., ed.), Marcel Dekker, New York, pp. 69–118.

Frizell, R. A., Rechkemmer, G., and Shoemaker, R. L. (1986) Altered regulation of airway epithelial cell chloride channels in cystic fibrosis. *Science* **233**, 558–560.

Frohler, M., Saito, Y., Ackenheil, M., Bak, R., Bondy, B., Feistenauer, E., Hoftscuster, E., Vakis, A., and Welter, D. (1985) Catecholaminergic binding sites of blood cells of healthy volunteers with special respect to circadian rhythms. *Pharmacopsychiatria* **18**, 147,148.

Galant, S.P. and Britt, S. (1984) Uncoupling of the beta-adrenergic receptor as a mechanism of in vitro neutrophil desensitization. *J. Lab. Clin. Med.* **103**, 322–332.

Galant, S. P., Duriseti, L., Underwood, S., and Insel, P. A. (1978a) Decreased beta-adrenergic receptor in polymorphonuclear leukocytes after adrenergic therapy. *New Engl. J. Med.* **299**, 933–936.

Galant, S. P., Underwood, S., Duriseti, L., and Insel, P. A. (1978b) Characterization of high affinity beta-2 adrenergic receptor binding of (-)[^{3}H] dihydroalprenolol to human polymorphonuclear cell particulates. *J. Lab.Clin. Med.* **92**, 613–618.

Galant, S. P., Morton, L., Herbst, J., and Wood, C. (1981) Impaired beta-adrenergic binding and function in cystic fibrosis neutrophils. *J. Clin. Invest.* **68**, 253–258.

Galant, S. P., Duriseti, L., Underwood, S., Allred, S., and Insel, P. A. (1980) Beta-adrenergic receptors of polymorphonuclear particulates in bronchial asthma. *J. Clin. Invest.* **65**, 577–586.

Galant, S. P., Underwood, S., Allred, S., and Hanifin, J. M. (1979) Beta- adrenergic receptor binding on polymorphonuclear leukocytes in atopic dermatitis. *J. Invest. Dermatol.* **72**, 330–332.

Gavras, H. (1986) How does salt raise blood pressure? A hypothesis. *Hypertension* **8**, 83–88.

Ginsberg, A. M., Clutter, W. E., Shah, S. D., and Cryer, P. E. (1981) Triiodothyronine-induced thyrotoxicosis increases mononuclear leukocyte beta-adrenergic receptor density in man. *J. Clin. Invest.* **67**, 1785–1791.

Ginsburg, R., Bristow, M. R., Billingham, M. E., Stinson, E. D., Schroeder, J. S., and Harrison, D. C. (1983) Study of the normal and failing isolated human heart: Decreased response of failing heart to isoproterenol. *Am. Heart J.* **106**, 535–540.

Glaubiger, G., Tsai, B. S., Lefkowitz, R. J., Weiss, B., and Johnson, E. M. (1978) Chronic guanethidine treatment increases cardiac beta-adrenergic receptors. *Nature* **273**, 246–242.

Goldie, R. G., Spina, D., Henry, P. J., Lulich, K. M., and Paterson, J. W. (1986) In vitro responsiveness of human asthmatic bronchus to carbachol, histamine, beta-adrenergic agonists and theophylline. *Br. J. Clin. Pharmacol.* **22**, 669–676.

Goldstein, D., Lake, C. R., Chernow, B., Ziegler, M. G., Coleman, M. D., Taylor, A. A., Mitchell, J. R., Kopin, I. J., and Keiser, H. R. (1983) Age-dependence of hypertensive-normotensive differences in plasma norepinephrine. *Hypertension* **5**, 100–104.

Goldstein, D. S. (1983) Plasma catecholamines and essential hypertension. An analytical review. *Hypertension* **5**, 86–99.

Golf, S. and Hansson, V. (1986) Effects of beta-blocking agents on the density of beta-adrenoceptors and adenylate cyclase response in human myocardium: Intrinsic sympathomimatic activity favours receptor upregulation. *Cardiovasc. Res.* **20**, 637–644.

Greenacre, L. K. and Conolly, M. E. (1978). Desensitization of the beta-adrenoceptor of lymphocytes from normal subjects and patients with pheochromocytoma: Studies in vivo. *Br. J. Clin. Pharmacol.* **5**, 191–197.

Griese, M., Korholz, U., Korholz, D., Seeger, K., Wahn, V., and Reinhardt, D. (1988) Density and agonist-promoted high and low affinity states of the beta-adrenoceptor on human B-and T-cells. *Eur. J. Clin. Invest.* **18**, 213–217.

Gross, G. and Lues, I. (1985) Thyroid-dependent alteration of myocardial adrenoceptor and adrenoceptor-mediated responses in the rat. *Nauyn Schmiedebergs Arch. Pharmacol.* **329**, 427–439.

Halper, J. P., Brown, R. P., Sweeney, J. A., Kocsis, J. H., Peters, A., and Mann, J. J. (1988) Blunted beta-adrenergic responsivity of peripheral blood mononuclear cells in endogenous depression. *Arch. Gen. Psychiatry.* **45**, 241–244.

Halper, J. P., Mann, J. J., Weksler, M. E., Bilezikian, J. P., Sweeney, J. A., Brown, R. P., and Golbourne, T. (1984) Beta-adrenergic receptor and cyclic AMP levels in intact human lymphocytes: Effects of age and gender. *Life Sci.* **35**, 855–863.

Hamet, P. (1983) Metabolic aspects of hypertension, in *Hypertension* (Genest, J., Kuchel, O., Hamet, P., and Cantin, M., eds.) McGraw Hill, New York, pp. 408–427.

Hayes, J. S., Pollack, G. D., and Fuller, R. W. (1984) In vivo cardiovascular responses to isoproterenol, dopamine, and tyramine after prolonged infusion of isoproterenol. *J. Pharmacol. Exp. Ther.* **231**, 633–639.

Healy, D., Carney, P. A., and Leonard, B. B. (1983) Monoamine-related markers of depression: Changes following treatment. *J. Psychiatr. Res.* **17**, 251–260.

Hedberg, A., Kempf, F., Josephson, M. E., and Molinoff, P. B. (1985) Coexistence of beta-1 and beta-2 adrenergic receptors in the human heart: Effects of treatment with receptor antagonists or calcium entry blockers. *J. Pharmacol. Exp. Ther.* **234**, 561–568.

Hedberg, A., Gerber, J. G., Nies, A. S., Wolfe, B. B., and Molinoff, P. B. (1987) Effects of pindolol and propranolol on beta-adrenergic receptors on human lymphocytes. *J. Pharmacol. Exp. Ther.* **239**, 117–123.

Heinsimer, J. A., Davies, A. S., Downs, R. W., Levine, M. A., Spiegel, A. M., Drezner, M. K., DeLean, A., Wreggett, K. A., Caron, M. G., and Lefkowitz,

R. J. (1984) Impaired formation of beta-adrenergic receptor-nucleotide regulatory protein complexes in pseudohypoparathyroidism. *J. Clin. Invest.* **73**, 1335–1343.

Hill, R., Keenan, A. K., and Neligan, M. (1986) Increased activation of cardiac beta-adrenoceptor linked adenylate cyclase in hypertension. *Br. J. Pharmacol.* **87**, 106P.

Hughes, B., Marshall, D. R., Sobel, B. E., and Bergmann, S. R. (1986) Characterization of beta-adrenergic receptors in vivo with iodine-131 pindolol and gamma scintigraphy. *J. Nucl. Med.* **27**, 660–667.

Hui, K. K. P. and Conolly, M. E. (1981) Increased numbers of beta-receptors in orthostatic hypotension due to autonomic dysfunction. *N. Engl. J. Med.* **304**, 1473–1476.

Hui, K. K. P., Conolly, M. E., and Tashkin, D. P. (1982) Reversal of human lymphocyte beta-adrenoceptor desensitization by glucocorticoids. *Clin. Pharmacol. Ther.* **32**, 566–571.

Hui, K. K. P., Wolfe, R. N., and Conolly, M. E. (1982) Lymphocyte beta-adrenergic receptors are not altered in hyperthyroidism. *Clin. PharmacoL Ther.* **32**, 161-165.

Ingebretsen, C. G., Hawelu-Johnson, C., and Ingebretsen, W. R. (1983) Alloxan-induced diabetes reduces beta-adrenergic receptor number without affecting adenylate cyclase in rat ventricular membranes. *J. Cardiovasc. Pharmacol.* **5**, 454–461.

Insel, P. A. (1985) Adrenergic receptors in human blood cells, in *Pharmacology of Adrenoceptors* (Szabadi, E., Bradshaw, C. M., and Nahorski, S. R., eds.) Macmillan, London, pp. 215–225.

Insel, P. A. and Motulsky, H. J. (1984) A hypothesis linking intracellular sodium, membrane receptors and hypertension. *Life Sci.* **34**, 1009–1113.

Insel, P. A. and Motulsky, H. J. (1987) Physiologic and pharmacologic regulation of adrenergic receptors, in *Adrenergic Receptors in Man* (Insel, P. A., ed.) Marcel Dekker, New York, pp. 201–236.

Insel, P. A. and Ransnäs, L. A. (1988) G proteins and cardiovascular disease. *Circulation* **78**, 1511–1513.

Insel, T. R. and Cohen, R. M. (1987) Adrenergic receptors in psychiatric disease, in *Adrenergic Receptors in Man* (Insel, P.A., ed.) Marcel Dekker, New York, pp. 353–375.

Jenni-Eiermann, S., Von Hahn, H. P., Honegger, C. G., and Ulrich, J. (1984) Studies in neurotransmitter binding in senile dementia: Comparison of Alzheimer's and mixed vascular-Alzheimer's dementias. *Gerontology* **30**, 350–358.

Kahn, M. M., Sansoni, P., Silverman, E. O., Engelman, E. G., and Melmon, K. L. (1986) Beta-adrenergic receptors in human suppressor, helper and cytolytic lymphocytes. *Biochem. Pharmacol.* **35**, 1137–1142.

Kajiyama, H., Obara, K., Nomura, Y., and Segawa, T. (1982) The increase of cardiac beta-1 subtype of beta-adrenergic receptors in adult rats following neonatal 6-hydroxydopa treatment. *Eur. J. Pharmacol.* **77**, 75–77.

Kaliner, M., Shelhamer, J. H., Davis, P. B., Smith, L. J., and Venter, J. C. (1982) Autonomic nervous system abnormalities and allergy. *Ann. Intern. Med.* **96**, 349- 357.

Kariman, K. (1980) Beta-adrenergic receptor binding in lymphocytes from patients with asthma. *Lung* **158**, 41–51.

Karliner, J. S., Simpson, P. C., Honbo, N., and Woloszyn, W. (1986) Mechanisms and time course of beta-adrenoceptor desensitization in mammalian cardiac myocytes. *Cardiovasc. Res.* **20**, 221–228.

Kawai, Y., Powell, A., and Arinze, I. J. (1986) Adrenergic receptors in human liver plasma membranes: Predominance of beta-2 and alpha-1 receptor subtypes. *J. Clin. Endocrinol. Metab.* **62**, 827–832.

Kerry, R., Scrutton, M. C., and Wallis, R. B. (1984) Mammalian platelet adrenoceptors. *Br. J. Pharamcol.* **81**, 91–102.

Khan, J. P., Gully, R. J., Cooper, T. B., Perumal, A. S., Smith, T. M., and Klein, D. F. (1987) Correlation of type A behaviour with adrenergic receptor density: Implications for coronary artery disease pathogenesis. *Lancet* **8565**, 937-939.

Kobilka, B. K., Dixon, R. A. F., Frielle, T., Dohlman, H. G., Bolanowski, M. A., Sigal, I. S., Yang-Feng, T. L., Francke, U., Caron, M. G., and Lefkowitz, R. J. (1987) cDNA for the human beta-2 adrenergic receptor: A protein with multiple membrane spanning domains and encoded by a gene whose chromosomal location is shared with that of the receptor for platelet-derived growth factor. *Proc. Natl. Acad. Sci. USA* **84**, 46–50.

Krall, J. F., Connelly, M., and Tuck, M. L. (1980) Acute regulation of beta-adrenergic catecholamine sensitivity in human lymphocytes. *J. Pharmacol. Exp. Ther.* **214**, 554–560.

Krawietz, W., Werdan, K., Schober, M., Erdmann, E., Rindfleisch, G. E., and Hannig, K. (1982) Different numbers of beta-receptors in human lymphocyte subpopulations. *Biochem. Pharmacol.* **31**, 133–136.

Kuchel, O. (1983) The autonomic nervous system and blood pressure regulation in human hypertension, in *Hypertension* (Genest, J., Kuchel, O., Hamet, P., and Cantin, M., eds.) McGraw Hill, New York, pp. 140–160.

Kumano, K., Upsher, M. E., and Khairalloh, P. (1983) Beta-adrenergic receptor response coupling in hypertrophied hearts. *Hypertension* **5** (Suppl I), I-175–I-183.

Landmann, R. M. A., Burgisser, E., Wesp, M., and Bühler, F. R. (1984) Beta-adrenergic receptors are different in subpopulations of human circulating lymphocytes. *J. Recept. Res.* **4**, 37–50.

Larner, A. C. and Ross, E. M. (1981) Alteration in the protein components of catecholamine-sensitive adenylate cyclase during maturation of rat reticulocytes. *J. Biol. Chem.* **256**, 9551–9557.

Lefkowitz, R. J., Stadel, J. M., and Caron, M. G. (1983) Adenylate cyclase-coupled beta-adrenergic receptor: Structure and mechanism of activation and desensitization. *Annu. Rev. Biochem.* **52**, 159–186.

Leimbach, W. N., Wallin, B. G., Victor, R. G., Aylward, P. T., Sudlöf, G., and Mark, A. L. (1986) Direct evidence from intraneural recordings for increased central sympathetic outflow in patients with heart failure. *Circulation* **73**, 913–919.

Lerer, B. (1984) Electroconvulsive shock and neurotransmitter receptors: Implications for mechanism of action and adverse effects of electroconvulsive therapy. *Biol. Psychiatry* **19**, 361–383.

Levine, M. A., Downs, R. W., Moses, A. M., Breslan, M. A., Marx, S. J., Lasker, R. D., Rizzoli, R. E., Aurbach, G. D., and Spiegel, A. M. (1983) Resistance to multiple hormones in patients with pseudohypoparathyroidism. Association with deficient activity of guanine nucleotide regulatory protein. *Am. J. Med.* **74**, 545–556.

Liggett, S. B., Marker, J. C., Shah, S. D., Roper, C. L., and Cryer, P. E. (1988) Direct relationship between mononuclear leukocyte and lung beta-adrenergic receptors and apparent reciprocal regulation of extravascular, and not intravascular, alpha- and beta-adrenergic receptors by the sympathochromaffin system in humans. *J. Clin. Invest.* **182**, 48–56.

Liggett, S. B., Shah, S. D., and Cryer, P. E. (1989a) Human tissue adrenergic receptors are not predictive of responses to epinephrine in vivo. *Am. J. Physiol.* **256**, E600–E609.

Liggett, S. B., Shah, S. D., and Cryer, P. E. (1989b) Increased fat and skeletal muscle beta-adrenergic receptors but unaltered metabolic and hemodynamic sensitivity to epinephrine in vivo in experimental human thyrotoxicosis. *J. Clin. Invest.* **83**, 803–809.

Limas, C. and Limas, C. J. (1978) Reduced number of beta-adrenergic receptors in the myocardium of spontaneously hypertensive rats. *Biochem. Biophys. Res. Commun.* **83**, 710–714.

Limas, C. and Limas, C. J. (1985) Cardiac beta-adrenergic receptor numbers in salt-dependent genetic hypertension. *Hypertension* **7**, 760–766.

Limas, C. J. and Limas, C. (1984) Decreased isoproterenol-induced "downregulation" of beta-adrenergic receptors in the myocardium of SHR. *Hypertension* (Supp. I), I31–I40.

Limas, C. J. and Limas, C. (1987) Altered intracellular adrenoceptor distribution in myocardium of spontaneously hypertensive rats. *Am. J. Physiol.* **253**, H904–H908.

Limas, C. J., Goldenberg, I. F., and Limas, C. (1989a) Autoantibodies against beta-adrenoreceptors in human idiopathic dilated cardiomyopathy. *Circ. Res.* **64**, 97–103.

Limas, C. J., Limas, C., and Goldenberg, I. F. (1989b) Intracellular distribution of adrenoceptors in the failing human myocardium. *Am. Heart J.* **117**, 131–126.

Linden, J., Patel, A., Spanier, A. M., and Weglicki, W. (1984) Rapid agonist-induced decrease of [^{125}I]-pindolol binding to beta-adrenergic receptors. Relationship to desensitization of cyclic AMP accumulation in intact heart cells. *J. Biol. Chem.* **259**, 15115–15122.

Lonati-Galligani, M. and Pirke, K. M. (1986) Beta-adrenergic receptor regulation in circulating mononuclear leukocytes in anorexia nervosa and bulimia. *Psychiatry Res.* **19**, 189–198.

Mack, J. A., Nielson, C. P., Stevens, D. L., and Vestal, R. E. (1986) Beta-adrenoceptor-mediated modulation of calcium ionophore activated polymorphonuclear leukocytes. *Br. J. Pharmacol.* **88**, 417–423.

Maisel, A. S., Motulsky, H. J., and Insel, P. A. (1985) Externalization of beta-adrenergic receptors promoted by myocardial ischemia. *Science* **230**, 183–186.

Maisel, A. S., Motulsky, H. J., and Insel, P. A. (1987a) Propranolol treatment externalizes beta-adrenergic receptors in guinea pig myocardium and prevents further externalization by ischemia. *Circ. Res.* **60**, 108–112.

Maisel, A. S., Motulsky, H. J., and Insel, P. A. (1987b) Life cycles of cardiac alpha$_1$- and beta-adrenergic receptors. *Biochem. Pharmacol.* **36**, 1–7.

Maisel, A. S., Fowler, P., Rearden, A., Motulsky, H. J., and Michel, M. C. (1989) A new method for isolation of human lymphocyte subsets reveals differential regulation of beta-adrenergic receptors by terbutoline treatment. *Clin. Pharmacol. Ther.* **46**, 429–439.

Maisel, A. S., Knowlton, K. U., Fowler, P., Rearden, A., Ziegler, M. G., Motulsky, H. J., Insel, P. A., and Michel, M. C. (1990) Adrenergic control of circulating lymphocyte subpopulations: Effects of congestive heart failure, dynamic exercise, and terbutaline treatment. *J. Clin. Invest.* **85**, 462–467.

Maisel, A. S., Lee, P., Michel, M. C., Carter, S., Ziegler, M. G., and Insel, P. A. (1989b) Altered lymphocyte beta-adrenergic receptors and cyclic AMP generation in congestive heart failure. Submitted.

Mäki, T., Kontula, K., Myllynen, P., and Härkönen, M. (1987) Beta-adrenergic receptors of human lymphocytes in physically active and immobilized subjects: Characterization by a polyethylene glycol precipitation assay. *Scand. J. Clin. Invest.* **47**, 261–268.

Malbon, C. C., Graziano, M. P., and Johnson, G. L. (1984) Fat cell beta-adrenergic receptor in the hypothyroid rat: Impaired interaction with the stimulatory regulatory component of adenylate cyclase. *J. Biol. Chem.* **259**, 3254–3260.

Malbon, C. C., Rapiejko, P. J., and Watkins, D. C. (1988) Permissive hormone regulation of hormone-sensitive effector systems. *Trends Pharmacol. Sci.* **9**, 33–36.

Mann, J. J., Brown, R. P., Halper, J. P., Sweeney, J. A., Kocsis, J. H., Stokes, P. E., and Bilezikian, J. P. (1985) Reduced sensitivity of lymphocyte beta-adrenergic receptors in patients with endogenous depression and psychomotor agitation. *N. Engl. J. Med.* **313**, 715–720.

Marinetti, G. V., Rosenfeld, S. I., Thiem, P. A., Condemi, J. J., and Leddy, J. P. (1983) Beta-adrenergic receptors of human leukocytes: Studies with intact mononuclear and polymorphonuclear cells and membranes comparing two radioligands in the presence and absence of chloroquine. *Biochem. Pharmacol.* **32**, 2033–2043.

Marsh, J. D., Barry, W. H., Meer, E. J., Alexander, R. W., and Smith, T. W. (1980) Desensitization of chick embryo ventricle to the physiological and biological effects of isoproterenol. *Circ. Res.* **47**, 493–501.

Martinsson, A., Larsson, K., and Hjemdahl, P. (1987) Studies in vivo and in vitro of terbutaline-induced beta-adrenoceptor desensitization in healthy subjects. *Clin. Sci.* **72**, 47–54.

Meurs, H., Koeter, G. H., de Vries, K., and Kauffman, H. F. (1982) The beta-adrenergic system and allergic bronchial asthma: Changes in lymphocyte beta-adrenergic receptor number and adenylate cyclase activity after an allergen-induced asthmatic attack. *J. Allergy Clin. Immunol.* **70**, 272–280.

Meurs, H., Kauffman, H. F., Koeter, G. H., Timmerman, A., and de Vries, K. (1987) Regulation of the beta-receptor-adenylate cyclase system in lymphocytes of allergic patients with asthma: Possible role for protein kinase C in allergen-induced nonspecific refractoriness of adenylate cyclase. *J. Allergy Clin. Immunol.* **80**, 329–339.

Meurs, H., Kauffman, H. F., Timmermans, A., Amsterdam, F. I., Van Koeter, G. H., and de Vries, K. (1986) Phorbol 12-myristate 13-acetate induced beta-adrenergic receptors uncoupling and nonspecific desensitization of adenylate cyclase in human mononuclear leukocytes. *Biochem. Pharmacol.* **35**, 4217–4222.

Michel, M. C., Brodde, O.-E., and Insel, P. A. (1990) Peripheral adrenergic receptors in hypertension. *Hypertension,* in press.

Michel, M. C., Beckeringh, J. J., Ikezono, K., Kretsch, R., and Brodde, O-E. (1986) Lymphocyte beta-2 adrenoceptors mirror precisely beta-2 adrenoceptor, but poorly beta-1 adrenoceptor changes in the human heart. *J. Hypertension* **4** (Suppl. 6), S215–S218.

Michel, M. C., Pingsmann, A., Beckeringh, J. J., Zerkowski, H. R., Doetsch, N., and Brodde, O.-E. (1988) Selective regulation of beta-1, and beta-2 adrenoceptors in the human heart by chronic beta-adrenoceptor antagonist treatment. *Br. J. Pharmacol.* **94**, 685–692.

Michel, M. C., Kanezik, R., Khamssi, M., Knorr, A., Siegl, H., Beckeringh, J. J., and Brodde, O.-E. (1989) Alpha- and beta-adrenoceptors in hypertension I: Cardiac and renal alpha-1, beta-1, and beta-2 adrenoceptors in rat models of acquired hypertension. *J. Cardiovasc. Pharmacol.* **13**, 421–431.

Minakuchi, K., Ogawa, K., Ban, M., and Satake, T. (1981) Decreased generation of cyclic AMP in lymphocytes by beta-adrenergic stimulation in heart failure. *Jpn. Heart J.* **22**, 585–592.

Minneman, K. P., Pittman, R. N., and Molinoff, P. B. (1981) Beta-adrenergic receptor subtypes: Properties, distribution and regulation. *Annu. Rev. Neurosci.* **4**, 419-461.

Motulsky, H. J. and Insel, P. A. (1982) Adrenergic receptors in man: Direct identification, physiological regulation and clinical alterations. *N. Engl. J. Med.* **307,** 18–29.

Motulsky, H. J., Cunningham, E. M. S., De Blasi, A., and Insel, P. A. (1986) Agonists promote rapid desensitization of redistribution of beta-adrenergic receptors on intact human mononuclear leukocytes. *Am. J. Physiol.* **250,** E583–E590.

Mukherjee, A., Wong, T. M., Buja, L. M., Lefkowitz, R. J., and Willerson, J. T. (1979) Beta-adrenergic and muscarinic cholinergic receptors in canine myocardium. Effects of ischemia. *J. Clin. Invest.* **64,** 1423–1428.

Mukherjee, A., Bush, L. R., McCoy, K. E., Duke, R. J., Hagler, H., Buja, L. M., and Willerson, J. T. (1982) Relationship between beta-adrenergic receptor numbers and physiological responses during experimental canine myocardial ischemia. *Circ. Res.* **50,** 735–741.

Murphee, S. S. and Saffitz, J. E. (1989) Distribution of beta-adrenergic receptors in failing human myocardium. Implications for mechanisms of downregulation. *Circulation* **79,** 1214–1225.

Nadel, J. A. and Barnes, P. J. (1984) Autonomic regulation of the airways. *Annu. Rev. Med.* **35,** 451–467.

Neve, K. A. and Molinoff, P. B. (1986) Effects of chronic administration of agonists and antagonists on the density of beta-adrenergic receptors. *Am. J. Cardiol.* **57,** 17F–22F.

Nielson, C. P. (1987) Beta-adrenergic modulation of the polymorphonuclear leukocyte respiratory burst is dependent upon the mechanism of cell activation. *J. Immunol.* **139,** 2392–2397.

Noji, T., Tashiro, M., Yagi, H., Nagashima, K., Suzuki, S., and Kurouwe, T. (1986) Adaptive regulation of beta-adrenergic receptors in children with insulin-dependent diabetes mellitus. *Horm. Metab. Res.* **18,** 604–606.

Ohman, E. M., Butler, J., Kelly, J., Horgan, J., and O'Malley, K. (1987) Beta-adrenoceptor adaptation to endurance training. *J. Cardiovasc. Pharmacol.* **10,** 728–731.

Olson, E. N., Kelley, D. A., and Smith, P. B. (1981) Characterization of rat skeletal muscle sarcolemma during the development of diabetes. *Exp. Neurol.* **73,** 154–172.

Oppenheim, G., Mintzer, J., Halperin, Y., Eliakim, R., Stessman, J., and Ebstein, R. P. (1984) Acute desensitization of lymphocyte beta-adrenergic stimulated adenylate cyclase in old age and Alzheimer's disease. *Life Sci.* **35,** 1795–1802.

Paietta, E. and Schwarzmeier, J. O. (1983) Differences in beta-adrenergic receptor density and adenylate cyclase activity between normal and leukaemic leukocytes. *Eur. J. Clin. Invest.* **13,** 339–346.

Pandey, G. N., Janicak, P. G., and Davis, J. M. (1987) Decreased beta-adrenergic receptors in the leukocytes of depressed patients. *Psychiatry Res.* **22,** 265–273.

Pandey, G. N., Heinze, W. J., Brown, B. D., and Davis, J. M. (1979) Electroconvulsive shock treatment decreases beta-adrenergic receptor sensitivity in rat brain. *Nature* **280,** 234–235.

Parker, C. W. and Smith, J. W. (1973) Alteration in cyclic adenosine monophosphate metabolism in human bronchial asthma. I. Leukocyte responsiveness to beta-adrenergic agents. *J. Clin. Invest.* **52,** 48–59.

Pochet, R., Delespesse, G., Gausset, P. W., and Collet, H. (1979) Distribution of beta-adrenergic receptors in human lymphocyte subpopulations. *Clin. Exp. Immunol.* **38,** 578–584.

Prichard, B. N. C., Tomlinson, B., Walden, R. J., and Bhattacharjee, P. (1983) The beta-adrenergic blockade withdrawal phenomenon. *J. Cardiovasc. Pharmacol.* **5,** S56–S62.

Rainbow, T. C., Parsons, B., and Wolfe, B. B. (1984) Quantitative autoradiography of beta-1 and beta-2 adrenergic receptors in rat brain. *Proc. Natl. Acad. Sci. USA* **81,** 1585–1589.

Ransnäs, L. A., Hammond, H. K., and Insel, P. A. (1988) Increased G_s in myocardial membranes from hyperthyroid pigs. *Clin. Res.* **36,** 52A.

Rasmussen, H., Lake, W., and Allen, J. E. (1975) The effect of catecholamines and prostaglandins upon human and rat erythrocytes. *Biochem. Biophys. Acta* **411,** 63–73.

Ratje, D. D. and Wisser, H. (1986) Alpha- and beta-adrenergic receptor activity in circulating blood cells of patients with pheochromocytoma: Effects of adrenalectomy. *Acta Endocrinol.* **111,** 80–58.

Reinhardt, D., Zehmisch, T., Becker, B., and Nagel-Hiemke, M. (1984) Age-dependency of alpha- and beta-adrenoceptors on thrombocytes and lymphocytes of asthmatic and nonasthmatic children. *Eur. J. Pediatr.* **142,** 111–116.

Reithman, C. and Werdan, K. (1989) Noradrenaline-induced desensitization in cultured heart cells as a model for the defects of the adenylate cyclase system in severe heart failure. *Nauyn. Schmiedebergs Arch. Pharmacol.* **339,** 128–144.

Richelsen, B. and Sørensen, M. S. (1987) Alpha-2 and beta-adrenergic receptor binding and action in gluteal adipocytes from patients with hypothyroidism and hyperthyroidism. *Metabolism* **36,** 1031–1039.

Roan, Y. and Galant, S. P. (1982) Decreased neutrophil beta-adrenergic receptors in the neonate. *Pediatric Res.* **16,** 591–593.

Robertson, D. and Hollister, A. S. (1987) Adrenergic receptors in neurologic disorders, in *Adrenergic Receptors in Man* (Insel, P. A., ed.) Marcel Dekker, New York, pp. 339–352.

Robertson, D., Hollister, A. S., Carey, E. L., Tung, C. S., and Goldberg, M. R. (1984) Increased vascular beta-2 adrenergic receptor responsiveness in autonomic dysfunction. *J. Am. Coll. Cardiol.* **3,** 850–856.

Rosen, S. G., Berk, M. A., Popp, D. A., Serusclat, P., Smith, E. B., Shah, S. D., Ginsberg, A. M., Clutter, W. E., and Cryer, P. E. (1984) Beta-2 and alpha-2-adrenergic receptors and receptor coupling to adenylate cyclase in human mononuclear leukocytes and platelets in relation to physiological variations of sex steroids. *J. Clin. Endo. and Metab.* **58**, 1068–1075.

Rosendorff, C., Susanni, E., Hurwitz, M. L., and Ross, L. P. (1985) Adrenergic receptors in hypertension: Radioligand binding studies. *J. Hypertension* **3**, 571–581.

Ruoho, A. E., DeClergue, L. L., and Busse, W. W. C. (1980) Characterization of granulocyte beta-adrenergic receptors in atopic eczema. *J. Allergy Clin. Immunol.* **66**, 46–51.

Sager, G. (1982) Receptor binding sites for beta-adrenergic ligands on human erythrocytes. *Biochem. Pharmacol.* **31**, 99–104.

Sager, G. (1983) Beta-2 adrenergic receptors in intact human erythrocytes. *Biochem. Pharmacol.* **32**, 1946–1949.

Samuels, H. H., Foreman, B. M., Horowitz, Z. D., and Ye, Z-S. (1988) Regulation of gene expression by thyroid hormone. *J. Clin. Invest.* **81**, 957–967.

Samuelson, W. M. and Davies, A. O. (1984) Hydrocortisone-induced reversal of beta-adrenergic receptor uncoupling. *Am. Rev. Resp. Dis.* **130**, 1023–1026.

Sandnes, D., Gjerde, I., Refsnes, M., and Jacobsen, S. (1987) Downregulation of surface beta-adrenoceptors on intact human mononuclear leukocytes. Time course and isoproterenol concentration dependence. *Biochem. Pharmacol.* **36**, 1303–1311.

Sano, Y., Watt, G., and Townley, R. G. (1983) Decreased mononuclear cell beta-adrenergic receptors in bronchial asthma: Parallel studies of lymphocyte and granulocyte desensitization. *J. Allergy Clin. Immunol.* **72**, 495–503.

Savarese, H. and Berkowitz, B. A. (1979) Beta-adrenergic receptor decrease in diabetic rat hearts. *Life Sci.* **25**, 2075–2078.

Schildkraut, J. J. (1965) The catecholamine hypothesis of affective disorders: A review of supporting evidence. *Am. J. Psychiatry* **122**, 509–520.

Schofield, P. R., Rhee, L. M., and Peralta, E. G. (1987) Primary structure of the human beta-receptor gene. *Nucleic Acids Res.* **15**, 3636.

Schopf, R. E. and Lemmel, E. M. (1983) Control of the reduction of oxygen intermediates of human polymorphonuclear leukocytes and monocytes by beta-adrenergic receptors. *J. Immunopharmacology* **5**, 203–216.

Serusclat, P., Rosen, S. G., Smith, E. B., Shah, S. D., Clutter, W. E., and Cryer, P. E. (1983) Mononuclear leukocyte beta-2 adrenergic receptors and adenylate cyclase sensitivity in insulin-dependent diabetes mellitus. *Diabetes* **32**, 825–829.

Shamoon, H., Hendler, R., and Sherwin, R. S. (1980) Altered responsiveness to cortisol, epinephrine and glucagon in insulin-infused juvenile onset diabetics. *Diabetes* **29**, 284–291.

Sheppard, J. R., Gormus, R., and Moldow, C. F. (1977) Catecholamine hormone receptors are reduced in chronic lymphocytic leukaemia. *Nature* **269**, 693–695.

Shimohama, S., Taniguchi, T., Fugiwara, M., and Kameyama, M. (1987) Changes in beta-adrenergic receptor subtypes in Alzheimer-type dementia. *J. Neurochem.* **48**, 1215–1221.

Sleight, P. C. (1986) Use of beta adrenoceptor blockade during and after acute myocardial infarction. *Annu. Rev. Med.* **37**, 415–425.

Smith, U., Sjostrom, L., Stenstrom, G., Isaksson, O., and Jacobsson, B. (1977) Studies on the catecholamine resistance in fat cells from patients with pheochromocytoma. *Eur. J. Clin. Invest.* **7**, 355–361.

Snavely, M. D., Ziegler, M. G., and Insel, P. A. (1985) Subtype selective down-regulation of rat renal cortical alpha- and beta-adrenergic receptor by catecholamines. *Endocrinology* **117**, 2182–2189.

Snavely, M. D., Mahan, L. C., O'Connor, D. T., and Insel, P. A. (1983) Selective downregulation of adrenergic receptor subtypes in rats with pheochromocytoma. *Endocrinology* **113**, 354–363.

Souchet, T., Bree, R., Baatard, R., Fontenaille, C., D'Athis, P., Tillement, J. P., Kiechel, J. R., and Lhoste, F. (1986) Impaired regulation of beta-2 adrenergic receptor density in mononuclear cells during chronic renal failure. *Biochem. Pharmacol.* **35**, 2513–2519.

Steer, M. L. and Atlas, D. (1982) Demonstration of human platelet beta-adrenergic receptors using ^{125}I-labeled cyanopindolol and ^{125}I-labeled hydroxybenzyl-pindolol. *Biochim. Biophys. Acta* **686**, 240–244.

Sundaresan, P., Guarnucci, M. M., and Izzo, J. L. (1987) Adrenal medullary regulation of rat renal cortical adrenergic receptors. *Am. J. Physiol.* **253**, F1063–F1067.

Szefler, S. J., Edwards, C. K., Haslett, C., Zahniser, N. R., Miller, J. A., and Henson, P. N. (1987) Effects of cell isolation procedures and radioligand selection on the characterization of human leukocyte beta-adrenergic receptors. *Biochem. Pharmacol.* **36**, 1589–1597.

Szentivanyi, A. (1968) The beta-adrenergic theory of the atopic abnormality in bronchial asthma. *J. Allergy* **42**, 203–232.

Tashkin, D. P., Conolly, M. E., Deutsch, R. I., Hui, K. K., Luttner, M., Scarpace, P., and Abrass, I. (1982) Subsensitization of beta-adrenergic receptors in airways and lymphocytes of healthy and asthmatic subjects. *Am. Rev. Resp. Dis.* **125**, 185–193.

Tecoma, E. S., Motulsky, H. J., Traynor, A. E., Omann, G. M., Muller, H., and Sklar, L. A. (1986) Transient catecholamine modulation of neutrophil activation: Kinetic and intracellular aspects of isoproterenol action. *J. Leukocyte Biol.* **40**, 629–644.

Tenner, T. E., Mukherjee, A., and Hester, R. K. (1982) Reserpine induced supersensitivity and the proliferation of cardiac beta-adrenoceptors. *Eur. J. Pharmacol.* **77**, 61–65.

Thomas, J. A. and Marks, B. H. (1978) Plasma norepinephrine in congestive heart failure. *Am. J. Cardiol.* **41,** 233–243.

Titinchi, S., Shamma, M. A., Patel, K. R., Kerr, J. W., and Clark, B. (1984) Circadian variation in number and affinity of beta-2 adrenoceptors in lymphocytes of asthmatic patients. *Clin. Sci.* **66,** 323–328.

Tohmeh, J. F. and Cryer P. E. (1980) Biphasic adrenergic modulation of beta-adrenergic receptors in man. Agonist-induced early increment and late decrement in beta-adrenergic receptor number. *J. Clin. Invest.* **65,** 836–840.

Tse, J., Powell, J. R., Baste, C. A., Priest, R. E., and Kuo, J. F. (1979) Isoproterenol-induced cardiac hypertrophy: Modification in characteristics of beta-adrenergic receptor, adenylate cyclase, and ventricular contraction. *Endocrinology* **105,** 246–255.

Tsujimoto, G., Manger, W. M., and Hoffman, B. B. (1984) Desensitization of beta-adrenergic receptors by pheochromocytoma. *Endocrinology* **114,** 1271–1278.

Unverferth, D.V., Blanford, M., Kates, R. E., and Leier, C. V. (1980) Tolerance to dobutamine after a 72-h continuous infusion. *Am. J. Med.* **69,** 262–266.

Van Dop, C. and Bourne, H. R. (1983) Pseudohypoparathyroidism. *Annu. Rev. Med.* **34,** 259–266.

Van Vliet, P. D., Burchell, H. B., and Titus, H. L. (1966) Focal myocarditis associated with pheochromocytoma. *N. Engl. J. Med.* **274,** 1102–1108.

Vanden Berg, W., Leferink, J. G., Fokkens, J. K., Kreukniet, J., Maes, R. A. A., and Bruynzeel, P. L. B. (1982) Clinical implication of drug induced desensitization of the beta-receptor after continuous oral use of terbutaline. *J. Allergy Clin. Immunol.* **69,** 410–417.

Vatner, D. E., Kirby, D. A., Homcy, C. J., and Vatner, S. F. (1985a) Beta-adrenergic and cholinergic receptors in hypertension-induced hypertrophy. *Hypertension* **7** (Supp. I), I-55–I-60.

Vatner, D. E., Lavallee, M., Amano, J., Finizola, A., Homcy, C. J., and Vatner, S. F. (1985b) Mechanisms of supersensitivity to sympathomimetic amines in the chronically denervated heart of the conscious dog. *Circ. Res.* **57,** 55–64.

Vedin, J. A. and Wilhelmsson, C. E. (1983) Beta-receptor blocking agents in the secondary prevention of coronary heart disease. *Annu. Rev. Pharmacol. Toxicol.* **23,** 29–44.

Venter, J. C., Fraser, C. M., and Harrison, L. C. (1980) Autoantibodies to beta-2 adrenergic receptors: A possible cause of adrenergic hyporesponsiveness in allergic rhinitis and asthma. *Science* **207,** 1361–1363.

Viquerat, C. E., Daly, P., Swedberg, K., Evers, C., Curran, D., Parmley, W. W., and Chatterjee, K. (1985) Endogenous catecholamine levels in chronic heart failure: Relation to the severity of hemodynamic abnormalities. *Am. J. Med.* **78,** 455–460.

Volicier, L., Green, I., and Sinex, F. M. (1985) Epinephrine-induced cyclic AMP production in skin fibroblasts from patients with dementia of Alzheimer type and controls. *Neurobiol. Aging* **6**, 35–38.

Wang, X. L. and Brodde, O-E. (1985) Identification of a homogenous class of beta-2 adrenoceptors in human platelets by (-)-[^{125}I]iodopinolol binding. *J. Cyclic Nucleotide Protein Phosphor. Res.* **10(5)**, 439–450.

Westfall, T.C. and Meldrum, M.J. (1985) Alterations in the release of norepinephrine at the vascular neuroeffector junction in hypertension. *Annu. Rev. Pharm. Toxicol.* **25**, 621–641.

Whitsett, J. A., Noguchi, A., and Moore, J. J. (1982) Developmental aspects of alpha- and beta-adrenergic receptors. *Semin. Perinatol.* **6**, 125–141.

Widdicombe, J. H. (1986) Cystic fibrosis and beta-adrenergic response of airway epithelial cell cultures. *Am. J. Physiol.* **251**, R818–R822.

Williams, L. T., Synderman, R., and Lefkowitz, R. J. (1976) Identification of beta-adrenergic receptors in human lymphocytes by (-)[^{3}H]alprenolol binding. *J. Clin. Invest.* **57**, 149–155.

Williams, R. S., Guthrow, L. E., and Lefkowitz, R. J. (1979) Beta-adrenergic receptors of human lymphocytes are unaltered by hyperthyroidism. *J. Clin. Endocrinol. Metab.* **48**, 503–505.

Winther, K., Klysner, R., Geisler, A., and Anderson, P. H. (1985) Characterization of human platelet beta-adrenoceptors. *Thromb. Res.* **40**, 757–767.

Wood, A. J. J. (1983) Beta-blocker withdrawal. *Drugs* **25** (Suppl. 2), 318–321.

Wood, K., Whiting, K., and Coppen, A. (1986) Lymphocyte beta-adrenergic receptor density of patient with recurrent affective illness. *J. Affective Disord.* **10**, 3–8.

Wright, A. F., Crichton, D. N., Loudon, J. B., Morten, J. E. N., and Steel, C. M. (1984) Beta-adrenoceptor binding defects in cell lines from families with manic-depressive disorder. *Annu. Human Genetic* **48**, 201–214.

Zohar, J., Bonnet, J., Drummer, D., Fisch, R., Epstein, R. P., and Belmaker, R. H. (1983) The response of lymphocyte beta-adrenergic receptors to chronic propranolol treatment in depressed patients, schizophrenic patients and normal controls. *Biol. Psychiatry* **18**, 553–560.

Zucchelli, P., Sturani, A., Zuccala, A., Santoro, A., Degli Eposti, E., and Chiarini C. (1985) Dysfunction of the autonomic nervous system in patients with end-stage renal failure. *Contrib. Nephrol.* **45**, 69–81.

CHAPTER 8

Drug and Hormonal Regulation of the Beta-Adrenergic Receptor–Adenylate Cyclase System

Gary L. Stiles

1. Introduction

Catecholamines, acting through a variety of membrane-bound receptors, produce a wide range of physiological effects. Most mammalian organ systems are affected by either circulating or locally released epinephrine or norepinephrine. These agents produce their effects by interacting with two major classes of receptors, termed alpha-adrenergic and beta-adrenergic receptors. In this chapter, I shall deal exclusively with the beta-adrenergic receptor system. In 1967, Lands provided physiologic evidence for the existence of two subtypes of beta-adrenergic receptors, which he termed beta-1 and beta-2 (Lands et al., 1967). Subsequently, radioligand binding and biochemical analysis has confirmed the existence of these two subtypes (Stiles et

The β-Adrenergic Receptors Ed.: J. P. Perkins © 1991 The Humana Press Inc.

al., 1984). This topic is covered in more detail in other chapters. It has become clear that both beta-1 and beta-2 adrenergic receptors stimulate the membrane-bound enzyme adenylate cyclase, thus increasing the intracellular concentration of cyclic AMP. Cyclic AMP appears to be the second messenger of beta-adrenergic agonists in all tissues examined (Sutherland and Rall, 1960).

With the advent of radioligand binding techniques in 1974, it was quickly realized that beta-adrenergic receptors were present in most tissues and were not static entities, but rather were under dynamic regulation (Stiles et al., 1984). Receptor number was found to increase (upregulation) or decrease (downregulation) in response to a wide variety of pathophysiologic interventions or hormonal manipulations (Stiles et al., 1984). Over the subsequent 15 y, our knowledge of the components of the beta-adrenergic receptor–adenylate cyclase system has increased dramatically, and this newly acquired information has indicated that there are multiple steps in the transmembrane signaling process that may be altered by pathophysiologic conditions. Thus, it is now abundantly clear that alterations in receptor number represent but a single mechanism utilized by the organism to alter sensitivity to catecholamines. In this chapter, I shall discuss a multiformity of mechanisms whereby the sensitivity of the beta-adrenergic receptor–adenylate cyclase system is altered by drugs and hormones.

The beta-adrenergic receptor–adenylate cyclase system consists of at least three major protein components. These are the hormone receptor, the catalytic moiety of the enzyme adenylate cyclase, which converts ATP to cyclic AMP, and a coupling protein known as the stimulatory guanine nucleotide regulatory protein, variously abbreviated as G_s or N_s. This latter protein is regulated by guanine nucleotides such as GTP and is involved in the "coupling" of the receptor to its effector, the enzyme adenylate cyclase. The G_s protein has been purified from a variety of tissues and is known to consist of a heterotrimer of molecular mass 45, 35, and 8 kDa (Sternweis et al., 1981; Codina et al., 1984). This G protein, by directly interacting with GTP, produces two effects: The first is to perturb the interaction of agonists with the beta-adrenergic receptor; the second is to activate the catalytic moiety of the enzyme adenylate cyclase. Using radioligand binding techniques coupled with computer modeling, it has been demonstrated that the beta-adrenergic receptor is capable of existing in two discrete states having either low or high affinity for an agonist (Stiles et al., 1984). Biochemical studies subsequently indicated that these two forms of the receptor likely correspond to the receptor alone (low affinity) and the receptor coupled to G protein (high affinity) (Limbird et al., 1980; Limbird and Lefkowitz, 1978). These two discrete forms of the receptor likely exist in an equilibrium in the membrane. Agonists, such as isoproterenol, promote the interaction of hormone receptor with the coupling protein, whereas antagonists, such as propranolol, do not favor the formation of this complex. Since antagonists display equal affinity for the two states of the receptor, their competition curves are uniphasic and conform to what is predicted by the law of mass action for binding to a single class of sites. In contrast, agonist competition curves are biphasic, are shifted to a lower affinity, and become uniphasic in the presence of guanine nucleotides (Stiles et al., 1984). Guanine nucleotides, therefore, convert all the high-affinity-state receptors to the low-affinity state coincident with hormone promoted activation of the enzyme. Thus, a simple model to explain the activation of adenylate cyclase by beta agonists would be as follows: A hormone interacts with a beta-adrenergic receptor and initially forms a low-affinity binary complex (H•R). If the hormone is an agonist, it can then interact with a G protein to form a high-affinity complex (H•R•G). The formation of this ternary complex then permits GTP to associate with the G protein. This association results in the dissociation of H•R•G, and the GTP-liganded α subunit of the G protein can then interact with and activate adenylate cyclase. The activation can be terminated when GTP is hydrolyzed to GDP by a GTPase that is associated with the α subunit of G_s (Stiles et al., 1984). Detailed models of the biochemical mechanisms involved in activation of the catalytic unit are described in other chapters of this book.

The complexity of the transmembrane signaling pathway, that results in the activation of adenylate cyclase briefly described above highlights the number of possible sites and interactions that could be perturbed in altered pathophysiologic states. Alterations in any of these steps might account for the apparent altered sensitivity to catecholamines associated with a variety of drug and hormone perturbations. In fact, qualitative and quantitative alterations in several of the components, as well as in the interactions between the components, have been described for different pathophysiological changes. Thus, changes in the number of beta-adrenergic receptors, the affinity of agonists for the beta-adrenergic receptors, and the concentrations of the G_s protein or the ability of the G_s protein and the beta-adrenergic receptor to interact may all be perturbed. Figure 1 is a schematic representation of the beta-adrenergic receptor–adenylate cyclase system and some of the methodologies available to quantify the components of the system and their interactions. Various aspects of the methodology, interpretation, and pitfalls of radioligand binding assays are described in detail in other chapters. The remainder of this chapter will discuss the pathophysiologic regulation of the beta-adrenergic receptor–adenylate cyclase system in animal models.

2. Regulation of the Beta-Adrenergic Receptor– Adenylate Cyclase System by Compounds that Interact Directly or Indirectly with the Beta-Adrenergic Receptor System

The investigation of the phenomenon of desensitization, or tachyphylaxis, has been discussed in another chapter and shall not be further commented upon here.

2.1. Denervation Hypersensitivity

The phenomenon of denervation hypersensitivity has long been described in the clinical and animal literature. It is characterized by a marked increase in the sensitivity of a tissue to a given stimulus following the chronic deprivation of the stimulus. For example, chronic decreases in the local tissue concentration of neurotransmitters at the neural synapses will produce enhanced responsiveness to catecholamines in these cells when reexposed to the neurotransmitter. Thus, if one chemically or surgically denervates an adrenergically innervated tissue, and later the sensitivity of that tissue to a catecholamine chal-

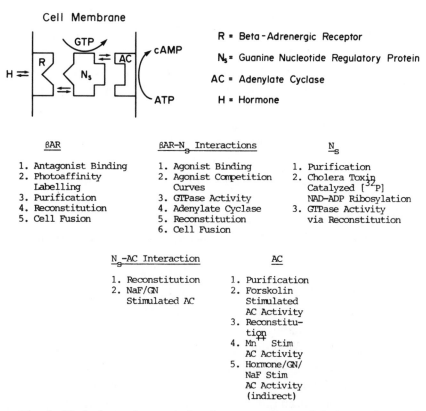

Fig. 1. Techniques for assessing the components of the beta-adrenergic receptor–adenylate cyclase system.

lenge is tested, the subsequent catecholamine effect would be much larger than that seen in tissue not denervated. The mechanisms responsible for this hypersensitive response have only recently begun to be elucidated. It appears from radioligand binding data that alterations in the beta-adrenergic receptor may play a role in the hypersensitive response seen following the depletion of catecholamines.

The neurotoxin 6-hydroxydopamine selectively destroys catecholaminergic neurons, and it has served as a useful tool to induce denervation hypersensitivity. This destruction of the nerve terminals leads to a depletion of norepinephrine. To study the effect of 6-hydroxydopamine on the central nervous system, Sporn et al. (1976) treated rats with intraventricular injections of 6-hydroxydopamine to destroy the nerve terminals. They found the density of beta-adrenergic receptors in the cerebral cortex to be 30% greater than that found in control animals. There was also an increase in the accumulation of cyclic AMP in the brain in response to maximally stimulating concentrations of isoproterenol in the treated animals. Thus, there appears to be a direct correlation between the increase of beta-adrenergic receptor number and increase in cyclic AMP accumulation. There was, however, no demonstration that it was the change in receptor number alone that accounted for all the changes in cyclic AMP. The authors do suggest, however, that the regulation of the postsynaptic response to catecholamines at central noradrenergic synapses is at least partially mediated by a dynamic regulation of the number of beta-adrenergic receptors in response to altered catecholamine levels. In this brain model, at least, depletion of catecholamines appears to upregulate the number of beta-adrenergic receptors, thus promoting an enhanced response to exogenously administered catecholamine derivatives.

Similar studies have been reported, using the rat heart as a model for hypersensitivity following denervation in animals systematically administered 6-hydroxydopamine (Chiu, 1978; Kajiyama et al., 1982; Nomura et al., 1980; Yamada et al., 1980). Nomura et al. (1980) found that when neonatal rats were given 6-hydroxydopamine, norepinephrine content was decreased by 55% in the atria and by 30% in the whole heart, whereas the norepinephrine content in the left ventricle did not significantly change. The fact that norepinephrine levels were lowest in the ventricle and did not change after 6-hydroxydopamine treatment is consistent with the idea that there are fewer nerve terminals in the ventricle. Following 6-hydroxydopamine treatment, the isoproterenol-induced stimulation of ornithine decarboxylase in whole-heart homogenates was markedly potentiated, the number of beta-adrenergic receptors in the whole heart homogenate was increased, and the positive inotropic response to norepinephrine in isolated atria was also potentiated. Nomura et al. noted that the chronotropic response to isoproterenol was not potentiated, even though the positive inotropic response to norepinephrine was potentiated. An explanation for this observation might relate to the fact that the atria of many species of animals contain mixtures of both beta-1 and beta-2 adrenergic receptors (Stiles et al., 1983b). The beta-2 adrenergic receptor appears to mediate, at least in part, the chronotropic response in the heart, whereas the beta-1 adrenergic receptors mediate both chronotropy and inotropy. Since norepinephrine is a very weak agonist at beta-2 adrenergic receptors, it might be that in the model of 6-hydroxydopamine treatment, beta-1 adrenergic receptors are significantly altered, but beta-2 adrenergic receptors are not. This would explain why there is no change in the chronotropic response to isoproterenol, whereas the inotropic response is markedly changed following denervation hypersensitivity.

This hypothesis was tested in a study by Kajiyama et al. (1982) in which they found that, after 6-hydroxydopamine treatment, the concentration of beta-adrenergic receptors in membranes prepared from the whole heart increased from 59–108 fmol/mg protein. This increase was caused by a change in the beta-1 adrenergic receptor concentration without any significant change in the beta-2 adrenergic receptors. All these findings are consistent with the notion that norepinephrine is the physiological catecholamine active at beta-1 adrenergic receptors, but probably has little if any effect on beta-2 adrenergic receptors in vivo.

Selective increases in beta-adrenergic receptor density in the rat cerebellum have also been found after 6-hydroxydopamine treatment (Wolfe et al., 1982). Investigators have also studied the effect of topically administered 6-hydroxydopamine on the rabbit eye. These studies have demonstrated increases in the beta-adrenergic receptor density in the cillary body of the rabbit eye following treatment, but no information was provided on changes in cyclic AMP content following denervation (Neufeld et al., 1978). Denervation hypersensitivity also has been studied by the chronic administration of reserpine, an agent capable of depleting norepinephrine stores in nerve terminals. The results in this type of study are strikingly similar to those that have been obtained with 6-hydroxydopamine treatment. Thus, in studies of rat submandibular gland (Bylund et al., 1981; Cutler et al., 1981) and in rabbit heart (Tenner et al., 1982) it was found that the number of beta-adrenergic receptors and the amount of cyclic AMP generated in response to beta-adrenergic agonists were both increased following treatment with reserpine.

The third pharmacological agent capable of producing depletion of catecholamines is guanethidine. Glaubiger et al. (1978) found that the number of cardiac beta-adrenergic receptors increased compared to control when rats were treated for 5 wk with guanethidine. An increased cyclic AMP accumulation also occurred in response to exogenous isoproterenol. These findings suggest that the increased responsiveness of cardiac tissue may be secondary to increased beta-adrenergic receptor number. This study was performed before detailed computer modeling was available to assess receptor–effector coupling. Finally, Canter and coworkers (Canter et al., 1981) investigated the effects of long-term changes of sympathetic nervous activity on the beta-adrenergic receptor–adenylate cyclase system in the rat pineal gland. It was demonstrated that either guanethidine or reserpine treatment led to a significant increase in the beta-adrenergic receptor number and the accumulation of cyclic AMP in response to norepinephrine.

Taken together, these studies strongly suggest that changes in the ambient concentration of catecholamines may induce a dynamic regulation of the beta-adrenergic receptor–adenylate cyclase system in a direction that would at least partially explain the increased sensitivity to catecholamines following prolonged deprivation of catecholamines. The fact that there is a concomitant increase in cyclic AMP accumulation to match the change in beta-adrenergic receptors suggests that these alterations may occur in a parallel fashion and may, in fact, be directly related. Presently, it is not known whether other components of this system are altered or if receptor G-protein coupling is perturbed under conditions of denervation hypersensitivity.

2.2. Propranolol Withdrawal Syndrome

As described in the preceding section, the chronic removal of catecholamines leads to a state of hypersensitivity. An analogous situation may occur if the pharmacological equivalent of catecholamine removal is obtained by chronically blocking the beta-adrenergic receptors with beta-adrenergic receptor blockers. Beta-adrenergic receptor antagonists have become one of the most commonly employed therapeutic agents. The beta-adrenergic blockers, particularly propranolol and now a whole host of selective and nonselective agents, are available and are used in the clinical practice of medicine for the treatment of angina pectoris, hypertension, and a variety of other conditions. It is, therefore, not surprising that investigators began to monitor the consequences of the abrupt withdrawal of these therapeutic agents and the subsequent effect on patients' symptoms and outcome. In 1974, Alderman et al. reported the sudden onset of unstable angina pectoris and myocardial infarction after the abrupt withdrawal of propranolol in patients who previously had stable angina pectoris. These early reports have subsequently been corroborated, suggesting that indeed an apparent withdrawal syndrome may exist (Miller et al., 1975). Boudoulas et al. (1977) first tested the possibility that a hypersensitivity to adrenergic stimulation after propranolol withdrawal might be responsible for the propranolol withdrawal syndrome. To study this phenomenon, they administered propranolol to normal subjects for 2 d and then abruptly withdrew it and tested the sensitivity of the patients to isoproterenol as delineated by their chronotropic and inotropic responses. They demonstrated that a hypersensitivity to isoproterenol was found 24–48 h after the abrupt withdrawal. Nattel et al. (1979) demonstrated similar findings in hypersensitive patients. In their studies, propranolol was abruptly withdrawn from nine patients who had been treated chronically for hypertension. These patients all displayed a transient hypersensitivity to the chronotropic effects of isoproterenol. The hypersensitivity began as early as 2 d after withdrawal and lasted up to 14 d. Two-thirds of these patients experienced excessive sweating and palpitations, all of which are consistent with a hyperadrenergic state. These findings, however, have not been universally corroborated. Using different protocols, Faulkner et al. (1973) could find no abnormality in the inotropic responsiveness of the human atria to norepinephrine 2 d after withdrawal of propranolol. Similarly, Lindenfeld et al. (1980) found no evidence for a hypersensitive adrenergic response to catecholamines either in normal subjects or in patients who have angina pectoris after sudden propranolol withdrawal. Thus, a direct correlation between the symptoms of the propranolol withdrawal phenomenon and hypersensitivity to catechol-amine responsiveness remains a tantalizing but as yet not thoroughly proven hypothesis. (*See also* Chapter 7).

Several attempts have been made to establish an animal model for the propranolol withdrawal syndrome. Webb and coworkers (Webb et al., 1981) chronically instrumented dogs with indwelling catheters and then gave propranolol for 14 d. Propranolol was then abruptly withdrawn, and measurements of inotropy and chronotropy were made serially. From 36–60 h after withdrawal, when propranolol levels had become undetectable, the isoproterenol dose–response curve for inotropy shifted to the left, consistent with a hypersensitive state. There was, however, no shift in the isoproterenol dose–response curve for chronotropy. This may relate to the differential regulation of beta-1 and beta-2 adrenergic receptors, although this possibility was not investigated in this study. The investigators did, however, measure the plasma norepinephrine levels and found no alterations either during drug treatment or after withdrawal. Using a rodent model, Kennedy and Donnelly (1982) found similar supersensitive responses in rats abruptly withdrawn from propranolol after either intraperitoneal or oral propranolol administration for 21 or 60 d. In addition, they found that the supersensitive response was extremely specific for adrenergic stimuli, because the chronotropic responsiveness to calcium, oubain, glucagon, acetylcholine, methoxamine, or 3-isobutyl 1-methylxanthine was unaltered during the withdrawal period. This suggested that only the beta-adrenergic receptors themselves were altered, and not other receptors, such as glucagon or alpha-adrenergic receptors, or the mechanisms responsible for calcium fluxes in the cell.

When taken in aggregate, the preceding physiological data suggest that following the abrupt withdrawal of propranolol, and presumably other beta-adrenergic blockers, a hyperadrenergic responsiveness may well ensue. This area of research has been controversial since its initial description, and further studies in normal humans as well as in animal models seem appropriate at the present time. If this putative syndrome does indeed exist, one must ask what mechanism might be responsible. In analogy with the denervation hypersensitivity model, it is not unreasonable to speculate that alterations in the beta-adrenergic receptor–adenylate cyclase system might be the responsible underlying mechanism. Boudoulas et al. (1977) suggested on theoretical grounds that an increase in the number of beta-adrenergic receptors could be responsible for the reported supersensitivity. They speculated that, if chronic exposure to catecholamines leads to down-regulation of beta-adrenergic receptor number, it is also plausible that occupancy of the receptors by propranolol would prevent them from being exposed to catecholamines. This might lead to increased density of beta-adrenergic receptors again by analogy with that seen in the denervation hypersensitivity model. They further reasoned that, if the beta-adrenergic antagonist were abruptly withdrawn, the excess beta-adrenergic receptors, which would then be coupled to the adenylate cyclase, would lead to an enhanced sensitivity to catecholamines with an increased production of cyclic AMP. To investigate the underlying mechanism, Glaubiger and Lefkowitz (1977) studied the effect of intraperitoneal injections of propranolol for 2 wk. They demonstrated that there was a doubling of the concentration of beta-adrenergic receptors in the ventricles of rats following the chronic propranolol therapy. They further demonstrated that there was no change in the dissociation constant for the antagonist radioligand [^{3}H]dihydro-alprenolol. This was the first report to suggest that an alteration in beta-adrenergic receptor number might provide an explanation for adrenergic hypersensitivity following propranolol withdrawal. Aarons and Molinoff (1982) used a rat model to study changes in the density of beta-adrenergic receptors in lymphocytes, heart, and lung after chronic treatment with propranolol. In this model, the rats underwent a chronic subcutaneous infusion of propranolol by osmotic minipumps. The propranolol levels achieved with these osmotic minipumps were in the same range as the levels found in many patients receiving propranolol therapy. The infusion of propranolol for 7 d resulted in significant increases in the density of beta-adrenergic receptors in the rat ventricle (28% increase) and in the rat lung (32% increase), and a 34% increase in the number of receptors on membranes derived from rat lymphocytes. They further demonstrated that the increase in beta-adrenergic receptors in the rat ventricle as well as the lungs was a consequence of increases in both beta-1 and beta-2 adrenergic receptors.

This demonstrated for the first time that both beta-1 and beta-2 adrenergic receptors were increased following propranolol treatment. In
addition, the changes seen on circulating lymphocytes seemed to parallel the changes seen in the heart and lung, suggesting to the authors
that the quantity of beta-adrenergic receptors on lymphocytes may
indeed be a good marker of changes that occur in the heart and lung,
making possible measurements in humans, where the relevant tissues
are not routinely available for assay. (*See also* Chapter 7).

Cooper et al. (1986) utilized a cat model to study the effect of
long-term beta-adrenergic blockade with propranolol. Chronically
beta-adrenergic-blocked cats demonstrated an enhanced contractility
in response to catecholamines following beta-adrenergic blocker
withdrawal compared to controls. Quantification of beta-adrenergic
receptors and the percentage of receptors in the agonist-specific
high-affinity state failed to disclose any difference between control
and propranolol treated animals. Yet, the catecholamine-stimulated
cyclic AMP accumulation was enhanced in treated animals, suggesting an enhanced "coupling" of hormone action to the biochemical
effector. Which specific step in the transmembrane signal is altered
remains a speculation.

Brodde et al. (1985) have studied the effect of a variety of beta-
adrenergic blockers both with and without intrinsic sympathomimetic
activity (ISA). ISA is a property of a subgroup of beta-adrenergic
blockers that display a partial agonism. The agonist properties are weak
compared to a full agonist, such as isoproterenol. In this study, three
different beta-adrenergic antagonists were given to healthy adult males
for 2 d: propranolol, a nonselective beta-adrenergic blocker with no
intrinsic sympathomimetic activity; mepindolol, which has significant
ISA; and alprenolol, a nonselective beta-adrenergic blocker with very
weak ISA. Following therapy, the following parameters were
assessed: heart rate, peripheral renin activity, and the number of beta-
adrenergic receptors on circulating lymphocytes. In the patients given
propranolol, it was found that the number of beta-adrenergic receptors
was increased by 25%, and plasma renin activity and heart rate were
reduced. Following withdrawal of propranolol, plasma renin activity
rose and rapidly reached peak levels, and heart rate significantly

increased. Of note was the fact that the density of beta-adrenergic receptors on lymphocytes (beta-2) declined slowly, still being significantly elevated at 3 d following withdrawal. Propranolol was not detectable in plasma after 24 h following withdrawal. In contrast, mepindolol treatment caused a 30% decrease in beta-adrenergic receptor density and plasma renin activity after 2 d. Following withdrawal, plasma renin activity rapidly reached control levels, whereas beta-2 adrenergic receptors were still significantly decreased for 4 d after withdrawal. Finally, alprenolol treatment led to a rapid fall in plasma renin activity, but did not significantly affect the number of beta-2 adrenergic receptors found on lymphocytes. The authors, therefore, conclude that intrinsic sympathomimetic activity (i.e., partial agonism) may modulate beta-2 adrenergic density and enhance tissue responsiveness to beta-adrenergic receptor stimulation. Further work is clearly warranted in this area in human studies.

Although these studies would appear to firmly establish alterations in beta-adrenergic receptors as the mechanism responsible for the putative increased sensitivity to catecholamines following abrupt withdrawal of beta-adrenergic blockers, not all studies have shown upregulation in beta-adrenergic receptor number after propranolol administration. Baker and Potter (1980) used a rodent model of propranolol withdrawal in which rats were given oral or intraperitoneal propranolol for 1–7 wk. Following this chronic therapy, the authors failed to find any alteration in receptor number or antagonist affinity following propranolol treatment. They did not, however, measure physiological responses, hence it is not known whether the animals actually developed a catecholamine supersensitivity. Of particular interest is the study of Kennedy and Donnelly (1982), wherein no evidence was found for upregulation of beta-adrenergic receptor number either during propranolol therapy or upon withdrawal. Several other studies also have found no alteration in beta-adrenergic receptor number following the administration of a variety of beta-adrenergic receptor antagonists in animal models.

In summary, it appears that chronic administration of beta-adrenergic blockers without intrinsic sympathomimetic activity to humans and animals may well be associated with a supersensitivity to catecholamines following the abrupt withdrawal of the therapeutic agents. How common and how clinically significant this withdrawal syndrome is remains to be determined. Although there is currently suggestive data that the mechanism involved in this syndrome may be secondary to alterations in receptor number, further work is warranted before this mechanism can be called fact, rather than hypothesis. It would seem that this continues to be an area for fruitful investigation.

2.3. Thyroid Hormone

There has long been an interest in the interrelationships among thyroid hormones, catecholamines, and the beta-adrenergic receptor–adenylate cyclase system. This interest has come about largely because many of the symptoms of hyperthyroidism in humans, such as tachycardia, increased thermogenesis, sweating, tremor, and increased cardiac output, are all suggestive of excessive catecholamine stimulation. Conversely, the symptoms and signs of hypothyroidism, such as bradycardia, are compatible with a reduced level of circulating catecholamines. Although the obvious implication is that catecholamine levels are altered during hyper- and hypothyroidism, a large body of evidence clearly documents that plasma and urinary catecholamine levels are not altered in a direction that could account for the signs and symptoms seen in hyper- and hypothyroidism. Thus, plasma norepinephrine levels are either low or normal in patients with hyperthyroidism and are usually elevated in patients with hypothyroidism (Coulombe et al., 1976). It has long been recognized that beta-adrenergic antagonists, such as propranolol, ameliorate the symptoms of hyperthyroidism even though beta-adrenergic blockers do not alter the biosynthesis, release, or peripheral metabolism of thyroid hormones. The accumulation of these data led to the realization that alterations in the sensitivity of various end organs to catecholamines might be responsible for the pathophysiologic changes seen in altered thyroid function. The obvious locus for this alteration in sensitivity was immediately thought to be the beta-adrenergic receptor-adenylate cyclase system. Since the advent of radioligand binding in 1974, the two most commonly studied pathophysiologic states related to beta-adrenergic receptors are the phenomenon of agonist-induced desensitization and the effect of thyroid hormones on the beta-adrenergic receptor–adenylate cyclase system.

2.3.1. Hyperthyroidism

In general, hyperthyroidism is associated with an increase in the number of beta-adrenergic receptors and, in some cases, an increase in the affinity of the beta-adrenergic receptor for agonists. There are exceptions to this generalization, however, that make it necessary to consider each tissue individually and each species separately. Table 1 summarizes results from several studies on the effects of thyroid hormone on the beta-adrenergic receptor–adenylate cyclase system. It should be noted at the outset that although the finding of increased beta-adrenergic receptors in hyperthyroidism has now been known for many years, the exact mechanism underlying this increase is still unclear. There are obviously multiple levels at which this regulation could occur, such as enhanced protein synthesis secondary to increases in transcription and/or translation, or decreased receptor degradation. As noted before, many of the most obvious manifestations of altered thyroid function are cardiovascular in nature. This is probably responsible for the large number of studies of beta-adrenergic receptor regulation in heart tissue. There is almost universal agreement that with hyperthyroidism the number of beta-adrenergic receptors in the heart increases. The original description of this increase was published by Williams et al. (1977). They demonstrated a doubling of the beta-adrenergic receptor number without any change in the K_d for [³H] DHA when rats were given thyroid hormone in vivo. More recent studies have suggested that in addition to the increases in beta-adrenergic receptor number, there is an apparent twofold increase in the affinity of the beta-adrenergic receptors for agonists, as manifested by a leftward-shifted agonist competition curve (Stiles and Lefkowitz, 1981). This shift presumably reflects an enhanced coupling of the receptor with the G_s. All these alterations would be expected to enhance the sensitivity of the system to the effects of catecholamines.

Tse et al. (1980) have further assessed the beta-adrenergic receptor–adenylate cyclase system in the rat heart. They found that with hyperthyroidism there is an increased number of beta-adrenergic

Table 1
β-Adrenergic Receptor–Adenylate Cyclase System in Hyperthyroidism

Tissue	βAR[a] number	Coupling process	Adenylate cyclase activity
Rat heart	Increase	Enhanced	Increase
Rat fat	No change	Enhanced	Increase
Rat lung	No change	ND[b]	ND
Turkey erythrocyte	No change	No change	Increase
Rat liver	Decrease	ND	Decrease

[a]βAR—β-adrenergic receptor.
[b]ND—Not determined.

receptors, an increased maximal stimulation of adenylate cyclase activity by isoproterenol, a decreased cyclic AMP phosphodiesterase activity, and an increase sensitivity to isoproterenol induced contraction in ventricular muscle strips. All these findings are compatible with an enhanced sensitivity of the heart to catecholamines, which may explain the tachycardia and increased force of contraction observed in hyperthyroidism. A study by Kempson et al. (1978) used rat heart ventricular slices incubated in vitro with thyroid hormones to study the time-course of the changes in the number of these receptors. They found that the increase in receptor number resulted from a two-phase process. The first phase was relatively rapid and reached a maximum by $1^1/_2$–2 hours. This phase was not blocked by protein synthesis inhibitors such as cycloheximide, but the process did require added amino acids for consistent results. A second phase of the increase in beta-adrenergic receptors occurred over a long period of time (approx 15 h), and this phase was completely blocked by the addition of protein synthesis inhibitors. Kempson et al. thus postulated that thyroid hormones can increase beta-adrenergic receptors by first permitting an already synthesized but dormant cohort of receptors to be expressed on the cell surface and then (as a secondary mechanism) by stimulating the synthesis of new beta-adrenergic receptors. To date there have been no further studies addressing this very interesting hypothesis. With the development of molecular biological techniques, it will be most important to study regulation of the gene coding for the beta-adrenergic receptor as well as the production of messenger RNA for the receptor.

In contrast to the increase in the number of receptors in the heart in hyperthyroidism, Scarpace and Abrass found that adult rats given thyroid hormone for 3 d exhibited no increase in the number of beta-adrenergic receptors in the lungs compared to control (Scarpace and Abrass, 1981). Whitsett et al. studied the effect of thyroid hormones on the developing rat lung (Whitsett et al., 1980). They found that the normal increase in beta-adrenergic receptor density that occurs in lung tissue between postnatal days 15 and 28 depends on the presence of thyroid hormones. If the animals were treated with propylthiouracil from the time of birth, the number of receptors at day 28 was markedly lower than the number in normal animals. This contrasts with the lack of effect of thyroid hormones in the adult lung. A similar lack of effect has been described in the number of beta-adrenergic receptors in whole-brain homogenates of rats who were treated with thyroid hormones (Smith et al., 1982).

Another model used for the study of hyperthyroidism is the turkey erythrocyte (Bilezikian et al., 1979; Furukawa et al., 1980). No alteration in the number of beta-adrenergic receptors or agonist affinity was found in erythrocyte membranes from hyperthyroid turkeys. There was, however, an increased sensitivity to submaximal concentrations of isoproterenol; the amount of cyclic AMP generated in whole cells from hyperthyroid animals was greater than that seen in control cells. The mechanism responsible for this enhanced cyclic AMP accumulation has not yet been determined.

In contrast to the enhanced sensitivity and responsiveness to catecholamines of many tissues in the hyperthyroid state, membranes derived from rat liver appear to demonstrate a hyporesponsiveness to catecholamines (Malbon and Greenberg, 1982; Malbon et al., 1978). In this model, the ability of beta-adrenergic agonists to stimulate adenylate cyclase activity is depressed, and this depressed sensitivity cannot be corrected with the use of phosphodiesterase inhibitors. In addition, there is an impaired activation of glycogen phosphorylase in parallel with the decreased adenylate cyclase activity. The decreased responsiveness in liver membranes derived from hyperthyroid animals is not, however, specific for the effects of catecholamines. There is also a decreased ability of glucagon, guanine nucleotides, and sodium fluoride to activate the adenylate cyclase system. In addition, both beta-adrenergic and glucagon receptor binding are significantly reduced in hyperthyroid liver membranes, a change consistent with generalized hyporesponsiveness of these membranes. Since sodium fluoride activation is depressed in these membranes, it would appear likely phenomenon the hyporesponsiveness is not secondary to an isolated change in receptor number. To assess whether there were also alterations in the quantity or functionality of the G_s, Malbon et al. quantified the G protein with the use of cholera toxin-mediated [^{32}P]ADP ribosylation with subsequent SDS-PAGE and autoradiography (Malbon and Greenberg, 1982). The quantity of G protein was not different in liver membranes from hyperthyroid animals and control animals. However, the functionality of the G_s protein, as assessed by reconstitution techniques, was reduced in terms of the ability of solubilized G_s protein of hyperthyroid liver membranes to promote sodium fluoride stimulated adenylate cyclase activity when reconstituted into membranes of S49 cyc⁻ lymphoma cells. These findings imply that in hyperthyroidism the diminished catecholamine activation of glycogen phosphorylase may be secondary to decreased numbers of beta-adrenergic receptors and to a diminished functionality of the G_s protein. Therefore, in this model, there appears to be a dual regulation of both beta-adrenergic receptors and the functionality of the G_s protein by the levels of thyroid hormone.

2.3.2. Hypothyroidism

To study the effects of hypothyroidism, animals are usually rendered hypothyroid by one of two mechanisms: Either through the administration of propylthiouracil or by thyroidectomy. In general, beta-adrenergic receptor number appears to be decreased in many tissues under hypothyroid conditions (Table 2). The earliest studies, which simply counted the number of beta-adrenergic receptors, were performed in the rat heart model. There appears to be universal agreement that beta-adrenergic receptor numbers are decreased in rat hearts from hypothyroid animals (Stiles et al., 1984). Brodde et al. (1980)

Table 2
β-Adrenergic Receptor–Adenylate Cyclase System in Hypothyroidism

Tissue	βAR[a] number	Coupling process	Adenylate cyclase activity
Rat heart	Decrease	No change	Decrease
Rat reticulocyte	Decrease	Decrease	Decrease
Turkey erythrocyte	Decrease	No change	Decrease
Rat brain	Decrease	ND[b]	ND
Rat skeletal muscle	Decrease	ND	Decrease
Rat fat	No change	Decrease	Decrease
Rat liver	Increase	ND	Increase

[a]βAR—β-adrenergic receptor.
[b]ND—Not determined.

found this change to be associated with decreased isoproterenol-stimulated contractility and decreased cyclic AMP accumulation with no alteration in calcium-stimulated contractility. In addition, they found that basal, isoproterenol-stimulated, and sodium fluoride stimulated adenylate cyclase activities were all significantly lower in membranes derived from hypothyroid animals than in control membranes. Since sodium fluoride activates adenylate cyclase through the G_s protein and not through a receptor-mediated process, these findings suggest that there may be alterations in multiple components of the beta-adrenergic receptor system in this pathophysiologic condition. In our study (Stiles and Lefkowitz, 1981) of the hypothyroid heart, we found no alteration in receptor–effector coupling as assessed by detailed agonist competition curves in membranes derived from hearts of hypothyroid rats. However, we did note a significant decrease in the number of beta-adrenergic receptors, as has been described by many other investigators.

There are now several examples of multiple component alterations in hypothyroidism. One such model is the rat reticulocyte (Stiles et al., 1981). First, the number of beta-adrenergic receptors was lower by approx 50% in reticulocytes of hypothyroid rats than in controls. Agonist competition curves revealed that the isoproterenol competition curve in "hypothyroid" reticulocyte membranes was shifted to the right (*see* Fig. 2). Computer modeling techniques indicated that this shift was primarily caused by the decreased ability of isoproterenol to stabilize the high-affinity state of the receptor. In contrast, the isoproterenol competition curve in the presence of Gpp (NH)p was not altered. The rightward-shifted isoproterenol competition curve would suggest that agonists might be less effective in activating adenylate cyclase. This was in fact found to be true when maximal isoproterenol stimulation produced 50% less cyclic AMP in reticulocyte membranes from hypothyroid animals than it produced in controls. Sodium fluoride stimulation also was diminished, suggesting that the G_s protein might be perturbed. When the quantity of G_s protein was assessed by [^{32}P]ADP ribosylation, it was found that the G protein was decreased by approx 40% in membranes derived from hypothyroid animals. These data are all compatible with the concepts that decreased levels of thyroid hormone can regulate the quantity of two components of the beta-adrenergic receptor–adenylate cyclase system and that these changes are in a direction that would explain the hyporesponsiveness of tissues to catecholamines in hypothyroidism.

Rat adipose tissue has provided another interesting model, wherein multiple components of the adenylate cyclase system have been altered in hypothyroidism. It has been known for a long time that the lipolytic response to administered epinephrine is markedly blunted in hypothyroidism. The preponderance of data suggest that beta-adrenergic receptor number is not significantly changed. However, agonist affinity appears to be decreased and catecholamine-stimulated adenylate cyclase activity is also reduced (Malbon, 1980a). Malbon and Gill (1979) assessed the G_s protein in fat cell membranes derived from normal and hypothyroid rats by means of cholera toxin induced [^{32}P]ADP ribosylation of the G_s protein. They found the quantity of G_s protein in fat from hypothyroid animals to be the same as that in controls, even though epinephrine-stimulated adenylate cyclase activity was diminished in the hypothyroid animals. In a followup study, Malbon et al. (1984) determined that membranes from hypothyroid-rat fat cells contain normal amounts of G_s and that peptide maps of the fragments generated by partial proteolysis of the radiolabeled peptide were essentially identical in the hypothyroid and euthyroid prepara-

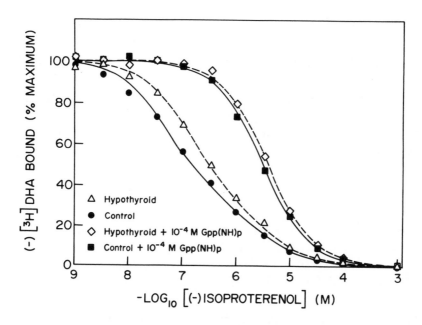

Fig. 2. Isoproterenol competition curves of isoproterenol with [³H]DHA in reticulocyte membranes from control and hypothyroid rats, in the presence and absence of guanine nucleotides, Gpp(NH)p. The reticulocytes were incubated with 2.5 n*M* [³H]DHA in competition with the indicated concentration of isoproterenol. Reproduced from *The Journal of Clinical Investigation*, 1981, vol. 68, p. 1452, by copyright permission of the American Society for Clinical Investigation.

tions. In addition, two-dimensional isoelectric focusing gels demonstrated no difference between the G_s from normal animals and that from hypothyroid animals. Using a variety of reconstitution and cell fusion techniques, Malbon et al. concluded that, although the number of receptors and the structure and function of the G_s protein were normal, there was a diminished ability of the receptor to effectively interact with G_s. They further concluded that this diminished interaction was likely to be the etiology of the decreased production of cyclic AMP in hypothyroid-rat fat tissue. A study by Malbon et al. (1985) demonstrated an enhanced ability of adenosine and adenosine analogs to inhibit cyclic AMP accumulation in hypothyroid compared to control animals. The mechanism appears to involve an enhanced concentration of the inhibitory G protein (G_i). Thus, both the diminished responsiveness to beta-adrenergic agonists and the enhanced responsiveness to adenosine agonists may be related to an increased quantity of G_i.

The beta-adrenergic receptor–adenylate cyclase system in hepatocytes again represents the exception rather than the rule in terms of its response to hypothyroidism. Malbon and colleagues (Malbon, 1980b; Malbon and Greenberg, 1982) have extensively studied the effects of hypothyroidism in the liver and found a potentiation of catecholamine-mediated accumulation of cyclic AMP and an activation of glycogen phosphorylase. There was no concomitant enhancement of glucagon-mediated glycogen phosphorylase activation, suggesting that hypothyroidism produced a specific alteration in beta-adrenergic agonist mediated processes. The same group extended their work to demonstrate that the beta-adrenergic receptor number increased two- to threefold and that maximal adenylate cyclase activity was enhanced threefold in response to isoproterenol. In contrast, sodium fluoride and glucagon-stimulated activities were no different in membranes from hypothyroid rats than in control rats. In addition, these authors demonstrated that there was no alteration in cyclic AMP phosphodiesterase activity and that the affinity of isoproterenol for the beta-adrenergic receptor was not altered. In this system, then, it appears that changes in receptor number in response to decreased thyroid hormone levels may be responsible for the increased catecholamine-stimulated activation of glycogen phosphorylase.

At least two studies have investigated the effect of hypothyroidism on the beta-adrenergic receptor–adenylate cyclase system in skeletal muscle. Sharma and Banerjee (1978) reported a lower number of beta-adrenergic receptors in skeletal muscle from thyroidectomized rats than in control rats. A study by Chu et al. (1985) demonstrated that, following 3 wk of therapy with propylthiouracil, the content of glycogen and glucose-6-phosphate in muscle was decreased. Basal and epinephrine-stimulated phosphorylase A and phosphorylase B kinase activities were also significantly reduced. Epinephrine-stimulated cyclic AMP accumulation and cyclic AMP dependent protein kinase activity were similarly decreased. Concomitant with all these changes, it was found that the decreased beta-adrenergic responsiveness of the enzymes involved in glycogen metabolism are likely secondary to increased activity of phosphoprotein phosphatases and to reduced beta-adrenergic receptor numbers and adenylate cyclase activity.

2.4. Adrenal Corticosteroids

Glucocorticoids, such as cortisol, have long been known to be essential for normal cellular metabolism. Studies over the past 25 y have indicated that normal circulating levels of cortisol must be present in order for other hormones to exert their normal influence upon the target cell. In 1953, Zweifach et al. demonstrated that corticosteroids could enhance the effects of epinephrine on the mesenteric vascular bed in the rat. In 1972, Exton et al. demonstrated that glucocorticoids had a "permissive role" in regulating the effects of cyclic AMP mediated hormone action in rat liver. They determined that adrenalectomized rats exhibited a markedly reduced catecholamine stimulation of gluconeogenesis and glycogenesis, even though the production of cyclic AMP was increased. Steroid replacement in these adrenalectomized animals restored the normal catecholamine responsiveness.

The mechanism whereby glucocorticoids produce this enhanced sensitivity to catecholamines has remained unclear. Several possible explanations have been proposed. These include:

1. Alterations in events distal to the enzyme adenylate cyclase, such as alterations in the levels or functionality of protein kinase;
2. Changes in the hormone receptor and its coupling to adenylate cyclase; or
3. Changes in the disposition of catecholamines, such as reduced neural uptake and the like.

Studies attempting to delineate the mechanisms whereby adrenal steroids or the lack thereof mediate changes in the beta-adrenergic receptor–adenylate cyclase system have been reported in both animals and humans, but only animal studies will be discussed here. Work on human tissues is described in Chapter 7 of this book.

In 1976, Wolfe et al. found that, after adrenalectomy in the rat, when catecholamine-induced effects on glucose metabolism are impaired, the number of hepatic beta-adrenergic receptors, the activity of the enzyme adenylate cyclase, and cyclic AMP production, are increased. This was the first demonstration of altered beta-adrenergic receptor number secondary to a loss in normal circulating corticosteroids. These data are quite consistent with the data mentioned above that demonstrated an increase in catecholamine-stimulated cyclic AMP accumulation in rat liver under similar conditions (Moylan et al., 1982). In this case, the change in receptor number is in a direction opposite to that which would explain the observed physiological effects. The suggestion was made that events distal to the beta-adrenergic receptor–adenylate cyclase system led to impaired gluconeogenesis and glycogenolysis. It was further proposed that the increased number of beta-adrenergic receptors and the increased cyclic AMP production were compensatory mechanisms to help maintain homeostasis. Although this teleological explanation is consistent with the data, there was no proof that the change in beta-adrenergic receptor number was indeed a compensatory mechanism. Guellaen et al. (1978) confirmed these results, that is, an increase in beta-adrenergic receptor number. They further demonstrated that, although there is a predominance of alpha-adrenergic receptors, rather than beta-adrenergic receptors, in the rat liver, there are no changes in the number of alpha-adrenergic receptors after adrenalectomy. This information suggests that the effects of epinephrine were not changing from an alpha-adrenergic-mediated mechanism to a beta-adrenergic receptor mechanism under the conditions studied.

Cortisol and some of its synthetic analogs are also known to exert a potentiating effect on catecholamine-stimulated bronchial relaxation. Mano et al. (1979) have investigated the mechanism by which this occurs. They administered hydrocortisone subcutaneously to male rats for 7 d and found that the number of pulmonary beta-adrenergic receptors increased by 70%, whereas there was no significant change in the K_d for the antagonist radioligand. By contrast, a 24% fall in beta-adrenergic receptors occurred after adrenalectomy. To demonstrate that these changes were a direct effect of cortisol depletion, they administered hydrocortisone, which led to an increase in beta-adrenergic receptor number back to control levels. Thus, the enhancement of catecholamine action by adrenal hormones may relate to an increase in beta-adrenergic receptor number. It should be noted, however, that these studies were done on whole-lung homogenates and not directly on vascular smooth muscle, which is presumably where the physiological effects are mediated. Fraser and Venter (1980) reported that hydrocortisone induced a dose-dependent increase in the rate of beta-adrenergic receptor synthesis in cultured human lung cells. After 24 h of treatment, the number of beta-adrenergic receptors was twice that found in control cells.

Corticosteroids have long been known to accelerate the maturation of the fetal lung. Cheng et al. (1980) found that when β-methozone was injected into rabbits at 25 d of pregnancy, the concentration of pulmonary beta-adrenergic receptors in the fetus increased 77% more than the concentration found in untreated fetuses. In contrast, no change was found in the number of beta-adrenergic receptors in treated fetal rabbit hearts. It is of interest that in the rabbit, pulmonary beta-adrenergic receptors are of the beta-1 subtype, not the beta-2 subtype that is found in many other mammalian lungs. This suggests that, although corticosteroids can increase the level of beta-1 adrenergic receptors in the rabbit lung, no changes are seen in the beta-1 adrenergic receptor of the rabbit heart. This difference may relate to maturation differences between the heart and the lung.

Two other examples of dexamethasone regulation of the beta-adrenergic receptor–adenylate cyclase system are worth noting. Lai et al. (1982) found that dexamethasone regulates beta-adrenergic receptor subtype in 3T3-L1 cells. In 3T3-L1 cells prior to differentiation, when phenotypically they display characteristics of fibrocytes or preadipocytes, the cells have predominately beta-2 adrenergic receptors. Following differentiation induced by dexamethasone, whereupon the cells acquire the characteristics of an adipocyte with a typical lipid droplet, the cells manifest predominantly beta-1 adrenergic receptors. Thus, as the cells undergo differentiation, the subtype of beta-adrenergic receptors changes almost completely from beta-2 to beta-1. This suggests that dexamethasone can alter transcription or translation such that either the gene encoding beta-1 adrenergic receptors becomes activated simultaneously with the beta-2 adrenergic receptor gene being inactivated or the translational efficiencies or stability of the mRNA transcripts for beta-1 and beta-2 are dramatically altered. Measurement of the transcripts for beta-1 and beta-2 will need to be made to differentiate between these two possibilities. The second study of interest is that of Rodan and Rodan (1986) wherein they found that treatment of ROS 17/28 cells (an osteosarcoma cell line) with dexamethasone increases the isoproterenol, guanine nucleotide, NaF, and forskolin-stimulated adenylate cyclase activity. It was determined that this enhanced sensitivity of adenylate cyclase to a variety of effectors was consequent to an increase in both beta-adrenergic receptors and G_s protein. This alteration in beta-adrenergic receptors and G_s was dependent on new protein synthesis. These findings also are consistent with changes in translation or transcription secondary to the dexamethasone treatment. Since the cDNA for both the beta-adrenergic receptors and the G proteins have been cloned, it will now be possible to address the molecular biological mechanisms responsible for these regulations.

Two studies have investigated the effect of adrenalectomy on the rat myocardial beta-adrenergic receptor–adenylate cyclase system. As noted above, corticosteroids can enhance the positive inotropic effect of catecholamines on the heart. In one study (Abrass and Scarpace, 1981) female rats underwent adrenalectomy and the number of beta-adrenergic receptors were quantified in animals postoperatively from 6 h to 7 d. These investigators found a significant (approx 50%) decrease in beta-adrenergic receptor number at all times after adrenalectomy. As an appropriate control, they treated a group of adrenalectomized animals with cortisol and demonstrated that this treatment reversed the effects of adrenalectomy. In contrast, Davies et al. (1981) used adrenalectomized male rats and found that the number of beta-adrenergic receptors did not change, but that the coupling mechanism between the receptor and the G protein was impaired compared with that of control animals. They demonstrated that the isoproterenol competition curve was shifted to the right and steepened after adrenalectomy compared to the control curve. In studying the effects of adrenalectomy on the adenylate cyclase system, they found that although basal, maximal isoproterenol-stimulated, and sodium fluoride stimulated activities were unaltered, the concentration of iso-proterenol needed for half maximal stimulation was increased significantly. These data would be consistent with an altered coupling of the receptor to the G protein. No further work has appeared to define more precisely the biochemical mechanisms responsible for these phenomenological observations.

The reasons for the differences in the two studies noted above are unclear but may relate to the different time-courses studied or the different sexes of the animals used. Further information will be needed to define the exact role corticosteroids play in maintaining the normal sensitivity of tissues with a variety of hormones at the biochemical and molecular biological level.

2.5. Gonadal Steroids

The studies to date on effects of progesterone, estrogen, and androgen on the beta-adrenergic receptor–adenylate cyclase system have been only preliminary in nature. Sex steroid hormones have long been known to influence catecholamine action, and catecholamines are known to be able to alter the function of ovaries. Jordan (1981) has investigated changes in ovarian beta-adrenergic receptors during the estrus cycle of the rat. During the progesterone phase, the concentration of receptors in ovarian membranes was twice the level seen during the estrogen phase. These results suggest that dynamic regulation of ovarian beta-adrenergic receptors occurs but the relation of this regulation to ovarian function remains mostly unknown. Other studies (Wagner et al., 1979) have shown that the chronic exposure of ovariectomized rats to estrogens decreases the catecholamine-mediated cyclic AMP accumulation by approx 10% in cerebral cortex. This change in cyclic AMP accumulation is associated with a concomitant decrease in the number of beta-adrenergic receptors.

Androgens, likewise, can regulate the number of beta-adrenergic receptors. Treatment of mice with testosterone leads to a time-dependent renal hypertrophy (Petrovic et al., 1981). A doubling of the number of beta-adrenergic receptors in renal membranes occurs at the same time. These studies also demonstrated that there is no change in the affinity of agonists or antagonists for the receptor, and no alterations in cyclic AMP accumulation were noted. Further work remains to be done on the interaction of the sex hormones with the beta-adrenergic receptor–adenylate cyclase system before any coherent picture of receptor regulation can emerge.

2.6. Antidepressant Drugs

Much work has gone into attempting to delineate the underlying pathophysiology responsible for the depressive illnesses. The realization that many of the antidepressant drugs could be classified as either tricyclic antidepressants or as monoamine oxidase inhibitors suggested that there might be a link between neurotransmitters and depression. The tricyclic antidepressants are known to inhibit the active reuptake of neurotransmitters, such as norepinephrine, released at the synapse. This would enhance norepinephrine's effect on postsynaptic beta-adrenergic receptors. On the other hand, monamine oxidase inhibitors prevent the intracellular degradation of norepinephrine thus increasing brain levels of catecholamines. A paradox has arisen, however, since treatment of patients with depression requires several weeks before any clinical response is noted, yet the above-noted biochemical effects of these drugs are known to occur quickly. Investigations into the beta-adrenergic receptor–adenylate cyclase system, primarily through the use of radioligand binding, has provided some insight into this seeming paradox.

Scores of experiments and papers have documented that the chronic adminstration of either tricyclic antidepressants or MAO inhibitors leads to a decrease in the number of cerebral beta-adrenergic receptors, whereas the affinity of isoproterenol for the receptor appears to be unaltered. The decreased number of beta-adrenergic receptors is associated with a diminished accumulation of cyclic AMP in response to isoproterenol or norepinephrine. The decreased cyclic AMP accumulation results not from enhanced phosphodiesterase activity, but rather from diminished isoproterenol-stimulated adenylate cyclase activity. These findings have been demonstrated in rat cerebral cortex by numerous investigators (Clements-Jewery, 1978;

Hertz et al., 1981; Rosenblatt et al., 1979; Sarai et al., 1978; Sellinger-Barnette, 1980; Sethy and Harris, 1981; Wirz-Justice et al., 1980; Wolfe et al., 1978; O'Donnell and Frazer, 1985; Duncan et al., 1985). Thus, the long-term effect of antidepressants may actually be a decreased rather than an enhanced, response to norepinephrine. The decreased number of beta-adrenergic receptors and decreased responsiveness to catecholamines may in fact be the result of a transient increase in local catecholamine levels followed by a compensatory downregulation of postsynaptic beta-receptors. It should be noted, however, that, although the changes in receptor number are associated with the antidepressant therapy, there is no evidence that this change is the direct mechanism by which antidepressants produce their clinical effectiveness. Recent reviews (Stahl, 1984) point out that there are several neurotransmitters and neurotransmitter receptors in the brain that appear to be regulated by antidepressant drugs. Thus, it would not be surprising if the mechanism of action of the antidepressant drugs were quite complex, with multiple receptor systems being altered simultaneously.

Studies on drugs that are effective as antidepressants but are not norepinephrine-uptake inhibitors or MAO inhibitors, have provided evidence that the increased levels of catecholamines brought about by these two classes of drugs may not be the only factor. Iprindole is an effective agent in the treatment of depression, but has little ability to block the uptake of norepinephrine or to inhibit monamine oxidase. Yet when rats were given this drug chronically, their cerebral cortices were found to contain fewer beta-adrenergic receptors than control animal cortices. These data suggest that a decrease in beta-adrenergic receptors may be a feature of all drug therapies that are effective for depression, whether or not the drug's action is to alter norepinephrine metabolism. Further work will certainly be required to dissect the exact mechanism by which antidepressants work and the underlying pathophysiologic derangement that produces depression.

2.7. Colchicine

Colchicine is the active alkaloid derived from various species of colchicum. It has been used for many years as a treatment for gout and is known to be a potent inhibitor of microtubular assembly in the cell. In addition, colchicine likely has many other modes of action that are unclear at this time. Colchicine appears to influence the function of the beta-adrenergic receptor adenylate cyclase system at several different levels. Insel and Kennedy (1978) have studied the effects of colchicine on the S49 lymphoma cell line growing in culture. They found that it enhanced the ability of beta-adrenergic agonists to increase cyclic AMP production. The mechanism responsible for this enhanced sensitivity remains largely unknown. The number of beta-adrenergic receptors was not changed and the affinity of the receptors, at least in whole cells, was unchanged in the presence of colchicine. More recent studies on the effects of colchicine have been performed by Limas and Limas (1984) and by Marsh et al. (1985). Their results suggest that colchicine alters the desensitization process in myocardial cells. Marsh et al. (1985) found that a 30-min exposure of intact cells to isoproterenol induced a loss of the high-agonist-affinity state of the receptor with no loss of receptor number. This decrease in high-affinity state was associated with a decreased contractile response to isoproterenol. If isoproterenol was removed, and the cells allowed to recover for 60 min, both high-affinity state and the contractile response returned toward normal. Longer incubations with isoproterenol did produce downregulation. If, however, the cells were preincubated with colchicine before exposure to isoproterenol for a 4-h period, the downregulation seen after 4 h was attenuated. In contrast, colchicine by itself had no effect on beta-adrenergic receptor number. Thus, receptor downregulation was inhibited in a concentration-dependent fashion by the inhibitor of microtubular function. These studies suggest a role for microtubules in receptor internalization during agonist exposure. Further work will be required to delineate the exact mechanism whereby internalization appears to be blunted.

2.8. Alcohol

Alcohol ingestion induces a wide range of alterations in the normal physiology of the body. Excessive alcohol intake can produce lethal effects. Research into the effects of ethanol have focused on two major areas. The first involves the effects of chronic ethanol ingestion on organ systems, such as the heart, the liver, and the brain.

The second involves the effects that abrupt withdrawal of alcohol has on these same systems. Chronic ethanol ingestion has long been associated with derangements in cardiac function, such as the induction of cardiac arrythmias and cardiomyopathies. These are manifested by symptoms of congestive heart failure in humans, secondary to a failure of the pumping function of the left ventricle, and biochemically by various alterations in cellular metabolism (Segel and Mason, 1982). There also have been reports that the inotropic effects of catecholamines are weaker in alcohol-treated rats than in control rats. The abrupt withdrawal of alcohol produces a constellation of symptoms that, in its most florid state, is termed delerium tremens. Less-severe manifestations include an apparent hyperadrenergic state associated with increased heart rate, increased blood pressure, tremulousness, and agitation. Because many of these symptoms are consistent with alterations in the sensitivity of various end organs to catecholamines, the possible involvement of the beta-adrenergic receptor–adenylate cyclase system was recognized early and was studied in experimental models. The first study to appear used a rat model and was performed by Banerjee and coworkers (Banerjee et al., 1978). They observed that chronic ethanol ingestion was associated with a fall in the number of beta-adrenergic receptors in both the whole brain and the heart of the rat. If the alcohol was suddenly withdrawn, a significant increase in the number of receptors in the brain was noted, without a change in the affinity for the antagonist radioligand. The increased beta-adrenergic receptor number provided an explanation for the symptoms of the hyperadrenergic state and provided a rationale for the common clinical practice of administering beta-adrenergic blockers such as propranolol, in the alcohol withdrawal syndrome. In contrast to this early study, Hunt et al. (1979) found no alterations in the density of beta-adrenergic receptors in the brain of rats either receiving a chronic alcohol therapy or following the cessation of alcohol treatment. It should be noted however, that, in this latter study, the alcohol was given for only 4 d, in contrast to the 60-d administration in the study of Banerjee et al. (1978). Another study, using a mouse model, by Kuriyama et al. (1981) found a significant increase in the beta-adrenergic receptor density in the cerebral cortices following the withdrawal of alcohol. In addition, they noted that short exposure to alcohol, such as 3–6 d, was not sufficient to induce the receptor number increase upon withdrawal. This study in mice may help explain the apparent disparity in the studies noted above using the rat model.

Several groups of investigators have studied beta-adrenergic receptor density in myocardium of animals given alcohol either by inhalation or in their diet (Sabourault et al., 1981; Segel and Mason, 1982). In these studies, there were no alterations in beta-adrenergic receptor density or antagonist affinity. In addition, in the study of Segel and Mason (1982) isoproterenol competition curves were analyzed, and no differences in agonist affinity were found between alcohol-treated animals and controls. This suggests that no alteration in the coupling of the beta-adrenergic receptor to G_s occurs. These data are consistent with the notion that an alteration in receptor number is not responsible for the manifestations of alcohol withdrawal in the heart.

2.9. Methylxanthines

The methylxanthines, such as aminophylline, isobutylmethyl-xanthine, and caffeine, have long been known to be inhibitors of the enzyme phosphodiesterase. It is only within the last few years that it has become apparent that methylxanthines are also potent inhibitors of adenosine receptors. It is now generally accepted that many of the effects of the methylxanthines are mediated through their antagonism of adenosine receptors, rather than through their ability to inhibit phos-phodiesterases (Fredholm, 1980). This is thought to be true because in order to inhibit phosphodiesterases millimolar concentrations of many of the methylxanthines are required, whereas micromolar concentra-tions are sufficient to block the effect of adenosine receptors, and the effects of methylxanthines are observed in the micromolar range.

Two subtypes of adenosine receptors are generally recognized, termed the A_1 and A_2 receptors. The subtypes can be defined by their relative affinities for adenosine agonist analogs such that the A_1 adenosine receptor displays a potency series of R-PIA (N^6-R-phenyl-isopropyl)adenosine > NECA > S-PIA, whereas the A_2 receptors' potency series is NECA > R-PIA > S-PIA (Stiles, 1985). Both subtypes are coupled to the enzyme adenylate cyclase. The A_1 adenosine receptor is coupled in an inhibitory manner, whereas the A_2 adenosine receptor is coupled in a stimulatory fashion.

The chronic administration of a methylxanthine, such as caffeine, followed by its abrupt withdrawal has long been known to produce a physiological condition characterized by an increased sensitivity to endogenous adenosine. This supersensitivity is likely analogous to that termed the propranolol withdrawal syndrome (Stiles et al., 1984). The caffeine withdrawal syndrome has been described in humans as having such symptoms as headache, myalgia, fatigue, and anxiety. The underlying mechanism responsible for this syndrome has remained largely unknown until recently (Green and Stiles, 1986). Because many of the symptoms of acute and chronic caffeine consumption and its withdrawal are catecholaminergic in nature, an obvious location to look for an effect was in the beta-adrenergic receptor adenylate cyclase system. In 1982, Goldberg et al. studied the effect of chronic caffeine ingestion on the number of beta-adrenergic receptors in the rat forebrain. Following chronic ingestion, there was a decrease in the number of beta-adrenergic receptors present. Similarly, Lowenstein et al. (1982) studied the effect of pentoxifylline, a methylxanthine derivative, and found a decrease in the number of beta-adrenergic receptors in the cerebral cortex of the rat following chronic ingestion. Both of these studies, therefore, provide evidence that methylxanthines are capable of regulating beta-adrenergic receptors in the brain.

One of the effects of A_1 adenosine receptor activation in the brain is to inhibit the release of norepinephrine (Green and Stiles, 1986). We therefore wondered whether chronic blockade of the A_1 adenosine receptors with caffeine might not lead to an enhanced release of norepinephrine. This would obviously increase local concentrations of catecholamines and hence downregulate beta-adrenergic receptors, as has been described in the studies mentioned above. Therefore, we chronically administered caffeine to rats and studied changes in both the beta-adrenergic receptor system and the A_1 adenosine receptor system. We found that chronic ingestion of the A_1 adenosine receptor antagonist, caffeine, leads to a sensitization of the A_1 adenosine receptor–adenylate cyclase system such that the equilibrium of high- and low-affinity-state receptors (which is exactly analogous to the high- and low-affinity-state model described above for beta-adrenergic receptors) is shifted to all high-affinity state. In addition, it was found that there was an enhanced sensitivity of the high-affinity state to guanine nucleotides. These changes were associated with a 35% enhancement of adenosine-mediated inhibition of adenylate cyclase. These data indicate that the adenosine receptor antagonist, caffeine, induces a compensatory sensitization of the A_1 adenosine receptor–adenylate cyclase system. In addition, we found that, in the same animals, there was a decrease in the number of beta-adrenergic receptors from 233 to 190 fmol/mg, without any change in antagonist affinity. The biochemical changes seen in the adenosine receptor–adenylate cyclase system following caffeine ingestion may provide a molecular mechanism for the caffeine withdrawal syndrome. In response to the caffeine's antagonism of the A_1 adenosine receptor, the high–affinity state of A_1 receptor is enhanced, resulting in a compensatory sensitization of the A_1 receptor-mediated inhibition of adenylate cyclase. The caffeine induces an increase in norepinephrine concentration, presumably by blocking the ability of the A_1 receptor to inhibit the release of norepinephrine, then induces a beta-adrenergic receptor down-regulation and probably an attenuation of beta-adrenergic receptor adenylate cyclase stimulation. Therefore, chronic caffeine ingestion, acting by A_1 adenosine receptor antagonism, can induce a compensatory alteration in both the A_1 and beta-adrenergic receptor–adenylate cyclase systems making production of cyclic AMP less favored. This study highlights the facts that interactions among different receptor systems can occur and that the mechanism of action of many drugs may be quite complex and involve multiple receptor systems. It is, therefore, the additive effect of these different systems that is responsible for their physiological and pathophysiological effects.

3. Summary

The last ten years have brought forth a dramatic increase in our understanding of hormone-sensitive adenylate cyclase systems and, in particular, the beta-adrenergic receptor–adenylate cyclase system. Hormone-sensitive adenylate cyclase systems are known to be composed of multiple membrane-bound protein components. First are the glycoprotein receptors that specifically recognize and bind catecholamines or other hormones. Second are the family of guanine nucleotide regulatory proteins that mediate either stimulation (G_s) or inhibition (G_i) of adenylate cyclase and act as the coupling protein to link the receptor with the effector. The third component is the enzyme adenylate cyclase itself that catalyzes the conversion of ATP to cyclic AMP. This brief review has attempted to highlight several important facts concerning the pathophysiologic regulation of the beta-adrenergic receptor–adenylate cyclase system. The first is that each of the components—the receptor, the G proteins, and the catalytic unit—appears to be regulated by a variety of interventions. Therefore, in order to understand the effect of a given drug or other intervention, it is necessary to study each component separately as well as to study how they interact with each other to promote the transmembrane signal that results in stimulation of the enzyme adenylate cyclase. Second, has been the attempt to highlight that the technologies are now available to study the system within the whole cell, within the membrane, and now in in vitro reconstituted systems. Finally, the great tissue variability that is seen in response to the same intervention or stimulus has been underscored. For example, hyperthyroidism is associated with an enhancement of catecholamine sensitivity in a variety of tissues, and this is associated with an increased number of beta-adrenergic receptors as well as an enhanced coupling mechanism, all resulting in an increase in the amount of cyclic AMP produced. In contrast, in the beta-adrenergic receptor system of liver membranes, there appears to be a hyporesponsiveness to catecholamines; this is associated with a decreased production of cyclic AMP.

Although dramatic strides have been made in the past ten years to understand this very important hormone-sensitive system, it is readily apparent that we are still at only the initial stages of understanding the structure and function of the system, and how it relates to homeostatic mechanisms in the body. The next several years will continue to bring forth a plethora of information about the genes that code for the components, how they are regulated by various interventions, and how the system might be perturbed and manipulated for therapeutic benefits in both animals and humans.

Acknowledgment

G. L. S. is an Established Investigator of the American Heart Association and is supported by NHLBI grant ROlHL35134 and a grant-in-aid from the A.H.A. (880662) with funds contributed in part by the North Carolina Affiliate.

References

Aarons, R. D. and Molinoff, P. B. (1982) Changes in the density of beta-adrenergic receptors in rat lymphocytes, heart and lung after chronic treatment with propranolol. *J. Pharmacol. Exp. Ther.* **221,** 439–443.

Abrass, I. B. and Scarpace, P. J. (1981) Glucocorticoid regulation of myocardial β-adrenergic receptors. *Endocrinology* **108,** 977–980.

Alderman, E. L., Coltart, D. J., Wettach, G. F., and Harrison, D. C. (1974) Coronary artery syndromes after sudden propranolol withdrawal. *Ann. Intern. Med.* **81,** 625–627.

Baker, S. P. and Potter, L. T. (1980) Effect of propranolol on β-adrenoceptors in rat hearts. *Br. J. Pharmacol.* **68,** 8–10.

Banerjee, S. P., Sharma, V. K., and Khanna, J. M. (1978) Alterations in β-adrenergic receptor binding during ethanol withdrawal. *Nature* **276,** 407–408.

Bilezikian, J. P., Loeb, J. N., and Gammon, D. E. (1979) The influence of hyperthyroidism and hypothyroidism on the β-adrenergic responsiveness of the turkey erythrocyte. *J. Clin. Invest.* **63,** 184–192.

Boudoulas, H., Lewis, R. P., Kates, R. E., and Dalamangas, G. (1977) Hypersensitivity to adrenergic stimulation after propranolol withdrawal in normal subjects. *Ann. Intern. Med.* **87,** 433–436.

Brodde, O. E., Schumann, H. J., and Wagner, J. (1980) Decreased responsiveness of the adenylate cyclase system on left atria from hypothyroid rats. *Mol. Pharmacol.* **17,** 180–186.

Brodde, O. E., Daul, A., Stuka, N., O'Hara, N., and Borchard, U. (1985) Effects of β-adrenoceptor antagonist administration on $β_2$-adrenoceptor density in human lymphocytes—The role of the "intrinsic sympathomimetic activity." *Naunyn-Schmiedebergs Arch. Pharmacol.* **328,** 417–422.

Bylund, D. B., Forte, L. R., Morgan, D. W., and Martinez, J. R. (1981) Effects of chronic reserpine administration on beta-adrenergic receptors, adenylate cyclase and phosphodiesterase of the rat submandibular gland. *J. Pharmacol. Exp. Ther.* **218,** 134–141.

Canter, E. H., Greenberg, L. H., and Weiss, B. (1981) Effects of long-term changes in sympathetic nervous activation on the beta-adrenergic receptor-adenylate cyclase complex of rat pineal gland. *Mol. Pharmacol.* **19,** 21–26.

Cheng, J. B., Goldfien, A., Ballodard, P. L., and Roberts, J. M. (1980) Gluco-corticoids increase pulmonary β-adrenergic receptors in fetal rabbit. *Endocrinology* **107**, 1646–1648.

Chiu, T. H. (1978) Chronic effects of 6-hydroxydopamine and reserpine on myocardial adenylate cyclase. *Eur. J. Pharmacol.* **52**, 385–388.

Chu, D. T. W., Shikama, H., Khatra, B. S., and Exton, J. H. (1985) Effects of altered thyroid status on β-adrenergic actions on skeletal muscle glycogen metabolism. *J. Biol. Chem.* **260**, 9994–10000.

Clements-Jewery, S. (1978) The development of corticol β-adrenoceptor sub-sensitivity in the rat by chronic treatment with trazodane, doxepin and mianserine. *Neuropharmacology* **17**, 779–781.

Codina, J., Holdebrandt, J. D., Sekura, R. D., Birnbaumer, M., Bryan, J., Manclark, R., Iyenger, R., and Birnbaumer, L. (1984) N_s and N_i, the stimulatory and inhibitory regulatory components of adenylyl cyclases—Purification of the human erythrocyte proteins without the use of activating regulatory ligands. *J. Biol. Chem.* **259**, 5871–5886.

Cooper, G., Kent, R. L., McGonigle, P., and Watanabe, A. M. (1986) Beta adrenergic receptor blockade of feline myocardium-cardiac mechanics, energetics, and beta adrenoceptor regulation. *J. Clin. Invest.* **77**, 441–445.

Coulombe, P., Dussault, J. H., and Walker, P. (1976) Plasma catecholamine concentrations in hyperthyroidism and hypothyroidism. *Metabolism* **25**, 973–979.

Cutler, L. S., Boccuzzi, J., Yaeger, C., Bottaro, B., Christian, C. P., and Martinez, J. R. (1981) Effects of reserpine treatment on β-adrenergic/adenylate cyclase modulated secretion and resynthesis by the rat submandibular gland. *Virchows Arch.* **A392**, 185–198.

Davies, A. O., DeLean, A., and Lefkowitz, R. J. (1981) Myocardial beta-adrenergic receptors from adrenalectomized rats: impaired formation of high affinity agonist-receptor complexes. *Endocrinology* **108**, 720–722.

Duncan, G. E., Paul, I. A., Harden, T. K., Mueller, R. A., Stumpf, W. E., and Breese, G. R. (1985) Rapid down regulation of beta adrenergic receptors by combining antidepressant drugs with forced swim: A model of antidepressant-induced neural adaptation. *J. Pharmacol. and Exp. Ther.* **234**, 402–408.

Exton, J. H., Friedmann, N., Wong, E. H., Brineaux, J. P., Corbin, J. D., and Park, C. R. (1972) Interaction of glucocorticoids with glucagon and epinephrine in the control of gluconeogenesis and glycogenolysis in liver and lipolysis in adipose tissue. *J. Biol. Chem.* **247**, 3579–3588.

Faulkner, S. L., Hopkins, J. T., Boerth, R. C., Young, J. L., Jellett, L. B., Nies, A. S., Bender, H. W., and Shand, D. G. (1973) Time required for complete recovery from chronic propranolol therapy. *N. Engl. J. Med.* **289**, 607–609.

Fraser, C. M. and Venter, J. C. (1980) The synthesis of β-adrenergic receptors in cultured human lung cells: Induction by glucocorticoids. *Biochem. Biophys. Res. Commun.* **94**, 390–397.

Fredholm, B. B. (1980) Are methylxanthine effects due to antagonism of endogenous adenosine? *Trends Pharmacol. Sci.* **1**, 129–132.

Furukawa, H., Loeb, J. N., and Bilezikian, J. P. (1980) Beta-adrenergic receptors and isoproterenol-stimulated potassium transport in erythrocytes from normal and hypothyroid turkeys. *J. Clin. Invest.* **66**, 1057–1064.

Glaubiger, G. and Lefkowitz, R. J. (1977) Elevated beta-adrenergic receptor number after chronic propranolol treatment. *Biochem. Biophys. Res. Comm.* **78**, 720–725.

Glaubiger, G., Tsai, B. S., Lefkowitz, R. J., Weiss, B., and Johnson, E. M. (1978) Chronic guanethidine treatment increases cardiac β-adrenergic receptors. *Nature* **273**, 240–242.

Goldberg, M., Curatolo, P., Tung, C., and Robertson, D. (1982) Caffeine down regulates β-adrenoceptors in rat forebrain. *Neurosci. Lett.* **31**, 47–52.

Green, R. M. and Stiles, G. L. (1986) Chronic caffeine ingestion sensitizes the A_1 adenosine receptor–adenylate cyclase system in rat cerebral cortex. *J. Clin. Invest.* **77**, 222–227.

Hanoune, J. (1978) Characterization of the binding of [^{3}H]dihydroergocryptine to the adrenergic receptor of hepatic plasma membranes: Comparison with the adrenergic receptor in normal and adrenalectomized rats. *J. Biol. Chem.* **253**, 1114–1120.

Hertz, L., Mukerji, S., and Richardson, J. S. (1981) Down regulation of adrenergic activity in astroglia by chronic treatment with an antidepressant drug. *Eur. J. Pharmacol.* **72**, 267–268.

Hunt, W. A., Harden, T. K., and Thurman, R. G. (1979) β-adrenergic receptor density in brains of ethanol-dependent rats. *Drug Alcohol Depend.* **4**, 327.

Insel, P. A. and Kennedy, M. S. (1978) Colchicine potentiates β-adrenoceptor-stimulated cyclic AMP in lymphoma cells by an action distal to the receptor. *Nature* **273**, 471–473.

Jordan, A. W. (1981) Changes in ovarian β-adrenergic receptors during the estrous cycle of the rat. *Biol. Reprod.* **24**, 245–248.

Kajiyama, H., Obara, K., Nomura, Y., and Segawa, T. (1982) The increase of cardiac B_1-subtype of β-adrenergic receptors in adult rats following neonatal 6-hydroxydopa treatment. *Eur. J. Pharmacol.* **77**, 75–77.

Kempson, S., Marinetti, G. V., and Shaw, A. (1978) Hormone action at the membrane level VII. Stimulation of dihydroalprenolol binding to beta-adrenergic receptors in isolated rat heart ventricle slices by triiodothyronine and thyroxine. *Biochim. Biophys. Acta* **540**, 320–329.

Kennedy, R. H. and Donnelly, Jr., T. E. (1982) Cardiac responsiveness after acute withdrawal of chronic propranolol treatment in rats. *Gen. Pharmacol.* **13**, 231–239.

Kuriyama, K., Muramatsu, M., Aiso, M., and Ueno, E. (1981) Alteration in β-adrenergic receptor binding in brain, lung, and heart during morphine and alcohol dependence and withdrawal. *Neuropharmacology* **20**, 659–666.

Lai, E., Rosen, O. M., and Raila, C. S. (1982) Dexamethasone regulates the β-adrenergic receptor subtype expressed by 3T3-L1 preadipocytes and adipocyte. *J. Biol. Chem.* **257**, 691–696.

Lands, A. M., Arnold, A., McAuliff, J. P., Cuduena, F. P., and Brown, T. G. (1967) Differentiation of receptor systems by sympathomimetic amines. *Nature* **214**, 597–598.

Limas, L. J. and Limas, C. (1984) Rapid recovery of cardiac β-adrenergic receptors after isoproterenol-induced "down"-regulation. *Circ. Res.* **55**, 524–531.

Limbird, L. E., Gill, D. M., and Lefkowitz, R. J. (1980) Agonist-promoted coupling of the β-adrenergic receptor with the guanine nucleotide regulatory protein of the adenylate cyclase system. *Proc. Natl. Acad. Sci. USA* **77**, 775–779.

Limbird, L. E. and Lefkowitz, R. J. (1978) Agonist-induced increase in apparent beta-adrenergic receptor size. *Proc. Natl. Acad. Sci. USA* **75**, 228–232.

Lindenfeld, J., Crawford, M. C., O'Rourke, R. A., Levine, S. P., Matiel, M. M., and Horwitz, L. D. (1980) Adrenergic responsiveness after abrupt propranolol withdrawal in normal subjects and patients with angina. *Circulation* **62**, 704–710.

Lowenstein, P. R., Vacas, M. I., and Cardinali, D. P. (1982) Effect of pentoxifylline on alpha- and beta-adrenoceptor sites in cerebral cortex, medial basal hypothalmus and pineal gland of the rat. *Neuropharmacology* **21**, 243–248.

Malbon, C. C. (1980a) The effects of thyroid status on the modulation of fat cell β-adrenergic receptor agonist affinity by guanine nucleotides. *Mol. Pharmacol.* **18**, 193–198.

Malbon, C. C. (1980b) Liver cell adenylate cyclase and β-adrenergic receptors increased β-adrenergic receptor number and responsiveness in the hypothyroid rat. *J. Biol. Chem.* **255**, 8692–8699.

Malbon, C. C. and Gill, D. M. (1979) ADP-ribosylation of membrane proteins and activation of adenylate cyclase by cholera toxin in fat cell ghosts from euthyroid and hypothyroid rats. *Biochim. Biophys. Acta* **586**, 518–527.

Malbon, C. C., Graziano, M. P., and Johnson, G. L. (1984) Fat cell β-adrenergic receptor in the hypothyroid rat: Impaired interaction with the stimulatory regulatory component of adenylate cyclase. *J. Biol. Chem.* **259**, 3254–3260.

Malbon, C. C. and Greenberg, M. L. (1982) 3,3',5-Triiodothyronine administration *in vivo* modulates the hormone-sensitive adenylate cyclase system of rat hepatocytes. *J. Clin. Invest.* **69**, 414–426.

Malbon, C. C., Li, S. Y., and Fain, J. N. (1978) Hormonal activation of glycogen phosphorylase in hepatocytes from hypothyroid rats. *J. Biol. Chem.* **253**, 8820–8825.

Malbon, C. C., Rapiejko, P. J., and Mangano, T. J. (1985) Fat cell adenylate cyclase system-enhanced inhibition by adenosine and GTP in the hypothyroid rat. *J. Biol. Chem.* **260**, 2558–2564.

Mano, K., Akbarzadeh, A., and Townley, R. G. (1979) Effect of hydrocortisone on beta-adrenergic receptors in lung membranes. *Life Sci.* **25**, 1925–1930.

Marsh, J. D., Lachance, D., and Kim, D. (1985) Mechanisms of β-adrenergic receptor regulation in cultured chick heart cells—role of cytoskeleton function and protein synthesis. *Circ. Res.* **57**, 171–181.

Miller, R. R., Olson, H. G., Amsterdam, E. A., and Mason, D. T. (1975) Propranolol withdrawal rebound phenomenon—Exacerbation of coronary events after abrupt cessation of antianginal therapy. *N. Engl. J. Med.* **293**, 416–418.

Moylan, R. D., Bariovsky, K., and Brooker, G. (1982) N^2,O^2-dibutyryl cyclic AMP and cholera toxin-induced β-adrenergic receptor loss in cultured cells. *J. Biol. Chem.* **257**, 4947–4950.

Nattel, S., Ragno, R. E., and Van Loon, G. (1979) Mechanism of propranolol withdrawal phenomenon. *Circulation* **59**, 1158–1164.

Neufeld, A. H., Zawistowski, K. A., Page, E. D., and Bromberg, B. B. (1978) Influences on the density of β-adrenergic receptors in the cornea and iris-ciliary body of the rabbit. *Invest. Opthalmol. Vis. Sci.* **17**, 1069–1075.

Nomura, Y., Kajiyama, H., and Segawa, T. (1980) Hypersensitivity of cardiac β-adrenergic receptors after neonatal treatment of rats with 6-hydroxydopa. *Eur. J. Pharmacol.* **66**, 225–232.

O'Donnell, J. M. and Frazer, A. (1985) Effects of clenbuterol and antidepressant drugs on beta adrenergic receptor/N-protein coupling in the cerebral cortex of the rat. *J. Pharmacol. Exp. Ther.* **234**, 30–36.

Petrovic, S. L., Stanic, M. A., Haugland, R. P., and Dowben, R. M. (1981) Increased β-adrenergic receptor complement in androgen-induced mouse kidney hypertrophy. *Biochim. Biophys. Acta* **676**, 329–337.

Rodan, S. B. and Rodan, G. A. (1986) Dexamethasone effects on β-adrenergic receptors and adenylate cyclase regulatory proteins G_s and G_i in ROS 17/28 cells. *Endocrinology* **118**, 2510–2518.

Rosenblatt, J. E., Pert, C. B., Tallman, J. F., Pert, A., and Bunney, W. E. (1979) The effect of imipramine and lithium on α- and β-receptor binding in rat brain. *Brain Res.* **160**, 186–191.

Sabourault, D., Bauche, F., Giudicelli, Y., Nordmann, J., and Nordmann, R. (1981) Alpha- and beta-adrenergic receptors in rat myocardium membranes after prolonged ethanol inhalation. *Experientia* **37**, 227–228.

Sarai, K., Frazer, A., Brunswick, D., and Mendels, J. (1978) Desmethylimipramine-induced decrease in β-adrenergic receptor binding in rat cerebral cortex. *Biochem. Pharmacol.* **27**, 2179–2181.

Scarpace, P. J. and Abrass, I. B. (1981) Thyroid hormone regulation of rat heart, lymphocyte, and lung β-adrenergic receptors. *Endocrinology* **108**, 1007–1011.

Segel, L. D. and Mason, D. T. (1982) Beta-adrenergic receptors in chronic alcoholic rat hearts. *Cardiovasc. Res.* **16**, 34–39.

Sharma, V. K. and Banerjee, S. P. (1978) Beta-adrenergic receptors in rat skeletal muscle: Effects of thyroidectomy. *Biochim. Biophys. Acta* **539**, 538–542.

Sellinger-Barnette, M. M., Mendels, J., and Frazer, A. (1980) The effect of psychoactive drugs on beta-adrenergic receptor binding sites in rat brain. *Neuropharmacology* **19**, 441–454.

Sethy, V. H. and Harris, D. W. (1981) Effect of norepinephrine uptake blockers on β-adrenergic receptors of the rat cerebral cortex. *Eur. J. Pharmacol.* **75**, 53–56.

Smith, R. M., Patel, A. J., Kingsbury, A. E., Hunt, A., and Balazs, R. (1982) Effects of thyroid state on brain development: β-adrenergic receptors and 5'-nucleotidase activity. *Brain Res.* **198**, 375–387.

Sporn, J. R., Harden, T. K., Wolfe, B. B., and Molinoff, P. B. (1976) β-adrenergic receptor involvement in 6-hydroxydopamine-induced supersensitivity in rat cerebral cortex. *Nature* **194**, 624–626.

Stahl, S. M. (1984) Regulation of neurotransmitter receptors by desipramine and other antidepressant drugs: The neurotransmitter receptor hypothesis of antidepressant action. *J. Clin. Psychiatry* **45**, 37–44.

Sternweis, P. C., Northup, J. K., Smigel, M. D., and Gilman, A. G. (1981) The regulatory component of adenylate cyclase—Purification and properties. *J. Biol. Chem.* **256**, 11517–11526.

Stiles, G. L., Caron, M. G. and Lefkowitz, R. J. (1984) β-adrenergic receptors: Biochemical mechanisms of physiological regulation. *Physiological Reviews* **64**, 661–743.

Stiles, G. L. (1985) The A_1 adenosine receptor: Solubilization of a guanine nucleotide sensitive form of the receptor. *J. Biol. Chem.* **260**, 6728–6732.

Stiles, G. L. (1986) Adenosine receptors: Structure function and regulation. *Trends Pharmacol. Sci.* **7**, 486–490.

Stiles, G. L. and Lefkowitz, R. J. (1981) Thyroid hormone modulation of agonist-beta adrenergic receptor interactions in the rat heart. *Life Sci.* **28**, 2529–2536.

Stiles, G. L., Stadel, J. M., DeLean, A., and Lefkowitz, R. J. (1981) Hypothyroidism modulates beta-adrenergic receptor–adenylate cyclase interactions in rat reticulocytes. *J. Clin. Invest.* **68**, 1450–1455.

Stiles, G. L., Strasser, R. H., Lavin, T. N., Jones, L. R., Caron, M. G., and Lefkowitz, R. J. (1983a) The cardiac beta-adrenergic receptor: structural similarities of $beta_1$ and $beta_2$ receptor subtypes demonstrated by photoaffinity labeling. *J. Biol. Chem.* **258**, 8443–8449.

Stiles, G. L., Taylor, S., and Lefkowitz, R. J. (1983b) Human cardiac beta-adrenergic receptors: Subtype heterogeneity delineated by direct radioligand binding. *Life Sci.* **33**, 467–473.

Sutherland, E. W. and Rall, T. W. (1960) The relation of adenosine-3',5' phosphate and phosphorylase to the actions of catecholamines and other hormones. *Pharmacol. Review* **12**, 265–299.

Tenner, T. E., Mukherjee, A., and Hester, R. K. (1982) Reserpine-induced super-sensitivity and the proliferation of cardiac β-adrenoceptors. *Eur. J. Pharmacol.* **77,** 61–65.

Tse, J., Wrenn, R. W., and Kuo, J. F. (1980) Thyroxine-induced changes in characteristics and activities of β-adrenergic receptors and adenosine 3', 5' monophosphate and guanosine 3', 5' monophosphate systems in the heart may be related to reported catecholamines supersensitivity in hyperthyroidism. *Endocrinology* **107,** 6–10.

Wagner, H. R., Crutcher, K. A., and Davis, J. N. (1979) Chronic estrogen treatment decreases β-adrenergic responses in rat cerebral cortex. *Brain Research* **171,** 147–151.

Webb, J. G., Newman, W. H., Walle, T., and Daniell, H. B. (1981) Myocardial sensitivity to isoproterenol following abrupt propranolol withdrawal in conscious dogs. *J. Cardiovasc. Pharmacol.* **3,** 622–635.

Whitsett, J. A., Darovec-Beckerman, C., Adams, K., Pollinger, J., and Needleman, H. (1980) Thyroid dependent maturation of β-adrenergic receptors in the rat lung. *Biochem. Biophys. Res. Comm.* **97,** 913–917.

Williams, L. T., Lefkowtiz, R. J., Watanabe, A. M., Hathaway, D. R., and Besch, H. R. (1977) Thyroid hormone regulation of β-adrenergic receptor number. *J. Biol. Chem.* **252,** 2767-2769.

Wirz-Justice, A., Kafka, M. S., Naber, D., and Weher, T. A. (1980) Circadian rhythms in rat brain alpha- and beta-adrenergic receptors are modified by chronic imipramine. *Life Sci.* **27,** 341–347.

Wolfe, B. B., Harden, T. K., and Mollnoff, P. B. (1976) β-adrenergic receptors in rat liver: Effects of adrenalectomy. *Proc. Natl. Acad. Sci. USA* **73,** 1343–1347.

Wolfe, B. B., Minnemann, K. P., and Molinoff, P. B. (1982) Selective increases in the density of cerebellar β_1-adrenergic receptors. *Brain Res.* **234,** 474–479.

Wolfe, B. B., Harden, T. K., Sporn, J. R., and Molinoff, P. B. (1978) Presynaptic modulation of beta-adrenergic receptors in rat cerebral cortex after treatment with antidepressants. *J. Pharmacol. Exp. Ther.* **207,** 446–457.

Yamada, S., Yamamura, H. I., and Roeske, W. R. (1980) Alterations in cardiac autonomic receptors following 6-hydroxydopamine treatment in rats. *Mol. Pharmacol.* **18,** 185–192.

Zweifach, B. W., Shorr, E., and Black, M. M. (1953) The influence of the adrenal cortex on behavior of terminal vascular bed. *Ann. NY Acad. Sci.* **56,** 626– 633.

Index

regulation of receptor density
in brain, 275–279
antidepressant
administration, 275
autoradiographic techniques,
278
autoradiography, 278
beta-1, 277
beta-2, 277, 278
beta-adrenergic receptors,
278
cerebral cortex, 275, 277
chronic administration, 277
chronic denervation, 278
clenbuterol, 277
decreases in density of beta-
adrenergic receptors,
277
denervation, 275
depression, 277
desipramine, 277
electroconvulsive shock
(ECS), 295
frontal cortex, 277
hormonal alterations, 275
6-hydroxydopamine, 278
in vivo treatment, 275
sensory cortex, 278
visual cortex, 279
"whisker barrel," 278, 279

C

cDNA for hamster-lung beta-
adrenergic cloning, 245, 249
Cell culture, 284, 285
[³H]DHA to Purkinje cells,
285
oligodendroglia, 285
polygonal astroglia, 285

primary cultures, 284
Characterization in reconstituted
prospholipid vesicles,
163–168
adenylate cyclase, 163, 164,
166, 167
antibodies, 167
chemical probes for protein
orientation, 167
criteria for determining
reproducibility and
relative activity of
preparations, 163
differential accessibility to
soluble and impermeant
probes, 166
distribution of proteins among
vesicles, 165
general problem, 165
[³H]dihydroalprenolol, 163
[¹²⁵I]iodocyanopindolol, 163
lipid:protein ration, 164
orientation of individual
protein molecules, 165
physical characterization of
reconstituted
phospholipid vesicles,
164
possible mechanisms for
inducing asymmetry,
168
receptor-sensitive or coupled
pool of G_s, 164
reconstituted beta-adrenergic
receptor, 163
right-side-in or inside-out, 166
size of vesicles, 165
spherical shell, 165
Chemical sequencing, 238